A PLUME BOOK

THE DESCENT OF MAN

CHARLES DARWIN was born in 1809 and in 1831 embarked on a five-year voyage around the world aboard HMS *Beagle* as the ship's unofficial naturalist. On his return he began writing up his theories and observations, culminating most famously in 1859 in *On the Origin of Species by Means of Natural Selection*. It was not until 1871 that Darwin published a detailed case for human evolution in *The Descent of Man, and Selection in Relation to Sex*. He died in 1882 and was buried in Westminster Abbey.

CARL ZIMMER's books include *Evolution: The Triumph of an Idea*, *Soul Made Flesh*, and *At the Water's Edge*. His articles about biology appear frequently in the *New York Times* and magazines such as *National Geographic* and *Scientific American*. He has won awards for his writing from the American Association for the Advancement of Science and the American Institute of Biological Sciences, and he has received fellowships from the John Guggenheim Memorial Foundation and the Alfred P. Sloan Foundation.

FRANS B. M. DE WAAL is a Dutch-born zoologist and ethologist and the author of the award-winning *Peacemaking among Primates, Good Natured,* and *Our Inner Ape*. He studies monkeys and ape social behavior. He is currently the director of the Living Links Center and a C. H. Candler Professor of Primate Behavior at Emory University.

THE
DESCENT OF MAN,
AND SELECTION IN
RELATION TO SEX

THE CONCISE EDITION

Charles Darwin

Selections and Commentary by CARL ZIMMER
Foreword by FRANS DE WAAL

A PLUME BOOK

PLUME
Published by Penguin Group
Penguin Group (USA) Inc., 375 Hudson Street, New York, New York 10014, U.S.A. • Penguin
Group (Canada), 90 Eglinton Avenue East, Suite 700, Toronto, Ontario, Canada M4P 2Y3 (a divi-
sion of Pearson Penguin Canada Inc.) • Penguin Books Ltd., 80 Strand, London WC2R 0RL, Eng-
land • Penguin Ireland, 25 St. Stephen's Green, Dublin 2, Ireland (a division of Penguin Books
Ltd.) • Penguin Group (Australia), 250 Camberwell Road, Camberwell, Victoria 3124, Australia
(a division of Pearson Australia Group Pty. Ltd.) • Penguin Books India Pvt. Ltd., 11 Community
Centre, Panchsheel Park, New Delhi – 110 017, India • Penguin Group (NZ), 67 Apollo Drive, Rose-
dale, North Shore 0632, New Zealand (a division of Pearson New Zealand Ltd.) • Penguin Books
(South Africa) (Pty.) Ltd., 24 Sturdee Avenue, Rosebank, Johannesburg 2196, South Africa

Penguin Books Ltd., Registered Offices: 80 Strand, London WC2R 0RL, England

First published by Plume, a member of Penguin Group (USA) Inc.

First Printing, November 2007
10 9 8 7 6 5 4 3 2 1

Commentary and foreword copyright © Penguin Group (USA) Inc., 2007
All rights reserved

 REGISTERED TRADEMARK—MARCA REGISTRADA

LIBRARY OF CONGRESS CATALOGING-IN-PUBLICATION DATA
Darwin, Charles, 1809–1882.
 [Descent of man. Selections]
 The descent of man, and selection in relation to sex / Charles Darwin ; selections and commentary by Carl Zimmer ;
foreword by Frans de Waal. — Concise ed.
  p. cm.
 Includes bibliographical references and index.
 ISBN 978-0-452-28888-1
 1. Evolution (Biology). 2. Sexual selection in animals. 3. Sexual dimorphism (Animals).
4. Sex differences. 5. Human beings—Origin. I. Zimmer, Carl, 1966– II. Title.
 QH365.D25Z56 2007
 576.8'2—dc22   2007022908

Printed in the United States of America
Set in Century Old Style and Bembo
Designed by Leonard Telesca

Contents

Foreword

Charles Darwin said he had wanted to keep his musings about "man" for himself. Publication of the revolutionary thoughts that he had written in his notebooks about the human animal would only reinforce the prejudices against his theory, he feared.

We should be immensely grateful that he overcame his reluctance, and gave us *The Descent of Man,* a book that continues to provoke new insights for many branches of science. *The Origin of Species* barely mentions our species at all, alluding to us only in the vaguest terms: "light will be thrown on the origin of man and his history."

Descent made up for this omission, in 1871, over a decade after *Origin.* The book starts out demonstrating the anatomical continuity between us and other animals. Nowadays, few readers will find this shocking, but then Darwin continues applying the same logic to mental continuity. In doing so, he took the opposite view from Alfred Wallace— his contemporary and codiscoverer of evolution by natural selection. In an 1864 essay, Wallace had argued that humans are physically a species of ape, but that our intelligence could not possibly have been produced by the same mechanism of natural selection that had produced our bodies. The noble human mind hints at intervention, Wallace suggested, by some higher being.

Echoes of his position can still be found today in the social sciences and humanities, where it is quite common to hear academics adhere to evolutionary logic while at the same time stressing a radical break between humans and other animals in the cognitive domain. Even the up-and-coming field of evolutionary psychology cannot resist this temptation, and manages to keep its textbooks mostly free of hairy cousins. Inasmuch as this approach tries to dissociate human evolution from continuity among all life forms, it is dramatically at odds with Darwin's message in *Descent*, which is that humans are animals in *both* body and mind.

As a primatologist, the parts that interest me most are those in which Darwin compares human and animal behavior. In one of his notebooks, he had claimed "He who understands baboon would do more toward metaphysics than Locke" (M Notebook, 16 August 1838). In *Descent*, Darwin spends considerable time reviewing whatever he knew about animal altruism and kindness, relating how primates are loyal to their friends (see the incident with the zookeeper who is saved by a small monkey, pp. 163–64), or how a blind pelican keeps a full belly thanks to its companions, p. 163. Not all of these stories should be taken literally, though, as Darwin often received them secondhand. But they helped Darwin make the case that animal social instincts may have provided the basis for the evolution of human morality. He stressed continuity, as in my favorite quote from *Descent:* ". . . any animal whatever, endowed with well-marked social instincts, the parental and filial affections being here included, would inevitably acquire a moral sense or conscience, as soon as its intellectual powers had become as well, or nearly as well developed, as in man."

This is typical of Darwin. Instead of shying away from the most challenging topic, he faces it head on. Morality is often proposed as that which sets humans apart, but such thinking was alien to Darwin. If we consider that much of the continuing opposition to evolutionary theory probably derives from people fearing "moral decay," and if we were to accept that we are mere animals, it is all the more relevant to read what Darwin himself had to say about morality. He did not see it as something coming from God, but from a need to survive. He saw humans as moral to the core of their being.

The parts of *Descent* in which Darwin speculates about moral evolution are often cited since they imply so-called group selection, a highly controversial topic. Darwin argued that if the members of certain tribes are endowed with loyalty to the group and a tendency to help each other, their tribe will likely supplant other tribes, resulting in these qualities being passed on to the next generation. Biologists have been debating group selection for the last few decades, but it is generally agreed that provided the people within a tribe are genetically related, Darwin's argument holds.

At the time of writing, Darwin did not know about genes, let alone DNA. He obviously knew about the inheritability of characteristics, and about the need for males and females to contribute, but not that maternal and paternal traits in fact stay segregated in the genome. A host of issues remained irresolvable because of the assumption that the characteristics of both parents are blended together in their descendants. But this lack of knowledge did not prevent him from proposing a second mechanism of evolution, sometimes considered equally important as natural selection, which is selection of sex-specific characteristics. Here the selection is not on how traits assist survival in a particular environment, but on how they affect mate choice.

Darwin had been worried about the fact that some animals possess traits that seemed extremely costly or risky. How does the male peacock get around with that enormous tail, and why do male Babirusa pigs have upper tusks that curve back over their snout and seem useless for digging or fighting? From a survival perspective, these traits were stupid. They seemed to mock Darwin's theory according to which every trait has its own utility—otherwise it would not be there. He told his children: "The sight of a feather in a peacock's tail, whenever I gaze at it, makes me sick!"

Darwin's deceptively simple solution that females may prefer males with the most extravagant ornamentation ran into immense opposition from scholars who did not wish to give females such an important role in the origin of male traits, and others who saw adornment and embellishment as morally corrupt, utterly frivolous, and perhaps even harmful for the species.

Darwin's ideas about humans as animals, about morality as an evolutionary product, and about bodily beauty being a turn-on for the other sex made *Descent* an incredibly radical book for its time. Even in modern times, not all of its implications have fully sunk in yet. For anyone with an interest in human evolution, there is only one logical starting point and it is this exposé of humanity's place in the larger scheme of nature. The current concise edition has made reading this otherwise somewhat lengthy book a much less daunting task, and will help readers absorb its central and most profound points.

—Frans de Waal

A Note on the Text

Darwin originally published *The Descent of Man* in 1871, correcting and revising it in subsequent editions. The 1879 edition is the basis of the Penguin Classics edition published in 2004. The editors of the Penguin edition, Darwin historians Adrian Desmond and James Moore, corrected many inconsistencies in fonts and spelling and translated many of the passages Darwin had quoted from other languages. They also added a magisterial introduction of their own. This abridged version is drawn from the Penguin edition, with separation marks indicating where portions of *The Descent of Man* have been omitted.

THE
DESCENT OF MAN,
AND SELECTION IN
RELATION TO SEX

I

Editor's Introduction

The history of science is rife with fateful meetings. The astronomer Tycho Brahe hires a young assistant named Johannes Kepler, who will go on to discover in Brahe's observations the law of planetary motion. A bright but aimless British physicist named Francis Crick is introduced to a boisterous young American biologist named James Watson. The two soon discover they share a curiosity about a strange molecule called DNA. And on a warm afternoon in the early spring of 1838, the young Charles Darwin climbed into an orangutan's cage.

The sight of a living ape was a new sensation at the time in England. Europeans only knew of apes through vague accounts from travelers returning from Africa and Indonesia. Beginning in 1835, however, chimpanzees and orangutans began to appear at London's Zoological Gardens. Face to face, the apes inspired uneasy fascination. Queen Victoria declared them "frightful, and painfully and disagreeably human."

The first ape to come to London, a chimpanzee named Tommy, was put in a sailor's suit. The second, a female orangutan named Jenny, was put in a dress. Both were taught to eat with spoons. Yet the humanlike qualities of these apes did not cause most observers to question the uniqueness of humans. An article about Jenny's humanlike behavior stressed that "in nothing does it trench upon the moral or mental provinces of man." After all, it was agreed, God had separately created man and orangutan and every other species on Earth in their current form.

He had endowed each species with its own combinations of traits, all of which displayed His handiwork. One need only look at the complex barbs of a feather or the muscles of the human hand to see the work of a Creator. An orangutan's ability to eat with a spoon was just a distraction from that great truth.

A few people, however, harbored some doubts. Charles Darwin was one of them. In 1838, those doubts had not yet flowered into a full-blown theory of life. Darwin was only twenty-nine at the time, and would not publish his first account of evolution for another twenty-one years. But his doubts were already deep enough to lure him into Jenny's cage.

Historians have long wondered how the seeds of evolution were planted in Darwin's mind. They were probably already there in his youth, but they remained dormant for years. His grandfather, the physician Erasmus Darwin, wrote a long poem called "Zoonomia" in the 1790s, in which he argued that life had changed over vast stretches of time, with new kinds of creatures emerging from old ones. This transformation would later become known as evolution. Charles Darwin learned more about evolution as a teenager, when he traveled to Edinburgh to study medicine. Surgery and autopsies appalled him. His fondness for nature was already becoming plain, as he explored tidal pools instead of sitting through lectures. Darwin soon found a mentor in the Scottish naturalist Robert Grant. Grant taught Darwin not only about natural history, but about evolution as well.

Grant had been deeply influenced by Darwin's grandfather, as well as by the more recent work of the French naturalist Jean Lamarck. Lamarck argued that two laws of nature gradually changed life over time. Simple life forms perpetually came into existence, and they were all driven toward greater degrees of perfection. Species also adapted to their particular environments through experience. Giraffes stretching to reach leaves, Lamarck suggested, might acquire longer necks. The giraffes could then pass down their longer necks to their offspring, Lamarck believed.

While Grant and a few other naturalists embraced Lamarck, most scholars viewed him with scorn. Lamarck was challenging a fundamental tenet of Christianity, and he even dared to suggest that human beings were also the product of evolution. Lamarck speculated how an ape much like orangutans or chimpanzees might gradually stand upright and begin to speak, and thus become human.

Despite the influence of his grandfather and Grant, Darwin doesn't appear to have taken evolution seriously as a young man. Rejecting medicine, he began preparing to join the clergy. At the University of Cambridge he read the influential 1802 book *Natural Theology* by the Reverend William Paley. Paley asked his readers to imagine walking across a heath and coming across a watch. By its sheer complexity, his readers would know someone had made it. Paley argued that eyes and feet and other traits showed similar signs of design. Darwin admired Paley's rhetoric, although he was not a terribly serious student of theology. In his free time, Darwin hunted for beetles in the countryside.

In 1831, as Darwin was finishing up at Cambridge, he received a rare invitation. Robert FitzRoy, the captain of HMS *Beagle*, was searching for a well-educated gentleman to join him on a voyage around the Earth. Darwin accepted, and became the *Beagle*'s unofficial naturalist. The voyage lasted five years, during which time Darwin explored Amazon jungles, climbed the Andes, and prowled Pacific islands few Europeans had ever seen. He gathered much of the raw material that he would later use to develop his own theory of evolution.

But Darwin was not yet an evolutionist. On the voyage of the *Beagle* he simply made observations—of mountains, of coral reefs, of giant tortoises—and sought natural explanations for how they came to be. One of his best guides was a newly published book he had brought along on the voyage, called *Principles of Geology*. The author, a British lawyer-turned-geologist named Charles Lyell, argued that mountains and valleys and other geological features were not the result of Noah's flood or some other sudden catastrophe. Instead, the surface of the planet was the product of a long series of gradual changes. Rain gradually wore down canyons. Mountains inched up out of the sea. Darwin saw ample evidence on his voyage to support Lyell's theory of an ancient, slowly changing Earth.

Lyell also discussed evolution in *Principles of Geology*. He sketched out Lamarck's arguments, describing how an orangutan supposedly "is made slowly to attain the attributes and dignity of man." But Lyell rejected Lamarck. He pointed out the recent discovery of ibises mummified for thousands of years in Egypt. If Lamarck were right, one would expect to see a difference between the mummified birds and living members of their species. No such difference had been found. Yet Lyell did not believe that all life was created at Earth's dawn. He proposed that new species were separately created over the course of Earth's history.

This argument did not make sense to Darwin. After his return to England in 1837, he began to carefully describe the fossils, birds, plants, and other specimens he had gathered on his travels. He recorded his ideas in a series of notebooks. The notebooks document his embrace of evolution. Just as geology showed evidence of a slowly changing Earth, living species showed evidence of a slow transformation. How else to explain fossil rodents and anteaters in South America that were giant versions of the mammals that Darwin saw on his visit? Darwin had discovered new species of finches on the remote Galapagos Islands, with beaks so different from one another he had not realized at first that they were finches at all. It was hard to reconcile those finches with the idea of special creation. Perhaps an ancestral finch had given rise to new species, Darwin thought, each with its own adaptations.

In his notebooks, Darwin sketched a tree, each branch a species joined in kinship with other species. Above the tree he wrote, "I think."

Humans, Darwin immediately recognized, might also belong on one of those branches. Hence his fascination with Jenny the orangutan. In the similarities between orangutans and humans Darwin saw signs of kinship, of a shared ancestry. On March 28, 1838, Darwin rode to the London Zoo and paid a visit to Jenny, who was weathering the British climate in the heated giraffe house. As a wealthy guest, Darwin was allowed to enter the cage itself. In a letter he wrote four days later to his sister Susan, he described what he saw:

> . . . the keeper showed her an apple, but would not give it [to] her, whereupon she threw herself on her back, kicked & cried, precisely like a naughty child.—She then looked very sulky & after two or three fits of pashion, the keeper said, "Jenny if you will stop bawling & be a good girl, I will give you the apple.—She certainly understood every word of this, &, though like a child, she had great work to stop whining, she at last succeeded, & then got the apple, with which she jumped into an arm chair & began eating it, with the most contented countenance imaginable.

Darwin watched Jenny gaze at herself in a mirror. She used bits of straw like tools. Her face contorted much as a child's would. Others might believe they were vastly different from an orangutan, but Darwin didn't. He decided that much of that difference was a superficial matter of clothes and manners. His mind raced back to the people he had encountered on his voyage aboard the *Beagle*. He had met naked Indians in Tierra del Fuego. But he had also met Fuegans who had traveled to England and taken up the customs of Western civilization.

"Let man visit Ourang-outang in domestication," he wrote in his notebook, "hear [its] expressive whine, see its intelligence when spoken [to]; as if it understand every word said—see its affection.—to those it knew.—see its passion & rage, sulkiness, & very actions of despair; let him look at savage, roasting his parent, naked, artless, not improving yet improvable & let him dare to boast of his proud preeminence."

Darwin kept his notebooks private. His dangerous thoughts about human origins would stew in his mind for more than three decades. He would finally share them with the world thirty-three years later, with the publication of his 1871 book, *The Descent of Man, and Selection in Relation to Sex.*

The Descent of Man is one of the most important books in the history of biology, but it is also one of the most baffling. A reader can be forgiven for wondering why the book exists at all. Twelve years earlier, Darwin had introduced the world to his theory of evolution with *On the Origin of Species by Means of Natural Selection*. An account of how evolution produced humans would have made a splendid final chapter. And yet Darwin scrupulously avoided almost any mention of humans in the *Origin of Species*. When Darwin finally did turn his attention to mankind and wrote *The Descent of Man*, he produced a sprawling book that seems arranged to frustrate any attempt to read it all the way through.

Darwin buries the skeleton of his argument under fleshy folds of esoterica—the length of European shinbones, the habit rabbits have of stamping their feet in fear, lizard snouts, peacock feathers, the Egyptian custom of knocking out their front teeth. In fact, even the title of the book is misleading, because most of *The Descent of Man* is not in fact about man at all. Darwin dedicates more than half of the book exclusively to the courtship of animals.

For all these frustrations, *The Descent of Man* marks a turning point in the history of science. It represents the first effort to trace the origin of human nature to a biologically realistic account of evolution. Darwin argued that the same natural processes that produced iris petals and scorpion tails also produced humanity's noblest features, such as language and morality. Of course, like any book, *The Descent of Man* was the product of its time. It is deeply tinted by the prejudices and assumptions of Victorian England. Its picture of humanity is in some ways disturbingly obsolete. And yet more than 130 years later it remains a living document. How many other books from 1871 appear regularly in the footnotes of papers published in the latest issues of scientific journals?

This abridged edition will, I hope, give readers new to *The Descent of Man* an appreciation of its importance. Each excerpt I have selected represents one of the book's major themes. To introduce them, I discuss the intellectual background to Darwin's arguments and then compare his ideas to the current understanding of human nature. The notion that Jenny the orangutan shared a common ancestor with a Victorian gentleman, for example, was outrageous in 1838, but today the evidence is overwhelming. Darwin made his original case by comparing the anatomy and behavior of humans and apes. Today scientists can compare humans and apes on the molecular level and put Darwin's hypothesis to the test. If orangutans were kin to humans, you'd expect that their DNA would preserve traces of that kinship. And indeed it does. Human DNA bears distinctive sequences shared only by orangutans, gorillas, chimpanzees, and bonobos. Jenny and Darwin were two cousins in a cage.

To understand the peculiar nature of *The Descent of Man*, it's essential to understand two things: Darwin's own view of human nature, and his struggle to present evolution to the public. Together, they made *The Descent of Man* such a paradox of a book.

Looking back from the twenty-first century, we must work hard to understand Darwin's conception of what it means to be human. Darwin viewed humanity with a mix of egalitarianism and elitism. Is this a contradiction? Perhaps, but it was one that Darwin shared with many of his peers.

Darwin's egalitarianism made him a passionate opponent of slavery. England outlawed trading in slaves in 1807, two years before Darwin was born, but slavery itself would not be outlawed in British colonies until 1833. Darwin grew up in an abolitionist household, and his experiences in South America during the voyage of the *Beagle* only hardened his disgust for slavery: In Brazil, slave families would be cruelly separated, and the ears of fugitives lopped off. This brutality was often justified on the basis of race: Europeans supposedly were inherently superior to other humans. Racism also underlay the vicious treatment of Indians in Argentina, where European settlers slaughtered families and took their land.

Evolution, to Darwin's mind, directly challenged justifications for this cruelty. Africans were not subhumans, fit only for slavery. All humans shared a common ancestry—not from Adam, but from apes. Africans, Fuegans, and Englishmen were no more different from one another than breeds of pigeons. In fact, many of the qualities that supposedly elevated humans above animals could be found among the beasts, such as Jenny.

Once Darwin recognized *that* humans and other species had evolved, he had to figure out *how* they did so. He did not reject Lamarck entirely, but he was soon captivated by a new concept, which he dubbed natural selection. It came to Darwin as he read an essay by the clergyman Thomas Malthus about the growth of human population. Malthus argued that the population of a country, if left unchecked, would rapidly explode, and warned that it would outstrip the country's food supply. The only reason that human history was not one long famine was because human populations were kept in check by diseases, catastrophes, and the like.

Darwin realized that Malthus's ideas had a huge importance with respect to animals and plants. The world could easily be overrun by flies or roses or toadstools or any other species if the population of that species could grow without check. Many individuals in each species must therefore die without offspring. Darwin realized that certain traits might determine whether an individual reproduced more than others. A thick coat of fur might allow one fox to weather a winter better than a more lightly attired one, and survive to produce more kits. Over the course of generations, the thick-furred foxes would become more and more common. Natural selection could not only alter life but also wipe it out. If two species competed for the same types of food, one might eventually drive the other to extinction.

Darwin's discovery of Malthus was a crucial moment in the history of science, one that would ultimately give rise to the modern understanding of biology. It's easy to forget, however, that Malthus was much more interested in people than in animals or plants. He championed a bleak sort of politics. Welfare was counterproductive, Malthus argued, because it only worsened the suffering from population growth. Darwin acquired a similarly bleak view of natural selection among humans. He was wary of welfare for the poor, since it removed the improving powers of hardship. And when two peoples met—such as the English and the Aborigines of Australia—they competed like two species. In Darwin's day, a number of cultures were dwindling in the face of European expansion, and Darwin fully expected them to disappear. Thus Darwin's view of human nature was a mix of abolitionist equality and a Victorian capitalist view of industrial society. He was hardly the only Victorian gentleman to hold both views at once.

In the early 1840s Darwin worked furiously to flesh out a theory of evolution by natural selection. He found that it could account for a vast range of patterns in nature, from the distribution of species to the relationship between fossils and living animals. He amassed evidence of the common ancestry of different species, evidence such as homologies—structural similarities between different organs: bat wings and human hands, for example.

Darwin translated his notebooks into a long formal description of the theory, but he did not immediately publish it. He knew that it would shock many of his colleagues, particularly in its implications for human origins. He shared the full details with only a few people, including his wife, Emma, and the botanist Joseph Hooker. Darwin's instincts proved sound. Shortly after Emma and Hooker learned about his theory, England began buzzing about a new book about evolution.

In 1844, a Scottish journalist named Robert Chambers anonymously published *Vestiges of the Natural History of Creation*. Chambers declared that life evolved according to natural laws. As lofty as humans might be, they had evolved just as any other animal had. *Vestiges* sold tens of thousands of copies, but it also triggered a furious backlash from England's scientific elite. Adam Sedgwick, Darwin's old geology teacher at Cambridge, declared that if the book were true, "religion is a lie; human law a mass of folly and a base injustice; morality is moonshine."

Vestiges's harsh reception did not force Darwin to abandon his theory. He set out to succeed where Chambers had failed. Instead of presenting a compelling mechanism by which life evolved, Chambers offered a fanciful stew of speculation. Darwin redoubled his efforts to make the case for evolution by natural selection. He gathered vast stores of information, sending letters around the world. He spent years studying barnacles, which had undergone extreme evolutionary transformation from their shrimplike ancestors. He published a massive two-volume monograph on barnacles that was both a zoological tour de force and a gold mine of insights about evolution. He also looked for signs of evidence of evolution in humans. In the expressions in the faces of his children, Darwin saw echoes of Jenny.

Darwin did not return to his natural selection manuscript for another decade. When he did, he was able to infuse it with all that he had learned in those years. His ideas about natural selection had become more subtle and complex. The reproductive success of an animal, he realized, also depended on one animal finding another animal. Here, too, there might be inequality. Strong males might be able to fight off weaker ones for the opportunity to mate with females. Females might be attracted to handsome males more than ugly ones. Attractive males would father more offspring, and their traits would become more common over the generations.

In 1856 Darwin dubbed this process sexual selection. He wondered if sexual selection might be the source of nature's extravagance, from warthog tusks to peacock tails. It might have even produced the differences between human races. Europeans became white and Africans became black thanks only to different conceptions of beauty. Sexual selection might have also produced the differences between men and women, as men competed for the opportunity to mate, and women chose from their prospective suitors.

By 1858 Darwin's book had reached 250,000 words, with no end in sight. But his glacial pace was interrupted in 1858 by a letter from the Far East. Alfred Russell Wallace, a naturalist who had supplied Darwin with some bird skins for his research, had been thinking about a theory of evolution. It was strikingly similar to Darwin's; Wallace even drew inspiration from Malthus as well. In his letter to Darwin, Wallace described his theory in detail. Once Darwin got over his shock, he arranged for Wallace's paper and a short summary of his own work to be read at a June meeting of the Linnean Society. Suddenly his ideas were public, but not in the way he had carefully planned. He struggled to write a scientific paper to expand on the Linnean talk, but it swelled up to book length. He decided that a book it would be: *The Origin of Species*.

The book ended up far shorter than Darwin's unpublished tome. But he still managed to squeeze in an awesome amount of natural history. His friend and public champion, Thomas Huxley, called *The Origin of Species* "a mass of facts crushed and pounded into shape, rather than held together by the ordinary medium of an obvious logical bond." Darwin had compiled that mass of facts to meet every objection he could imagine, and also to show how universal a process evolution was—to show how deeply evolution had left its mark on every corner of life.

One of those corners, Darwin knew, was our own species. His storehouse of natural history included a great deal of information on human evolution. But in *The Origin of Species*, he constrained himself to just a few fleeting mentions of humans. Darwin hinted at the evolution of human races and then coyly announced, "I may add that some little light can apparently be thrown on the origin of these differences, chiefly through sexual selection of a particular kind, but without here entering on copious details my reasoning would appear frivolous." Near the end of the book, he cryptically predicts, "Light will be thrown on the origin of man and his history."

Darwin's omission was a conscious, tactical maneuver in a campaign to win the public over to evolution. It would be hard enough to persuade his readers that animals or plants evolved. To add humans to evolution's list of accomplishments might instantly turn them away. "I thought that I should thus only add to the prejudices against my views," he later wrote.

Darwin was probably right. Most of the attacks launched against *The Origin of Species* in the 1860s sooner or later came around to the question of man's place in nature. Richard Owen, the greatest British anatomist of the nineteenth century, tried to refute evolution by finding a part of the human brain with no counterpart in apes. In 1860, Owen attended the most famous debate over Darwin, at Oxford University. Bishop Samuel Wilberforce delivered a furious attack on *The Origin of Species* and ended it by turning to Thomas Huxley and asking him whether he descended from an ape on his mother's or father's side. Jenny's ghost cast a long, worrisome shadow.

While others argued over man and ape, Darwin planned his next project after *The Origin of Species*. It would be a sprawling trilogy, packed with all the information he'd had to leave out of his first book on evolution. Originally, he hoped to include his material on human biology. But his plans, as they often did, went awry. The material on domesticated animals and plants proved massive enough to warrant a book of its own, which Darwin published in 1868. He became distracted by orchids, which proved a hothouse illustration of evolution's remarkable power to reshape life. And then his health took a serious turn for the worse, leaving him exhausted and unable to work for long stretches.

While Darwin was distracted from human evolution, others gave it their full attention. Archaeologists were discovering arrowheads and pottery shards and other evidence of an ancient period of human existence that they dubbed "prehistory." Huxley was organizing devastating attacks against Owen, showing that human brains were not so profoundly different from primate brains after all. Anthropologists tried to quantify the differences between the races. Some argued that the races had a single human origin, while others argued that each race must have evolved separately from primate ancestors. To Darwin's horror, the multiple-origins camp proved willing to use evolution to defend slavery.

Darwin kept abreast of these developments, but it was Wallace who finally pulled him into the debate over human evolution. In the late 1860s Darwin began to discuss his theory of sexual selection with Wallace. He explained how it could have driven the evolution of features in both animals and humans. Wallace only saw the need for one kind of selection: natural. Darwin argued that in birds, bright males and drab females were evidence of the choosiness of females. Wallace argued that the plumage of the two sexes helped each to survive—drab females, for example, being better camouflaged from predators as they sat on their nests. Over the course of their exchange, Darwin searched for more evidence he could provide to Wallace for sexual selection—evidence he would later present in *The Descent of Man*.

Darwin and Wallace debated sexual selection in the spirit of friendly competition. But suddenly the argument turned ugly. Wallace decided that once the very simplest features of humans had emerged—crude communication, tool-making, and so on—natural selection could not have driven those faculties to become as powerful as they are today. There would simply be no advantage to full-blown language, morality, and all the other things humans prided themselves in. Wallace was growing increasingly interested in séances and the supernatural, and he became convinced that human nature transcended nature itself. While our distant primate ancestors might have evolved by natural selection, Wallace concluded, supernatural forces took control of humanity's progress.

Darwin told Wallace that he was "dreadfully disappointed." Unlike Wallace, Darwin saw nothing in human nature beyond the scope of evolution. Morality had its counterpart in the selfless behavior of animals. Human language was just a particularly elaborate form of communication. Humans were different in degree from animals, not in kind.

Darwin could no longer rely on others to put man in his proper place. He would have to do it himself. By 1867 Darwin was referring to his "book" on man. Its final form reflected the complex history of Darwin's ideas about humans. He began by assembling evidence that humans had evolved, showing all the anatomical links to other species. He then argued that man's mental faculties also had precedents in the natural world. Darwin offered explanations for the differences between human races, but in order to do so, he had to introduce readers to sexual selection. Like so many of his other projects, this introduction got the better of him, and it came to dominate the book. At the end of *The Descent of Man*, Darwin combined all of his arguments into a single vision of humanity as the product of evolution, through both natural and sexual selection.

In March 1870 Darwin sent the manuscript to his long-time publisher, John Murray. Murray fretted that the book would stir up controversy and convinced Darwin to abandon his original title, *The Origin of Man*. Such tweaks notwithstanding, it remained dense with provocation, taking on all of religion's most cherished assumptions. The book certainly attracted some tough criticisms, but nothing compared to the uproar that had met *The Origin of Species*. It sold well and deeply penetrated the public discourse. Darwin began to appear in cartoons as half man, half ape.

Today thousands of scientists follow Darwin's path, investigating the evolutionary origins of human nature. They dig up fossils, they hunt for traces of history in our DNA, they search for humanlike behavior in primates and other animals. Yet human evolution remains a deeply controversial subject. In a 2007 survey, 48 percent of Americans stated that God created humans in their present form in the past 10,000 years. Opponents of evolution claim, like Adam Sedgwick did 163 years ago, that believing in evolution corrupts morality. Human evolution also underlies fierce debates over what is "natural" about human nature. In 2006, for example, Harvard University president Lawrence Summers aroused a storm of protest by suggesting that there were more male engineering professors than female ones because of biological differences between the sexes. Shortly afterward, Summers resigned. Evolutionary biologists are shedding light on such controversial issues, although they are a long way from providing definitive answers. In the biological world, nothing surpasses human nature in mystery. But scientists do not react to the mystery by throwing up their hands; they think of new experiments to run.

The Descent of Man is important not just as a scientific milestone, but as a cultural artifact. Darwin was a man of his time, and as controversial as his theory of evolution was, he left certain assumptions of Victorian England unchallenged. Men were intellectually superior to women, he assumed; Victorian monogamy was natural for our species, despite the wide range of other marital systems found in other cultures; and Europeans were in many ways the pinnacle of human development. One cannot ignore these aspects of *The Descent of Man*. Yet one must also bear in mind that science has always been an imperfect, all-too-human endeavor. But science also allows humans to reach deeper and deeper insights about the world, and about ourselves.

Introduction

THE NATURE OF the following work will be best understood by a brief account of how it came to be written. During many years I collected notes on the origin or descent of man, without any intention of publishing on the subject, but rather with the determination not to publish, as I thought that I should thus only add to the prejudices against my views. It seemed to me sufficient to indicate, in the first edition of my 'Origin of Species', that by this work 'light would be thrown on the origin of man and his history', and this implies that man must be included with other organic beings in any general conclusion respecting his manner of appearance on this earth. Now the case wears a wholly different aspect. When a naturalist like Carl Vogt ventures to say in his address as President of the National Institution of Geneva (1869), 'personne, en Europe au moins, n'ose plus soutenir la création indépendante et de toutes pièces, des espèces,' ['nobody, in Europe at any rate, would nowadays dare to support the idea of the separate and independent creation of every species'] it is manifest that at least a large number of naturalists must admit that species are the modified descendants of other species; and this especially holds good with the younger and rising naturalists. The greater number accept the agency of natural selection; though some urge, whether with justice the future must decide, that I have greatly overrated its importance. Of the older and honoured chiefs in natural science, many unfortunately are still opposed to evolution in every form.

In consequence of the views now adopted by most naturalists, and which will ultimately, as in every other case, be followed by others who are not scientific, I have been led to put together my notes, so as to see how far the general conclusions arrived at in my former works were applicable to man. This seemed all the more desirable, as I had never deliberately applied these views to a species taken singly. When we confine our attention to any one form, we are deprived of the weighty arguments derived from the nature of the affinities which connect together whole groups of organisms—their geographical distribution in past and present times, and their geological succession. The homological structure, embryological development, and rudimentary organs of a species remain to be considered, whether it be man or any other animal, to which our attention may be directed; but these great classes of facts afford, as it appears to me, ample and conclusive evidence in favour of the principle of gradual evolution. The strong support derived from the other arguments should, however, always be kept before the mind.

The sole object of this work is to consider, firstly, whether man, like every other species, is descended from some preexisting form; secondly, the manner of his development; and thirdly, the value of the differences between the so-called races of man. As I shall confine myself to these points, it will not be necessary to describe in detail the differences between the several races—an enormous subject which has been fully discussed in many valuable works. The high antiquity of man has recently been demonstrated by the labours of a host of eminent men, beginning with M. Boucher de Perthes; and this is the indispensable basis for understanding his origin. I shall, therefore, take this conclusion for granted, and may refer my readers to the admirable treatises of Sir Charles Lyell, Sir John Lubbock, and others. Nor shall I have occasion to do more than to allude to the amount of difference between man and the anthropomorphous apes; for Prof. Huxley, in the opinion of most competent judges, has conclusively shewn that in every visible character man differs less from the higher apes, than these do from the lower members of the same order of Primates.

This work contains hardly any original facts in regard to man; but as the conclusions at which I arrived, after drawing up a rough draft, appeared to me interesting, I thought that they might interest others. It has often and confidently been asserted, that man's origin can never be known: but ignorance more frequently begets confidence than does knowledge: it is those who know little, and not those who know much, who so positively assert that this or that problem will never be solved by science. The conclusion that man is the co-descendant with other species of some ancient, lower, and extinct form, is not in any degree new. Lamarck long ago came to this conclusion, which has lately been maintained by several eminent naturalists and philosophers; for instance, by Wallace, Huxley, Lyell, Vogt, Lubbock, Büchner, Rolle, &c.,[1] and especially by Häckel. This last naturalist, besides his great work, 'Generelle Morphologie' (1866), has recently (1868, with a second edit. in 1870), published his 'Natürliche Schöpfungsgeschichte', in which he fully discusses the genealogy of man. If this work had appeared before my essay

1. As the works of the first-named authors are so well known, I need not give the titles; but as those of the latter are less well known in England, I will give them:— 'Sechs Vorlesungen über die Darwin'sche Theorie': zweite Auflage, 1868, von Dr L. Büchner; translated into French under the title 'Conférences sur la Théorie Darwinienne', 1869. 'Der Mensch, im Lichte der Darwin'sche Lehre', 1865, von Dr F. Rolle. I will not attempt to give references to all the authors who have taken the same side of the question. Thus G. Canestrini has published ('Annuario della Soc. d. Nat', Modena, 1867, p. 81) a very curious paper on rudimentary characters, as bearing on the origin of man. Another work has (1869) been published by Dr Francesco Barrago, bearing in Italian the title of 'Man, made in the image of God, was also made in the image of the ape'.

had been written, I should probably never have completed it. Almost all the conclusions at which I have arrived I find confirmed by this naturalist, whose knowledge on many points is much fuller than mine. Wherever I have added any fact or view from Prof. Häckel's writings, I give his authority in the text; other statements I leave as they originally stood in my manuscript, occasionally giving in the foot-notes references to his works, as a confirmation of the more doubtful or interesting points.

During many years it has seemed to me highly probable that sexual selection has played an important part in differentiating the races of man; but in my 'Origin of Species' (first edition, p. 199) I contented myself by merely alluding to this belief. When I came to apply this view to man, I found it indispensable to treat the whole subject in full detail.[2] Consequently the second part of the present work, treating of sexual selection, has extended to an inordinate length, compared with the first part; but this could not be avoided.

I had intended adding to the present volumes an essay on the expression of the various emotions by man and the lower animals. My attention was called to this subject many years ago by Sir Charles Bell's admirable work. This illustrious anatomist maintains that man is endowed with certain muscles solely for the sake of expressing his emotions. As this view is obviously opposed to the belief that man is descended from some other and lower form, it was necessary for me to consider it. I likewise wished to ascertain how far the emotions are expressed in the same manner by the different races of man. But owing to the length of the present work, I have thought it better to reserve my essay for separate publication.

---

2. Prof. Häckel was the only author who, at the time when this work first appeared, had discussed the subject of sexual selection, and had seen its full importance, since the publication of the 'Origin'; and this he did in a very able manner in his various works.

2

Evidence for Human Evolution

The human body works, and works very well. Our eyes can capture exquisitely detailed images; our immune systems can fight off a practically infinite number of pathogens; our hands are nimble enough that with enough training surgeons can repair ruptured blood vessels and broken brains. In the early 1800s, when Charles Darwin was first learning about biology, such finely adapted traits were seen as evidence of God's handiwork. Only a supernatural Creator could produce structures so well-suited to their tasks.

This view of life was not an ignorant, prescientific delusion. In fact, many of the greatest figures of the scientific revolution in the 1600s believed that their discoveries about the workings of the human body were proof of God's design. With the scientific revolution God became a chemical and mechanical engineer. Natural theology, as this school of thought was known, survived into the early 1800s and was embraced by leading researchers such as Charles Bell, one of the great neurologists of the nineteenth century. In addition to his pathbreaking work on the anatomy and diseases of the nervous system, Bell also published *The Hand, Its Mechanism and Vital Endowments as Evincing Design.*

Only by recognizing the power of previous explanations of the human body can we appreciate how important Darwin's breakthrough was. Darwin's assault on natural theology was like a pincer movement, describing the mechanism of natural selection and offering example after example of human biology that made little sense except as evidence of evolution. In *The Descent of Man*, the evidence comes first. Darwin wants to show the many ways in which man, in his words, "still bears in his bodily frame the indelible stamp of his lowly origin."

In order to persuade his readers of this genealogical connection, Darwin must overcome the powerful notion that the human body is exquisitely well-engineered. To do so, he points out the many rudimentary features of the human body—the stump of a tail at the bottom of the spine, for example, or the lanugo, the hair that grows over the entire human fetus before falling out before birth. These traits make eminent sense if seen as vestiges of primate ancestors. They make little sense if seen as the handiwork of a designer.

In the twenty-first century, the human body continues to reveal more of its imperfections. Along with vestigial tails and fleeting lanugos, scientists are discovering molecular rudiments in the human genome. The most striking examples come from the human nose. Darwin noted that humans have a weak sense of smell compared to other mammals, but in his day scientists knew little about the mechanisms that make smell possible. Scientists now know that each olfactory nerve in the nose is studded with receptors that can grab odor molecules. Each kind of receptor is encoded by a separate gene. Its unique structure allows it to grab certain molecules. The combined information gathered by olfactory receptors lets us discriminate among a vast range of subtle smells.

Other animals carry their own olfactory receptor genes, and by comparing ours to theirs, scientists have reconstructed the history of smell. Over millions of years, mutations have accidentally produced extra copies of many olfactory receptor genes. Additional mutations altered the structure of those copies, changing their responses to odor molecules. As a result, we carry families of related olfactory receptor genes. Genes belonging to those same families can be found in chimpanzees, mice, and other mammals.

These families alone are powerful evidence of our common heritage with other animals. But even more powerful is the fact that many of the olfactory receptor genes in the human genome don't work. They have been hit by crippling mutations that prevent an olfactory receptor gene from being successfully translated into a working protein. These disabled genes are called pseudogenes. The human genome contains 388 working genes for olfactory receptors, and 414 pseudogenes. By contrast, a mouse has 1,037 working genes, and 354 pseudogenes.

By comparing the human and mouse genomes, scientists have reconstructed the rise and fall of olfactory receptor genes. All of the genes, broken or functional, in both mice and men, originated in an ancestral set of 754 functional genes. Those genes made receptors in the nose of a small mammal that lived approximately 100 million years ago. In the lineage that produced today's mice, many of those original genes were duplicated, and most of those duplicates still work today. In the lineage leading to humans, on the other hand, many more of the ancestral genes became pseudogenes. We carry some new duplicates of the ancestral genes, but far fewer than mice do. Many pseudogenes have disappeared altogether from the human genome.

The way scientists today understand evolutionary loss is different in some important ways from the way Darwin thought of it. Darwin envisioned a Lamarckian process by which disuse caused a structure to atrophy and then gradually shrivel away over generations. But careful studies of human genes have helped make today's biologists more Darwinian about evolutionary loss than Darwin himself. As species evolve new habits, some of their genes become unnecessary. Mutations that disable them are not eliminated by natural selection, because animals with those mutations suffer no penalty. In the case of smell, our loss of genes occurred as our primate ancestors made a profound sensory shift. Instead of relying mainly on smell for finding food or picking up signals from other members of their species, they came to rely more on eyesight. As they became more visual, they probably no longer needed a big repertoire of olfactory receptors. A mutation that struck one of the genes did not affect their reproductive fitness, and so it was not eliminated by natural selection. For mice, evolution took a very different path. Smell has remained important for mice over tens of millions of years. Losing a gene could make a mouse less fit. New olfactory receptor genes made them more fit and were thus favored by natural selection. Thus mice and men ended up with their very different collections of olfactory receptor genes.

In Darwin's day, many people scoffed at the notion that humans descended with other primates from a common ancestor. Many still do today. Olfactory receptor pseudogenes are a powerful antidote to illusions of human uniqueness. Humans' olfactory receptor genes are much more similar to those of chimpanzees than to those of mice. By the time the ancestors of humans and chimpanzees split about six million years ago, many olfactory receptor genes had turned to pseudogenes. The same pseudogenes can be found in both humans and chimpanzees. It would be hard enough for natural theologians to explain why most of the genes for smelling in humans are broken. It would be impossible to explain why chimpanzees and humans would be broken in the same way.

CHAPTER I:
The Evidence of the Descent of Man from some Lower Form

HE WHO WISHES to decide whether man is the modified descendant of some pre-existing form, would probably first enquire whether man varies, however slightly, in bodily structure and in mental faculties; and if so, whether the variations are transmitted to his offspring in accordance with the laws which prevail with the lower animals. Again, are the variations the result, as far as our ignorance permits us to judge, of the same general causes, and are they governed by the same general laws, as in the case of other organisms; for instance, by correlation, the inherited effects of use and disuse, &c.? Is man subject to similar malconformations, the result of arrested development, of reduplication of parts, &c., and does he display in any of his anomalies reversion to some former and ancient type of structure? It might also naturally be enquired whether man, like so many other animals, has given rise to varieties and sub-races, differing but slightly from each other, or to races differing so much that they must be classed as doubtful species? How are such races distributed over the world; and how, when crossed, do they react on each other in the first and succeeding generations? And so with many other points.

The enquirer would next come to the important point, whether man tends to increase at so rapid a rate, as to lead to occasional severe struggles for existence; and consequently to beneficial variations, whether in body or mind, being preserved, and injurious ones eliminated. Do the races or species of men, whichever term may be applied, encroach on and replace one another, so that some finally become extinct? We shall see that all these questions, as indeed is obvious in respect to most of them, must be answered in the affirmative, in the same manner as with the lower animals. But the several considerations just referred to may be conveniently deferred for a time: and we will first see how far the bodily structure of man shows traces more or less plain, of his descent from some lower form. In succeeding chapters the mental powers of man, in comparison with those of the lower animals, will be considered.

The Bodily Structure of Man—It is notorious that man is constructed on the same general type or model as other mammals. All the bones in his skeleton can be compared with corresponding bones in a monkey, bat, or seal. So it is with his muscles, nerves, blood-vessels and internal viscera. The brain, the most important of all the organs, follows the same law, as shewn by Huxley and other anatomists. Bischoff,[1] who is a hostile witness, admits that every chief fissure and fold in the brain of man has its analogy in that of the orang; but he adds that at no period of development do their brains perfectly agree; nor could perfect agreement be expected, for otherwise their mental powers would have been the same. Vulpian[2] remarks: 'Les différences réelles qui existent entre l'encéphale de l'homme et celui des singes supérieurs, sont bien minimes. Il ne faut pas se faire d'illusions à cet égard. L'homme est bien plus près des

---

1. 'Grosshirnwindungen des Menschen', 1868, s. 96. The conclusions of this author, as well as those of Gratiolet and Aeby, concerning the brain, will be discussed by Prof. Huxley in the Appendix alluded to in the Preface to this edition.
2. 'Lec. sur la Phys,' 1866, p. 890, as quoted by M. Dally, 'L'Ordre des Primates et le Transformisme', 1868, p. 29.

singes anthropomorphes par les caractères anatomiques de son cerveau que ceux-ci ne le sont non-seulement des autres mammifères, mais même de certains quadrumanes, des guenons et des macaques.' ['The real differences that exist between the brain of man and that of the apes are very minimal. We should not have any illusions about this. In the anatomical characteristics of his brain, man is much closer to anthropoid apes than these are to other mammals, and not only to them but even to some quadrumanes, guenons and macaques.'] But it would be superfluous here to give further details on the correspondence between man and the higher mammals in the structure of the brain and all other parts of the body.

It may, however, be worth while to specify a few points, not directly or obviously connected with structure, by which this correspondence or relationship is well shewn.

Man is liable to receive from the lower animals, and to communicate to them, certain diseases, as hydrophobia, variola, the glanders, syphilis, cholera, herpes, &c.;[3] and this fact proves the close similarity[4] of their tissues and blood, both in minute structure and composition, far more plainly than does their comparison under the best microscope, or by the aid of the best chemical analysis. Monkeys are liable to many of the same non-contagious diseases as we are; thus Rengger,[5] who carefully observed for a long time the *Cebus Azarae* in its native land, found it liable to catarrh, with the usual symptoms, and which, when often re-

3. Dr W. Lauder Lindsay has treated this subject at some length in the 'Journal of Mental Science', July 1871; and in the 'Edinburgh Veterinary Review', July 1858.
4. A Reviewer has criticised ('British Quarterly Review', Oct. 1st, 1871, p. 472) what I have here said with much severity and contempt; but as I do not use the term identity, I cannot see that I am greatly in error. There appears to me a strong analogy between the same infection or contagion producing the same result, or one closely similar, in two distinct animals, and the testing of two distinct fluids by the same chemical reagent.
5. 'Naturgeschichte der Säugethiere von Paraguay', 1830, s. 50.

current, led to consumption. These monkeys suffered also from apoplexy, inflammation of the bowels, and cataract in the eye. The younger ones when shedding their milk-teeth often died from fever. Medicines produced the same effect on them as on us. Many kinds of monkeys have a strong taste for tea, coffee, and spirituous liquors: they will also, as I have myself seen, smoke tobacco with pleasure.[6] Brehm asserts that the natives of north-eastern Africa catch the wild baboons by exposing vessels with strong beer, by which they are made drunk. He has seen some of these animals, which he kept in confinement, in this state; and he gives a laughable account of their behaviour and strange grimaces. On the following morning they were very cross and dismal; they held their aching heads with both hands, and wore a most pitiable expression: when beer or wine was offered them, they turned away with disgust, but relished the juice of lemons.[7] An American monkey, an Ateles, after getting drunk on brandy, would never touch it again, and thus was wiser than many men. These trifling facts prove how similar the nerves of taste must be in monkeys and man, and how similarly their whole nervous system is affected.

Man is infested with internal parasites, sometimes causing fatal effects; and is plagued by external parasites, all of which belong to the same genera or families as those infesting other mammals, and in the case of scabies to the same species.[8] Man is subject, like other mammals, birds, and even insects,[9] to that

6. The same tastes are common to some animals much lower in the scale. Mr A. Nicols informs me that he kept in Queensland, in Australia, three individuals of the *Phaseolarctus cinereus*; and that, without having been taught in any way, they acquired a strong taste for rum, and for smoking tobacco.

7. Brehm, 'Thierleben', B. i. 1864, s. 75, 86. On the Ateles, s. 105. For other analogous statements, see s. 25, 107.

8. Dr W. Lauder Lindsay, 'Edinburgh Vet. Review', July 1858, p. 13.

9. With respect to insects see Dr Laycock, 'On a General Law of Vital Periodicity', 'British Association', 1842. Dr Macculloch, 'Silliman's North American Journal of Science', vol. xvii. p. 305, has seen a dog suffering from tertian ague. Hereafter I shall return to this subject.

mysterious law, which causes certain normal processes, such as gestation, as well as the maturation and duration of various diseases, to follow lunar periods. His wounds are repaired by the same process of healing; and the stumps left after the amputation of his limbs, especially during an early embryonic period, occasionally possess some power of regeneration, as in the lowest animals.[10]

The whole process of that most important function, the reproduction of the species, is strikingly the same in all mammals, from the first act of courtship by the male,[11] to the birth and nurturing of the young. Monkeys are born in almost as helpless a condition as our own infants; and in certain genera the young differ fully as much in appearance from the adults, as do our children from their full-grown parents.[12] It has been urged by some writers, as an important distinction, that with man the young ar-

---

10. I have given the evidence on this head in my 'Variation of Animals and Plants under Domestication', vol. ii. p. 15, and more could be added.

11. 'Mares e diversis generibus Quadrumanorum sine dubio dignoscunt feminas humanas a maribus. Primum, credo, odoratu, postea aspectu. Mr Youatt, quidiu in Hortis Zoologicis (Bestiariis) medicus animalium erat, vir in rebus observandis cautus et sagax, hoc mihi certissime probavit, et curatores ejusdem loci et alii e ministris confirmaverunt. Sir Andrew Smith et Brehm notabant idem in Cynocephalo. Illustrissimus Cuvier etiam narrat multa de hâc re, quâ ut opinor, nihil turpius potest indicari inter omnia hominibus et Quadrumanis communia. Narrat enim Cynocephalum quendam in furorem incidere aspectu feminarum aliquarum, sed nequaquam accenditanto furore ab omnibus. Semper eligebat juniores, et dignoscebat in turbâ, et advocabat voce gestúque.' ['Males from various species of mammals clearly distinguish the anthropomorphous females from the males. First, I believe, by smell, then by appearance. Mr Youatt, who served for a long time as veterinarian at a zoo, and was cautious and astute in his observations, gave me a very certain demonstration of this, which was corroborated by the curators and other attendants at the same place. Sir Andrew Smith and Brehm noted the same thing amongst baboons. The famous Cuvier also has much to say on this matter and, in my opinion, we can point to nothing more unseemly than this amongst all human and mammal communities. For what he says is that a certain baboon fell into a frenzy at the sight of particular females, but was in no way equally incensed by all of them. He always chose the younger ones, and picked them out from the crowd, calling out to them with voice and gesture.']

12. This remark is made with respect to Cynocephalus and the anthropomorphous apes by Geoffroy Saint-Hilaire and F. Cuvier, 'Hist. Nat. des Mammifères', tom. i. 1824.

rive at maturity at a much later age than with any other animal: but if we look to the races of mankind which inhabit tropical countries the difference is not great, for the orang is believed not to be adult till the age of from ten to fifteen years.[13] Man differs from woman in size, bodily strength, hairiness, &c., as well as in mind, in the same manner as do the two sexes of many mammals. So that the correspondence in general structure, in the minute structure of the tissues, in chemical composition and in constitution, between man and the higher animals, especially the anthropomorphous apes, is extremely close.

Embryonic Development—Man is developed from an ovule, about the 125th of an inch in diameter, which differs in no respect from the ovules of other animals. The embryo itself at a very early period can hardly be distinguished from that of other members of the vertebrate kingdom. At this period the arteries run in arch-like branches, as if to carry the blood to branchiae which are not present in the higher vertebrata, though the slits on the sides of the neck still remain (*f, g,* fig. 1), marking their former position. At a somewhat later period, when the extremities are developed, 'the feet of lizards and mammals', as the illustrious Von Baer remarks, 'the wings and feet of birds, no less than the hands and feet of man, all arise from the same fundamental form'. It is, says Prof. Huxley,[14] 'quite in the later stages of development that the young human being presents marked differences from the young ape, while the latter departs as much from the dog in its developments, as the man does. Startling as this last assertion may appear to be, it is demonstrably true.'

As some of my readers may never have seen a drawing of an embryo, I have given one of man and another of a dog, at about

---

13. Huxley, 'Man's Place in Nature', 1863, p. 34.
14. 'Man's Place in Nature', 1863, p. 67.

the same early stage of development, carefully copied from two works of undoubted accuracy.[15]

After the foregoing statements made by such high authorities, it would be superfluous on my part to give a number of borrowed details, shewing that the embryo of man closely resembles that of other mammals. It may, however, be added, that the human embryo likewise resembles certain low forms when adult in various points of structure. For instance, the heart at first exists as a simple pulsating vessel; the excreta are voided through a cloacal passage; and the os coccyx projects like a true tail, 'extending considerably beyond the rudimentary legs'.[16] In the embryos of all air-breathing vertebrates, certain glands, called the corpora Wolffiana, correspond with, and act like the kidneys of mature fishes.[17] Even at a later embryonic period, some striking resemblances between man and the lower animals may be observed. Bischoff says that the convolutions of the brain in a human foetus at the end of the seventh month reach about the same stage of development as in a baboon when adult.[18] The great toe, as Prof. Owen remarks,[19] 'which forms the fulcrum when standing or walking, is perhaps the most characteristic peculiarity in the human structure', but in an embryo, about an inch in length, Prof. Wyman[20] found 'that the great toe was shorter than the others; and, instead of being parallel to them, projected

15. The human embryo (upper fig.) is from Ecker, 'Icones Phys.', 1851–1859, tab. xxx. fig. 2. This embryo was ten lines in length, so that the drawing is much magnified. The embryo of the dog is from Bischoff, 'Entwicklungsgeschichte des Hunde-Eies', 1845, tab. xi. fig. 42 B. This drawing is five times magnified, the embryo being twenty-five days old. The internal viscera have been omitted, and the uterine appendages in both drawings removed. I was directed to these figures by Prof. Huxley, from whose work, 'Man's Place in Nature', the idea of giving them was taken. Häckel has also given analogous drawings in his 'Schöpfungsgeschichte'.

16. Prof. Wyman in 'Proc. of American Acad. of Sciences', vol. iv. 1860, p. 17.

17. Owen, 'Anatomy of Vertebrates', vol. i. p. 533.

18. 'Die Grosshirnwindungen des Menschen', 1868, s. 95.

19. 'Anatomy of Vertebrates', vol. ii. p. 553.

20. 'Proc. Soc. Nat. Hist.', Boston, 1863, vol. ix. p. 185.

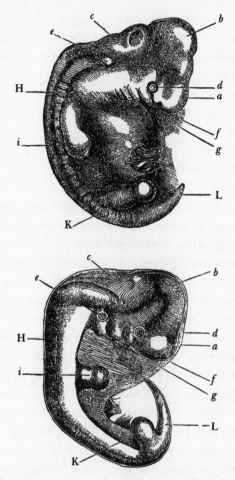

Fig. 1. Upper figure human embryo, from Ecker. Lower figure that of a dog, from Bischoff.

a. Fore-brain, cerebral hemispheres, &c.

b. Mid-brain, corpora quadrigemina.

c. Hind-brain, cerebellum, medulla oblongata.

d. Eye.

e. Ear.

f. First visceral arch.

g. Second visceral arch.

H. Vertebral columns and muscles in process of development.

i. Anterior ⎱ extremities.
K. Posterior ⎰

L. Tail or os coccyx.

at an angle from the side of the foot, thus corresponding with the permanent condition of this part in the quadrumana'. I will conclude with a quotation from Huxley,[21] who after asking, does man originate in a different way from a dog, bird, frog or fish? says, 'the reply is not doubtful for a moment; without question, the mode of origin, and the early stages of the development of man, are identical with those of the animals immediately below him in the scale: without a doubt in these respects, he is far nearer to apes than the apes are to the dog'.

Rudiments—This subject, though not intrinsically more important than the two last, will for several reasons be treated here more fully.[22] Not one of the higher animals can be named which does not bear some part in a rudimentary condition; and man forms no exception to the rule. Rudimentary organs must be distinguished from those that are nascent; though in some cases the distinction is not easy. The former are either absolutely useless, such as the mammae of male quadrupeds, or the incisor teeth of ruminants which never cut through the gums; or they are of such slight service to their present possessors, that we can hardly suppose that they were developed under the conditions which now exist. Organs in this latter state are not strictly rudimentary, but they are tending in this direction. Nascent organs, on the other hand, though not fully developed, are of high service to their possessors, and are capable of further development. Rudimentary organs are eminently variable; and this is partly intelligible, as they are useless, or nearly useless, and consequently

21. 'Man's Place in Nature', p. 65.
22. I had written a rough copy of this chapter before reading a valuable paper, 'Caratteri rudimentali in ordine all' origine del uomo' ('Annuario della Soc. d. Nat', Modena, 1867, p. 81), by G. Canestrini, to which paper I am considerably indebted. Häckel has given admirable discussions on this whole subject, under the title of Dysteleology, in his 'Generelle Morphologie' and 'Schöpfungsgeschichte'.

are no longer subjected to natural selection. They often become wholly suppressed. When this occurs, they are nevertheless liable to occasional reappearance through reversion—a circumstance well worthy of attention.

The chief agents in causing organs to become rudimentary seem to have been disuse at that period of life when the organ is chiefly used (and this is generally during maturity), and also inheritance at a corresponding period of life. The term 'disuse' does not relate merely to the lessened action of muscles, but includes a diminished flow of blood to a part or organ, from being subjected to fewer alternations of pressure, or from becoming in any way less habitually active. Rudiments, however, may occur in one sex of those parts which are normally present in the other sex; and such rudiments, as we shall hereafter see, have often originated in a way distinct from those here referred to. In some cases, organs have been reduced by means of natural selection, from having become injurious to the species under changed habits of life. The process of reduction is probably often aided through the two principles of compensation and economy of growth; but the later stages of reduction, after disuse has done all that can fairly be attributed to it, and when the saving to be effected by the economy of growth would be very small,[23] are difficult to understand. The final and complete suppression of a part, already useless and much reduced in size, in which case neither compensation nor economy can come into play, is perhaps intelligible by the aid of the hypothesis of pangenesis. But as the whole subject of rudimentary organs has been discussed and illustrated in my former works,[24] I need here say no more on this head.

Rudiments of various muscles have been observed in many

---

23. Some good criticisms on this subject have been given by Messrs. Murie and Mivart, in 'Transact. Zoolog. Soc.', 1869, vol. vii. p. 92.
24. 'Variation of Animals and Plants under Domestication', vol. ii. pp. 317 and 397. See also 'Origin of Species', 5th edit. p. 535.

parts of the human body;[25] and not a few muscles, which are regularly present in some of the lower animals can occasionally be detected in man in a greatly reduced condition. Every one must have noticed the power which many animals, especially horses, possess of moving or twitching their skin; and this is effected by the *panniculus carnosus*. Remnants of this muscle in an efficient state are found in various parts of our bodies; for instance, the muscle on the forehead, by which the eyebrows are raised. The *platysma myoides*, which is well developed on the neck, belongs to this system. Prof. Turner, of Edinburgh, has occasionally detected, as he informs me, muscular fasciculi in five different situations, namely in the axillae, near the scapulae, &c., all of which must be referred to the system of the *panniculus*. He has also shewn[26] that the *musculus sternalis* or *sternalis brutorum*, which is not an extension of the *rectus abdominalis*, but is closely allied to the *panniculus*, occurred in the proportion of about three per cent. in upwards of 600 bodies: he adds, that this muscle affords 'an excellent illustration of the statement that occasional and rudimentary structures are especially liable to variation in arrangement'.

Some few persons have the power of contracting the superficial muscles on their scalps; and these muscles are in a variable and partially rudimentary condition. M. A. de Candolle has communicated to me a curious instance of the long-continued persistence or inheritance of this power, as well as of its unusual development. He knows a family, in which one member, the present head of the family, could, when a youth, pitch several heavy books from his head by the movement of the scalp alone; and he won wagers by performing this feat. His father, uncle,

25. For instance M. Richard ('Annales des Sciences Nat.', 3rd series, Zoolog. 1852, tom. xviii. p. 13) describes and figures rudiments of what he calls the 'muscle pédieux de la main', which he says is sometimes 'infiniment petit'. Another muscle, called 'le tibial postérieur', is generally quite absent in the hand, but appears from time to time in a more or less rudimentary condition.

26. Prof. W. Turner, 'Proc. Royal See. Edinburgh', 1866–67, p. 65.

grandfather, and his three children possess the same power to the same unusual degree. This family became divided eight generations ago into two branches; so that the head of the above-mentioned branch is cousin in the seventh degree to the head of the other branch. This distant cousin resides in another part of France; and on being asked whether he possessed the same faculty, immediately exhibited his power. This case offers a good illustration how persistent may be the transmission of an absolutely useless faculty, probably derived from our remote semi-human progenitors; since many monkeys have, and frequently use the power, of largely moving their scalps up and down.[27]

The extrinsic muscles which serve to move the external ear, and the intrinsic muscles which move the different parts, are in a rudimentary condition in man, and they all belong to the system of the *panniculus*; they are also variable in development, or at least in function. I have seen one man who could draw the whole ear forwards; other men can draw it upwards; another who could draw it backwards;[28] and from what one of these persons told me, it is probable that most of us, by often touching our ears, and thus directing our attention towards them, could recover some power of movement by repeated trials. The power of erecting and directing the shell of the ears to the various points of the compass, is no doubt of the highest service to many animals, as they thus perceive the direction of danger; but I have never heard, on sufficient evidence, of a man who possessed this power, the one which might be of use to him. The whole external shell may be considered a rudiment, together with the various folds and prominences (helix and anti-helix, tragus and anti-tragus, &c.) which in the lower animals strengthen and support the ear when erect, without adding much to its weight. Some authors,

27. See my 'Expression of the Emotions in Man and Animals', 1872, p. 144.
28. Canestrini quotes Hyrtl. ('Annuario della Soc. dei Naturalisti', Modena, 1867, p. 97) to the same effect.

however, suppose that the cartilage of the shell serves to transmit vibrations to the acoustic nerve; but Mr Toynbee,[29] after collecting all the known evidence on this head, concludes that the external shell is of no distinct use. The ears of the chimpanzee and orang are curiously like those of man, and the proper muscles are likewise but very slightly developed.[30] I am also assured by the keepers in the Zoological Gardens that these animals never move or erect their ears; so that they are in an equally rudimentary condition with those of man, as far as function is concerned. Why these animals, as well as the progenitors of man, should have lost the power of erecting their ears, we cannot say. It may be, though I am not satisfied with this view, that owing to their arboreal habits and great strength they were but little exposed to danger, and so during a lengthened period moved their ears but little, and thus gradually lost the power of moving them. This would be a parallel case with that of those large and heavy birds, which, from inhabiting oceanic islands, have not been exposed to the attacks of beasts of prey, and have consequently lost the power of using their wings for flight. The inability to move the ears in man and several apes is, however, partly compensated by the freedom with which they can move the head in a horizontal plane, so as to catch sounds from all directions. It has been asserted that the ear of man alone possesses a lobule; but 'a rudiment of it is found in the gorilla',[31] and, as I hear from Prof. Preyer, it is not rarely absent in the negro.

The celebrated sculptor, Mr Woolner, informs me of one little peculiarity in the external ear, which he has often observed both in men and women, and of which he perceived

29. 'The Diseases of the Ear', by J. Toynbee, FRS, 1860, p. 12. A distinguished physiologist, Prof. Preyer, informs me that he had lately been experimenting on the function of the shell of the ear, and has come to nearly the same conclusion as that given here.
30. Prof. A. Macalister, 'Annals and Mag. of Nat. History', vol. vii. 1871, p. 342.
31. Mr St George Mivart, 'Elementary Anatomy', 1873, p. 396.

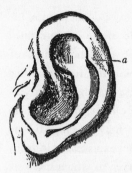

Fig. 2. Human Ear, mod-
elled and drawn by Mr
Woolner.

a. The projecting point.

the full significance. His attention was first called to the subject whilst at work on his figure of Puck, to which he had given pointed ears. He was thus led to examine the ears of various monkeys, and subsequently more carefully those of man. The peculiarity consists in a little blunt point, projecting from the inwardly folded margin, or helix. When present, it is developed at birth, and, according to Prof. Ludwig Meyer, more frequently in man than in woman. Mr Woolner made an exact model of one such case, and sent me the accompanying drawing. (Fig. 2.) These points not only project inwards towards the centre of the ear, but often a little outwards from its plane, so as to be visible when the head is viewed from directly in front or behind. They are variable in size, and somewhat in position, standing either a little higher or lower; and they sometimes occur on one ear and not on the other. They are not confined to mankind, for I observed a case in one of the spider-monkeys (*Ateles beelzebuth*) in our Zoological Gardens; and Dr E. Ray Lankester informs me of another case in a chimpanzee in the gardens at Hamburg. The helix obviously consists of the extreme margin of the ear folded inwards; and this folding appears to be in some manner connected with the whole external ear being permanently pressed backwards. In many monkeys, which do not stand high in the order, as baboons and some species of macacus,[32] the upper portion of the ear is slightly pointed, and the margin is not at all folded inwards; but if the margin were to be thus folded, a slight

32. See also some remarks, and the drawings of the ears of the Lemuroidea, in Messrs Murie and Mivart's excellent paper in 'Transact. Zoolog. Soc.', vol. vii. 1869, pp. 6 and 90.

point would necessarily project inwards towards the centre, and probably a little outwards from the plane of the ear; and this I believe to be their origin in many cases. On the other hand, Prof. L. Meyer, in an able paper recently published,[33] maintains that the whole case is one of mere variability; and that the projections are not real ones, but are due to the internal cartilage on each side of the points not having been fully developed. I am quite ready to admit that this is the correct explanation in many instances, as in those figured by Prof. Meyer, in which there are several minute points, or the whole margin is sinuous. I have myself seen, through the kindness of Dr L. Down, the ear of a microcephalous idiot, on which there is a projection on the outside of the helix, and not on the inward folded edge, so that this point can have no relation to a former apex of the ear. Nevertheless in some cases, my original view, that the points are vestiges of the tips of formerly erect and pointed ears, still seems to me probable. I think so from the frequency of their occurrence, and from the general correspondence in position with that of the tip of a pointed ear. In one case, of which a photograph has been sent me, the projection is so large, that supposing, in accordance with Prof. Meyer's view, the ear to be made perfect by the equal development of the cartilage throughout the whole extent of the margin, it would have covered fully one-third of the whole ear. Two cases have been communicated to me, one in North America, and the other in England, in which the upper margin is not at all folded inwards, but is pointed, so that it closely resembles the pointed ear of an ordinary quadruped in outline. In one of these cases, which was that of a young child, the father compared the ear with the drawing which I have given[34] of the ear of a monkey, the *Cynopithecus niger*, and says that their out-

33. Ueber das Darwin'sche Spitzohr, Archiv für Path. Anat. und Phys. 1871, p. 485.
34. 'The Expression of the Emotions', p. 136.

lines are closely similar. If, in these two cases, the margin had been folded inwards in the normal manner, an inward projection must have been formed. I may add that in two other cases the outline still remains somewhat pointed, although the margin of the upper part of the ear is normally folded inwards—in one of

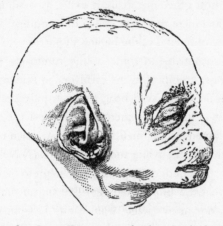

Fig. 3. Foetus of an Orang. Exact copy of a photograph, shewing the form of the ear at this early age.

them, however, very narrowly. The following woodcut (Fig. 3.) is an accurate copy of a photograph of the foetus of an orang (kindly sent me by Dr Nitsche), in which it may be seen how different the pointed outline of the ear is at this period from its adult condition, when it bears a close general resemblance to that of man. It is evident that the folding over of the tip of such an ear, unless it changed greatly during its further development, would give rise to a point projecting inwards. On the whole, it still seems to me probable that the points in question are in some cases, both in man and apes, vestiges of a former condition.

The nictitating membrane, or third eyelid, with its accessory muscles and other structures, is especially well developed in birds, and is of much functional importance to them, as it can be rapidly drawn across the whole eye-ball. It is found in some reptiles and amphibians, and in certain fishes, as in sharks. It is fairly well developed in the two lower divisions of the mammalian series, namely, in the monotremata and marsupials, and in some few of the higher mammals, as in the walrus. But in man, the quadrumana, and most other mammals, it exists, as is admitted by all anatomists, as a mere rudiment, called the semilunar fold.[35]

The sense of smell is of the highest importance to the greater number of mammals—to some, as the ruminants, in warning them of danger; to others, as the carnivora, in finding their prey; to others, again, as the wild boar, for both purposes combined. But the sense of smell is of extremely slight service, if any, even to the dark coloured races of men, in whom it is much more highly developed than in the white and civilised races.[36] Nevertheless it does not warn them of danger, nor guide them to their food; nor does it prevent the Esquimaux from sleeping in the most fetid atmosphere, nor many savages from eating half-putrid meat. In Europeans the power differs greatly in different individuals, as I am assured by an eminent naturalist who possesses this sense highly developed, and who has attended to the subject. Those who believe in the principle of gradual evolution, will not readily admit that the sense of smell in its present state was originally acquired

35. Müller's 'Elements of Physiology', Eng. translat., 1842, vol. ii. p. 1117. Owen, 'Anatomy of Vertebrates', vol. iii. p. 260; ibid on the Walrus, 'Proc. Zoolog. Soc.', November 8th, 1854. See also R. Knox, 'Great Artists and Anatomists', p. 106. This rudiment apparently is somewhat larger in Negroes and Australians than in Europeans, see Carl Vogt, 'Lectures on Man', Eng. translat. p. 129.

36. The account given by Humboldt of the power of smell possessed by the natives of South America is well known, and has been confirmed by others. M. Houzeau ('Etudes sur les Facultés Mentales,' &c., tom. i. 1872, p. 91) asserts that he repeatedly made experiments, and proved that Negroes and Indians could recognise persons in the dark by their odour. Dr W. Ogle has made some curious observations on the connection between the power of smell and the colouring matter of the mucous membrane of the olfactory region, as well as of the skin of the body. I have, therefore, spoken in the text of the dark-coloured races having a finer sense of smell than the white races. See his paper, 'Medico-Chirurgical Transactions', London, vol. liii. 1870, p. 276.

by man, as he now exists. He inherits the power in an enfeebled and so far rudimentary condition, from some early progenitor, to whom it was highly serviceable, and by whom it was continually used. In those animals which have this sense highly developed, such as dogs and horses, the recollection of persons and of places is strongly associated with their odour; and we can thus perhaps understand how it is, as Dr Maudsley has truly remarked,[37] that the sense of smell in man 'is singularly effective in recalling vividly the ideas and images of forgotten scenes and places'.

Man differs conspicuously from all the other Primates in being almost naked. But a few short straggling hairs are found over the greater part of the body in the man, and fine down on that of the woman. The different races differ much in hairiness; and in the individuals of the same race the hairs are highly variable, not only in abundance, but likewise in position: thus in some Europeans the shoulders are quite naked, whilst in others they bear thick tufts of hair.[38] There can be little doubt that the hairs thus scattered over the body are the rudiments of the uniform hairy coat of the lower animals. This view is rendered all the more probable, as it is known that fine, short, and pale-coloured hairs on the limbs and other parts of the body, occasionally become developed into 'thickset, long, and rather coarse dark hairs', when abnormally nourished near old-standing inflamed surfaces.[39]

I am informed by Sir James Paget that often several members of a family have a few hairs in their eyebrows much longer than the others; so that even this slight peculiarity seems to be inherited. These hairs, too, seem to have their representatives; for in the chimpanzee, and in certain species of Macacus, there are scattered hairs of considerable length rising from the naked skin

37. 'The Physiology and Pathology of Mind', 2nd edit. 1868, p. 134.
38. Eschricht, Ueber die Richtung der Haare am menschlichen Körper, 'Müller's Archiv für Anat. und Phys.', 1837, s. 47. I shall often have to refer to this very curious paper.
39. Paget, 'Lectures on Surgical Pathology', 1853, vol. i. p. 71.

above the eyes, and corresponding to our eyebrows; similar long hairs project from the hairy covering of the superciliary ridges in some baboons.

The fine wool-like hair, or so-called lanugo, with which the human foetus during the sixth month is thickly covered, offers a more curious case. It is first developed, during the fifth month, on the eyebrows, and face, and especially round the mouth, where it is much longer than that on the head. A moustache of this kind was observed by Eschricht[40] on a female foetus; but this is not so surprising a circumstance as it may at first appear, for the two sexes generally resemble each other in all external characters during an early period of growth. The direction and arrangement of the hairs on all parts of the foetal body are the same as in the adult, but are subject to much variability. The whole surface, including even the forehead and ears, is thus thickly clothed; but it is a significant fact that the palms of the hands and the soles of the feet are quite naked, like the inferior surfaces of all four extremities in most of the lower animals. As this can hardly be an accidental coincidence, the woolly covering of the foetus probably represents the first permanent coat of hair in those mammals which are born hairy. Three or four cases have been recorded of persons born with their whole bodies and faces thickly covered with fine long hairs; and this strange condition is strongly inherited, and is correlated with an abnormal condition of the teeth.[41] Prof. Alex. Brandt informs me that he has compared the hair from the face of a man thus characterised, aged thirty-five, with the lanugo of a foetus, and finds it quite similar in texture; therefore, as he remarks, the case may be attributed to an arrest of development in the hair, together with its continued growth.

40. Eschricht, ibid. s. 40, 47.
41. See my 'Variation of Animals and Plants under Domestication', vol. ii. p. 327. Prof. Alex. Brandt has recently sent me an additional case of a father and son, born in Russia, with these peculiarities. I have received drawings of both from Paris.

Many delicate children, as I have been assured by a surgeon to a hospital for children, have their backs covered by rather long silky hairs; and such cases probably come under the same head.

It appears as if the posterior molar or wisdom-teeth were tending to become rudimentary in the more civilised races of man. These teeth are rather smaller than the other molars, as is likewise the case with the corresponding teeth in the chimpanzee and orang; and they have only two separate fangs. They do not cut through the gums till about the seventeenth year, and I have been assured that they are much more liable to decay, and are earlier lost than the other teeth; but this is denied by some eminent dentists. They are also much more liable to vary, both in structure and in the period of their development, than the other teeth.[42] In the Melanian races, on the other hand, the wisdom-teeth are usually furnished with three separate fangs, and are generally sound; they also differ from the other molars in size, less than in the Caucasian races.[43] Prof. Schaaffhausen accounts for this difference between the races by 'the posterior dental portion of the jaw being always shortened' in those that are civilised,[44] and this shortening may, I presume, be attributed to civilised men habitually feeding on soft, cooked food, and thus using their jaws less. I am informed by Mr Brace that it is becoming quite a common practice in the United States to remove some of the molar teeth of children, as the jaw does not grow large enough for the perfect development of the normal number.[45]

With respect to the alimentary canal, I have met with an

---

42. Dr Webb, 'Teeth in Man and the Anthropoid Apes', as quoted by Dr C. Carter Blake in 'Anthropological Review', July 1867, p. 299.
43. Owen, 'Anatomy of Vertebrates', vol. iii. pp. 320, 321, and 325.
44. 'On the Primitive Form of the Skull', Eng. translat. in 'Anthropological Review', Oct. 1868, p. 426.
45. Prof. Montegazza writes to me from Florence, that he has lately been studying the last molar teeth in the different races of man, and has come to the same conclusion as that given in my text, viz., that in the higher or civilised races they are on the road towards atrophy or elimination.

account of only a single rudiment, namely the vermiform appendage of the caecum. The caecum is a branch or diverticulum of the intestine, ending in a cul-de-sac, and is extremely long in many of the lower vegetable-feeding mammals. In the marsupial koala it is actually more than thrice as long as the whole body.[46] It is sometimes produced into a long gradually-tapering point, and is sometimes constricted in parts. It appears as if, in consequence of changed diet or habits, the caecum had become much shortened in various animals, the vermiform appendage being left as a rudiment of the shortened part. That this appendage is a rudiment, we may infer from its small size, and from the evidence which Prof. Canestrini[47] has collected of its variability in man. It is occasionally quite absent, or again is largely developed. The passage is sometimes completely closed for half or two-thirds of its length, with the terminal part consisting of a flattened solid expansion. In the orang this appendage is long and convoluted: in man it arises from the end of the short caecum, and is commonly from four to five inches in length, being only about the third of an inch in diameter. Not only is it useless, but it is sometimes the cause of death, of which fact I have lately heard two instances: this is due to small hard bodies, such as seeds, entering the passage, and causing inflammation.[48]

In some of the lower Quadrumana, in the Lemuridae and Carnivora, as well as in many marsupials, there is a passage near the lower end of the humerus, called the supra-condyloid foramen, through which the great nerve of the fore limb and often the great artery pass. Now in the humerus of man, there is generally a trace of this passage, which is sometimes fairly well devel-

---

46. Owen, 'Anatomy of Vertebrates', vol. iii. pp. 416, 434, 441.
47. 'Annuario della Soc. d. Nat.', Modena, 1867, p. 94.
48. M. C. Martins ('De l'Unité Organique', in 'Revue des Deux Mondes', June 15, 1862, p. 16), and Häckel ('Generelle Morphologie', B, ii. s. 278), have both remarked on the singular fact of this rudiment sometimes causing death.

oped, being formed by a depending hook-like process of bone, completed by a band of ligament. Dr Struthers,[49] who has closely attended to the subject, has now shewn that this peculiarity is sometimes inherited, as it has occurred in a father, and in no less than four out of his seven children. When present, the great nerve invariably passes through it; and this clearly indicates that it is the homologue and rudiment of the supra-condyloid foramen of the lower animals. Prof. Turner estimates, as he informs me, that it occurs in about one per cent of recent skeletons. But if the occasional development of this structure in man is, as seems probable, due to reversion, it is a return to a very ancient state of things, because in the higher Quadrumana it is absent.

There is another foramen or perforation in the humerus, occasionally present in man, which may be called the intercondyloid. This occurs, but not constantly, in various anthropoid and other apes,[50] and likewise in many of the lower animals. It is remarkable that this perforation seems to have been present in man much more frequently during ancient times than recently. Mr Busk[51] has collected the following evidence on this head: Prof. Broca 'noticed the perforation in four and a half per cent. of the arm-bones collected in the "Cimetière du Sud", at Paris; and in the Grotto of Orrony, the contents of which are referred to the Bronze period, as many as eight humeri out of thirty-two were perforated; but this extraordinary proportion, he thinks,

---

49. With respect to inheritance, see Dr Struthers in the 'Lancet', Feb. 15, 1873, and another important paper, ibid., Jan. 24, 1863, p. 83. Dr Knox, as I am informed, was the first anatomist who drew attention to this peculiar structure in man; see his 'Great Artists and Anatomists', p. 63. See also an important memoir on this process by Dr Gruber, in the 'Bulletin de l'Acad. Imp. de St. Pétersbourg', tom. xii. 1867, p. 448.

50. Mr St George Mivart, 'Transact. Phil. Soc.', 1867, p. 310.

51. 'On the Caves of Gibraltar', 'Transact. Internat. Congress of Prehist. Arch.', Third Session, 1869, p. 159. Prof. Wyman has lately shewn (Fourth Annual Report, Peabody Museum, 1871, p. 20), that this perforation is present in thirty-one per cent of some human remains from ancient mounds in the Western United States, and in Florida. It frequently occurs in the negro.

might be due to the cavern having been a sort of "family vault".
Again, M. Dupont found thirty per cent of perforated bones in
the caves of the Valley of the Lesse, belonging to the Reindeer
period; whilst M. Leguay, in a sort of *dolmen* at Argenteuil, observed twenty-five per cent to be perforated; and M. Pruner-Bey
found twenty-six per cent in the same condition in bones from
Vauréal. Nor should it be left unnoticed that M. Pruner-Bey
states that this condition is common in Guanche skeletons.' It
is an interesting fact that ancient races, in this and several other
cases, more frequently present structures which resemble those
of the lower animals than do the modern. One chief cause seems
to be that the ancient races stand somewhat nearer in the long
line of descent to their remote animal-like progenitors.

In man, the os coccyx, together with certain other vertebrae
hereafter to be described, though functionless as a tail, plainly
represent this part in other vertebrate animals. At an early embryonic period it is free, and projects beyond the lower extremities; as may be seen in the drawing (Fig. 1.) of a human embryo.
Even after birth it has been known, in certain rare and anomalous
cases,[52] to form a small external rudiment of a tail. The os coccyx
is short, usually including only four vertebrae, all anchylosed together: and these are in a rudimentary condition, for they consist,
with the exception of the basal one, of the centrum alone.[53] They
are furnished with some small muscles; one of which, as I am
informed by Prof. Turner, has been expressly described by Theile
as a rudimentary repetition of the extensor of the tail, a muscle
which is so largely developed in many mammals.

---

52. Quatrefages has lately collected the evidence on this subject. 'Revue des Cours
    Scientifiques', 1867–1868, p. 625. In 1840 Fleischmann exhibited a human foe-
    tus bearing a free tail, which, as is not always the case, included vertebral bod-
    ies; and this tail was critically examined by the many anatomists present at the
    meeting of natualists at Erlangen (see Marshall in Niederländischen Archiv für
    Zoologie, December 1871).
53. Owen, 'On the Nature of Limbs', 1849, p. 114.

The spinal cord in man extends only as far downwards as the last dorsal or first lumbar vertebra; but a thread-like structure (the *filum terminale*) runs down the axis of the sacral part of the spinal canal, and even along the back of the coccygeal bones. The upper part of this filament, as Prof. Turner informs me, is undoubtedly homologous with the spinal cord; but the lower part apparently consists merely of the *pia mater*, or vascular investing membrane. Even in this case the os coccyx may be said to possess a vestige of so important a structure as the spinal cord, though no longer enclosed within a bony canal. The following fact, for which I am also indebted to Prof. Turner, shews how closely the os coccyx corresponds with the true tail in the lower animals: Luschka has recently discovered at the extremity of the coccygeal bones a very peculiar convoluted body, which is continuous with the middle sacral artery; and this discovery led Krause and Meyer to examine the tail of a monkey (Macacus), and of a cat, in both of which they found a similarly convoluted body, though not at the extremity.

The reproductive system offers various rudimentary structures; but these differ in one important respect from the foregoing cases. Here we are not concerned with the vestige of a part which does not belong to the species in an efficient state, but with a part efficient in the one sex, and represented in the other by a mere rudiment. Nevertheless, the occurrence of such rudiments is as difficult to explain, on the belief of the separate creation of each species, as in the foregoing cases. Hereafter I shall have to recur to these rudiments, and shall shew that their presence generally depends merely on inheritance, that is, on parts acquired by one sex having been partially transmitted to the other. I will in this place only give some instances of such rudiments. It is well known that in the males of all mammals, including man, rudimentary mammae exist. These in several instances have become well developed, and have yielded a copious supply of milk. Their essential identity in the two sexes is likewise shewn by their occasional sympathetic enlargement in both during an attack of the measles. The *vesicula prostatica*, which has been observed in many male mammals, is now universally acknowledged to be the homologue of the female uterus, together with the connected passage. It is impossible to read Leuckart's able description of this organ, and his reasoning, without admitting the justness of his conclusion. This is especially clear in the case of those mammals in which the true female uterus bifurcates, for in the males of these the vesicula likewise bifurcates.[54] Some other rudimentary structures belonging to the reproductive system might have been here adduced.[55]

The bearing of the three great classes of facts now given is unmistakeable. But it would be superfluous fully to recapitulate the line of argument given in detail in my 'Origin of Species'. The homological construction of the whole frame in the members of the same class is intelligible, if we admit their descent from a common progenitor, together with their subsequent adaptation to diversified conditions. On any other view, the similarity of pattern between the hand of a man or monkey, the foot of a horse, the flipper of a seal, the wing of a bat, &c., is utterly inexplicable.[56] It is no scientific explanation to assert that they

54. Leuckart, in Todd's 'Cyclop. of Anat.', 1849–52, vol. iv. p. 1415. In man this organ is only from three to six lines in length, but, like so many other rudimentary parts, it is variable in development as well as in other characters.

55. See, on this subject, Owen, 'Anatomy of Vertebrates', vol. iii. pp. 675, 676, 706.

56. Prof. Bianconi, in a recently published work, illustrated by admirable engravings ('La Théorie Darwinienne et la création dite indépendante', 1874), endeavours to show that homological structures, in the above and other cases, can be fully explained on mechanical principles, in accordance with their uses. No one has shewn so well, how admirably such structures are adapted for their final purpose; and this adaptation can, as I believe, be explained through natural selection. In considering the wing of a bat, he brings forward (p. 218) what appears to me (to use Auguste Comte's words) a mere metaphysical principle, namely, the preservation 'in its integrity of the mammalian nature of the animal'. In only a few cases does he discuss rudiments, and then only those parts which are partially rudimentary, such as the

have all been formed on the same ideal plan. With respect to development, we can clearly understand, on the principle of variations supervening at a rather late embryonic period, and being inherited at a corresponding period, how it is that the embryos of wonderfully different forms should still retain, more or less perfectly, the structure of their common progenitor. No other explanation has ever been given of the marvellous fact that the embryos of a man, dog, seal, bat, reptile, &c., can at first hardly be distinguished from each other. In order to understand the existence of rudimentary organs, we have only to suppose that a former progenitor possessed the parts in question in a perfect state, and that under changed habits of life they became greatly reduced, either from simple disuse, or through the natural selection of those individuals which were least encumbered with a superfluous part, aided by the other means previously indicated.

Thus we can understand how it has come to pass that man and all other vertebrate animals have been constructed on the same general model, why they pass through the same early stages of development, and why they retain certain rudiments in common. Consequently we ought frankly to admit their community of descent; to take any other view, is to admit that our own structure, and that of all the animals around us, is a mere snare laid to entrap our judgment. This conclusion is greatly strengthened, if we look to the members of the whole animal series, and consider the evidence derived from their affinities or classification, their geographical distribution and geological succession. It is only our natural prejudice, and that arrogance which made our fore-

---

little hoofs of the pig and ox, which do not touch the ground; these he shews clearly to be of service to the animal. It is unfortunate that he did not consider such cases as the minute teeth, which never cut through the jaw in the ox, or the mammae of male quadrupeds, or the wings of certain beetles, existing under the soldered wing-covers, or the vestiges of the pistil and stamens in various flowers, and many other such cases. Although I greatly admire Prof. Bianconi's work, yet the belief now held by most naturalists seems to me left unshaken, that homological structures are inexplicable on the principle of mere adaptation.

fathers declare that they were descended from demi-gods, which leads us to demur to this conclusion. But the time will before long come, when it will be thought wonderful that naturalists, who were well acquainted with the comparative structure and development of man, and other mammals, should have believed that each was the work of a separate act of creation.

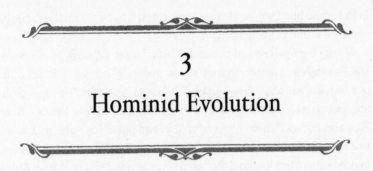

3

Hominid Evolution

Having pointed out some of the imperfections and peculiarities of our anatomy, Darwin moves to the heart of his argument: Humans evolved from an apelike ancestor, primarily by means of natural selection. Darwin may have gotten some of the details wrong, but those shortcomings do not take away from an astonishingly visionary chapter.

Darwin maps out a path by which apes that walked on all fours might have evolved into the human form. Here Darwin has in mind Wallace's arguments against natural selection in humans. Darwin did not see any insuperable barrier. A species of ape might have shifted from life in the trees to a life on the ground, and natural selection would have triggered changes throughout its entire body. The shift would also have opened the way to the use of tools and an increasing level of intelligence.

Darwin does not consider fossil evidence as he sketches out this hypothesis. When he was writing *The Descent of Man*, scientists were only beginning to discover the prehistoric record of our lineage. Spears and skeletons were being unearthed. But these prehistoric humans seemed to be no different in anatomy from living people. If Darwin was right, fossils of more apelike ancestors must be waiting to be found. In 1864 scientists uncovered fragments from a humanlike skeleton in the Neander Valley in Germany. Neanderthal man, as the skeleton came to be known, was hard to fathom. It had a massive browridge and stout limb bones. But Thomas Huxley considered it to be within the range of variation found in living humans.

Today the fossil record of humans is far larger, although by necessity still incomplete. Neanderthals, it's now generally agreed, split off from our own lineage about five hundred thousand years ago. The split probably occurred in Africa, after which Neanderthals spread into the Near East and Europe. There they endured in and out of Ice Ages until about twenty-eight thousand years ago. Our own species, *Homo sapiens*, evolved in Africa about two hundred thousand years ago and emerged from Africa at some point around one hundred thousand years ago, arriving in Europe forty thousand years ago. The two species may have interbred to a limited extent, but the fossil record shows Neanderthals retreating, century after century, to a few refuges before disappearing altogether.

Humans and Neanderthals belong to a group of species called hominids. In other words, they both belong to the lineage that split off from other apes and eventually gave rise to us. Neanderthals represent a relatively young twig on the hominid branch. The oldest fossils of hominids date back about six million years, and scientists have found a number of other hominid fossils between then and now. While they leave many questions unanswered, they do offer a chance to test Darwin's ideas.

Bipedalism may have evolved, to some extent, in the earliest hominids. Their femurs have the right shape for supporting an upright torso. Their skulls appear to have sat atop the spine, rather than projecting forward as they do in gorillas and other apes. Yet these early hominids probably did not walk quite as upright as we do. The oldest hominid for which scientists have a relatively complete skeleton is *Australopithecus afarensis*, which existed from about three to four million years ago. It stood only three or four feet tall and had long, hooklike hands that may have been adapted for moving through trees. The oldest fossils of hominids that look as if they were capable of modern walking (and running) are from less than two million years ago.

Darwin turns out to have been wrong in his suggestion that the shift to an upright stance was accompanied by a major shift in cognitive ability. The oldest stone tools—simple blades probably used for butchering meat—appeared about 2.6 million years ago. That's four million years after the earliest signs of bipedalism. The hominids that lived when the oldest stone tools were made had small brains—about a third the size of our own and roughly on par with that of a chimpanzee.

Hominid brains got larger about two million years ago as hominids got taller. But there might be less to that change than meets the eye. A cow has a bigger brain than a dog, but that does not make the cow smarter. Larger mammals seem to need larger brains simply to control their bodies. Only about half a million years ago did hominids begin to break this rule in a significant way. Hominid brains rapidly became much larger than one would expect from a mammal of our size. By about one hundred thousand years ago this rapid swelling came to a stop. Like Darwin, paleoanthropologists argue from the evidence that this evolutionary change occurred as hominids adapted to an ecological niche unlike the kind occupied by chimpanzees and other apes. They relied increasingly on cognition for survival—technological cognition for building stone tools they could use for scavenging meat, for example, and social cognition to cooperate in the search for food. We are, in other words, adapted to the cognitive niche.

Darwin, by necessity, could only offer indirect evidence for natural selection in humans. For one thing, we have all the raw ingredients necessary for natural selection to take place. Humans, he pointed out, vary in endless ways, both physically and mentally. Following Malthus, Darwin argued that freely reproducing would leave the planet covered many times over with human beings. Humans were limited in their reproduction, and some were more limited than others. If a trait could increase a person's reproductive success, it could become more common over the generations.

As Darwin recognized, some of these variations were not hereditary—and thus could not serve as the raw material for natural selection. In Darwin's day, scientists had not yet begun to learn how to separate the effect of the environment on an organism from the effects of hereditary differences. It is extremely easy to confuse the two.

In the early 1970s, for example, the anthropologist Barry Bogin began to study the short stature of Maya Indians in Guatemala. Some scholars called them the pygmies of Central America, since the men averaged only five feet two, the women four feet eight. The other major ethnic group in Guatemala is the Ladinos, who are of Spanish descent. Ladinos are of average height. It seemed obvious that the difference between the Mayans and the Ladinos must be genetic. But Bogin demonstrated that the difference was a matter of culture: The Maya Indians were kept in severe poverty compared to the Ladinos, and had less food and medicine. That poverty helped spark a civil war in the country, which sent a million Guatemalans to the United States. Bogin found some Mayan refugee families and began to study the growth of their children. By the year 2000 American Mayans were four inches taller than Guatemalan Mayans—and the same height as Guatemalan Ladinos. The so-called pygmies only needed a decent diet to grow much taller.

Natural selection cannot act on this sort of variation. It can only act on hereditary variations that have their origins in different genetic codes. Over the past six million years, mutations have cropped up in the genomes of our ancestors, and natural selection has favored some of them. Together, those positively selected mutations have helped to make us what we are today.

Scientists are just beginning to reconstruct the history of natural selection in our ancestors. To begin with, they must first pinpoint the parts of our DNA that were altered. One way to do this is to scan the human genome for genes, then search for their counterparts in other species. The versions of a gene found in humans, chimpanzees, and mice all descend from an ancestral gene that was carried by a tiny mammal that lived one hundred million years ago. In some cases, the gene has not changed significantly in all that time and is nearly identical in us and other mammals. In other words, they encode an identical protein. In the past, animals must have regularly acquired mutations to that gene, but natural selection weeded out the mutations that caused harmful changes to the protein.

In some cases, however, humans have a version of a gene that encodes a protein with a different structure. By statistically comparing these differences, scientists have pinpointed specific genes that have experienced significant natural selection in the time since our ancestors branched off from other apes. They identified hundreds of such genes. Now the challenge is to figure out what those genes do in humans, as opposed to other animals. In many cases, the genes do not have much to do with what we like to think of as our unique human nature. Some provide resistance to malaria and other diseases. These genes have undergone drastic remodeling because the genes of pathogens can evolve as well, trapping us in a host-parasite arms race. A few genes, however, offer some intriguing hints. Some of them, for example, are known to be expressed in neurons in the human brain. But simply pinpointing where a gene is active in the body is a tiny step toward finding out what it actually does in the human body.

CHAPTER 2:

On the Manner of Development of Man from some Lower Form

IT IS MANIFEST that man is now subject to much variability. No two individuals of the same race are quite alike. We may compare millions of faces, and each will be distinct. There is an equally great amount of diversity in the proportions and dimensions of the various parts of the body; the length of the legs being one of the most variable points.[1] Although in some quarters of the world an elongated skull, and in other quarters a short skull prevails, yet there is great diversity of shape even within the limits of the same race, as with the aborigines of America and South Australia—the latter a race 'probably as pure and homogeneous in blood, customs, and language as any in existence'—and even with the inhabitants of so confined an area as the Sandwich Islands.[2] An eminent dentist assures me that there is nearly as much diversity in the teeth as in the features. The chief arteries so frequently run in abnormal courses, that it has been found useful for surgical purposes to calculate from 1040 corpses how often each course prevails.[3] The muscles

---

1. 'Investigations in Military and Anthropolog. Statistics of American Soldiers', by B. A. Gould, 1869, p. 256.
2. With respect to the 'Cranial forms of the American aborigines', see Dr Aitken Meigs in 'Proc. Acad. Nat. Sci.', Philadelphia, May, 1868. On the Australians, see Huxley, in Lyell's 'Antiquity of Man', 1863, p. 87. On the Sandwich Islanders, Prof. J. Wyman, 'Observations on Crania', Boston, 1868, p. 18.
3. 'Anatomy of the Arteries', by R. Quain. Preface, vol. i. 1844.

are eminently variable: thus those of the foot were found by Prof. Turner[4] not to be strictly alike in any two out of fifty bodies; and in some the deviations were considerable. He adds, that the power of performing the appropriate movements must have been modified in accordance with the several deviations. Mr J. Wood has recorded[5] the occurrence of 295 muscular variations in thirty-six subjects, and in another set of the same number no less than 558 variations, those occurring on both sides of the body being only reckoned as one. In the last set, not one body out of the thirty-six was 'found totally wanting in departures from the standard descriptions of the muscular system given in anatomical text books'. A single body presented the extraordinary number of twenty-five distinct abnormalities. The same muscle sometimes varies in many ways; thus Prof. Macalister describes[6] no less than twenty distinct variations in the *palmaris accessorius*.

The famous old anatomist, Wolff,[7] insists that the internal viscera are more variable than the external parts: *Nulla particula est quae non aliter et aliter in aliis se habeat hominibus*. [There is no part whose condition does not differ in different men.] He has even written a treatise on the choice of typical examples of the viscera for representation. A discussion on the beau-ideal of the liver, lungs, kidneys, &c., as of the human face divine, sounds strange in our ears.

The variability or diversity of the mental faculties in men of the same race, not to mention the greater differences between the men of distinct races, is so notorious that not a word need here be said. So it is with the lower animals. All who have had charge of menageries admit this fact, and we see it plainly in

4. 'Transact. Royal Soc. Edinburgh', vol. xxiv. pp. 175, 189.
5. 'Proc. Royal Soc.', 1867, p. 544; also 1868, pp. 483, 524. There is a previous paper, 1866, p. 229.
6. 'Proc. R. Irish Academy', vol. x. 1868, p. 141.
7. 'Act. Acad. St Petersburg', 1778, part ii. p. 217.

our dogs and other domestic animals. Brehm especially insists that each individual monkey of those which he kept tame in Africa had its own peculiar disposition and temper: he mentions one baboon remarkable for its high intelligence; and the keepers in the Zoological Gardens pointed out to me a monkey, belonging to the New World division, equally remarkable for intelligence. Rengger, also, insists on the diversity in the various mental characters of the monkeys of the same species which he kept in Paraguay; and this diversity, as he adds, is partly innate, and partly the result of the manner in which they have been treated or educated.[8]

I have elsewhere[9] so fully discussed the subject of Inheritance, that I need here add hardly anything. A greater number of facts have been collected with respect to the transmission of the most trifling, as well as of the most important characters in man, than in any of the lower animals; though the facts are copious enough with respect to the latter. So in regard to mental qualities, their transmission is manifest in our dogs, horses, and other domestic animals. Besides special tastes and habits, general intelligence, courage, bad and good temper, &c., are certainly transmitted. With man we see similar facts in almost every family; and we now know, through the admirable labours of Mr Galton,[10] that genius which implies a wonderfully complex combination of high faculties, tends to be inherited; and, on the other hand, it is too certain that insanity and deteriorated mental powers likewise run in families.

With respect to the causes of variability, we are in all cases very ignorant; but we can see that in man as in the lower animals, they stand in some relation to the conditions to which each species has been exposed, during several generations. Domesticated animals vary more than those in a state of nature; and this

8. Brehm, 'Thierleben', B. i. s. 58, 87. Rengger, 'Säugethiere von Paraguay', s. 57.
9. 'Variation of Animals and Plants under Domestication', vol. ii. chap. xii.
10. 'Hereditary Genius: an Inquiry into its Laws and Consequences', 1869.

is apparently due to the diversified and changing nature of the conditions to which they have been subjected. In this respect the different races of man resemble domesticated animals, and so do the individuals of the same race, when inhabiting a very wide area, like that of America. We see the influence of diversified conditions in the more civilised nations; for the members belonging to different grades of rank, and following different occupations, present a greater range of character than do the members of barbarous nations. But the uniformity of savages has often been exaggerated, and in some cases can hardly be said to exist.[11] It is, nevertheless, an error to speak of man, even if we look only to the conditions to which he has been exposed, as 'far more domesticated'[12] than any other animal. Some savage races, such as the Australians, are not exposed to more diversified conditions than are many species which have a wide range. In another and much more important respect, man differs widely from any strictly domesticated animal; for his breeding has never long been controlled, either by methodical or unconscious selection. No race or body of men has been so completely subjugated by other men, as that certain individuals should be preserved, and thus unconsciously selected, from somehow excelling in utility to their masters. Nor have certain male and female individuals been intentionally picked out and matched, except in the well-known case of the Prussian grenadiers; and in this case man obeyed, as might have been expected, the law of methodical selection; for it is asserted that many tall men were reared in the villages inhabited by the grenadiers and their tall wives. In Sparta, also, a form of selection was followed, for it was enacted that all

11. Mr Bates remarks ('The Naturalist on the Amazons', 1863, vol. ii. p. 159), with respect to the Indians of the same South American tribe, 'no two of them were at all similar in the shape of the head; one man had an oval visage with fine features, and another was quite Mongolian in breadth and prominence of cheek, spread of nostrils, and obliquity of eyes.'

12. Blumenbach, 'Treatises on Anthropolog.', Eng. translat., 1865, p. 205.

children should be examined shortly after birth; the well-formed and vigorous being preserved, the others left to perish.[13]

If we consider all the races of man as forming a single species, his range is enormous; but some separate races, as the Americans and Polynesians, have very wide ranges. It is a well-known law that widely-ranging species are much more variable than species with restricted ranges; and the variability of man may with more truth be compared with that of widely-ranging species, than with that of domesticated animals.

Not only does variability appear to be induced in man and the lower animals by the same general causes, but in both the same parts of the body are affected in a closely analogous manner. This, has been proved in such full detail by Godron and Quatrefages, that I need here only refer to their works.[14] Monstrosities,

13. Mitford's 'History of Greece', vol. i. p. 282. It appears also from a passage in Xenophon's 'Memorabilia,' B. ii. 4 (to which my attention has been called by the Rev. J. N. Hoare), that it was a well recognised principle with the Greeks, that men ought to select their wives with a view to the health and vigour of their children. The Grecian poet, Theognis, who lived 550 bc, clearly saw how important selection, if carefully applied, would be for the improvement of mankind. He saw, likewise, that wealth often checks the proper action of sexual selection. He thus writes:

> With kine and horses, Kurnus! we proceed
> By reasonable rules, and choose a breed
> For profit and increase, at any price;
> Of a sound stock, without defect or vice.
> But, in the daily matches that we make,
> The price is everything: for money's sake,
> Men marry: women are in marriage given;
> The churl or ruffian, that in wealth has thriven,
> May match his offspring with the proudest race:
> Thus everything is mix'd, noble and base!
> If then in outward manner, form, and mind,
> You find us a degraded, motley kind,
> Wonder no more, my friend! the cause is plain,
> And to lament the consequence is vain.

(The Works of J. Hookham Frere, vol. ii. 1872, p. 334.)

14. Godron, 'De l'Espèce', 1859, tom. ii. livre 3. Quatrefages, 'Unite de l'Espèce Humaine', 1861. Also Lectures on Anthropology, given in the 'Revue des Cours Scientifiques', 1866–1868

which graduate into slight variations, are likewise so similar in man and the lower animals, that the same classification and the same terms can be used for both, as has been shewn by Isidore Geoffroy St-Hilaire.[15] In my work on the variation of domestic animals, I have attempted to arrange in a rude fashion the laws of variation under the following heads:—The direct and definite action of changed conditions, as exhibited by all or nearly all the individuals of the same species, varying in the same manner under the same circumstances. The effects of the long-continued use or disuse of parts. The cohesion of homologous parts. The variability of multiple parts. Compensation of growth; but of this law I have found no good instance in the case of man. The effects of the mechanical pressure of one part on another; as of the pelvis on the cranium of the infant in the womb. Arrests of development, leading to the diminution or suppression of parts. The reappearance of long-lost characters through reversion. And lastly, correlated variation. All these so-called laws apply equally to man and the lower animals; and most of them even to plants. It would be superfluous here to discuss all of them;[16] but several are so important, that they must be treated at considerable length.

The direct and definite action of changed conditions—This is a most perplexing subject. It cannot be denied that changed conditions produce some, and occasionally a considerable effect, on organisms of all kinds; and it seems at first probable that if sufficient time were allowed this would be the invariable result. But I have failed to obtain clear evidence in favour of this conclusion; and valid reasons may be urged on the other side, at least as far as

15. 'Hist. Gén. et Part. des Anomalies de l'Organisation', in three volumes, tom. i. 1832.
16. I have fully discussed these laws in my 'Variation of Animals and Plants under Domestication', vol. ii. chap. xxii, and xxiii. M. J. P. Durand has lately (1868) published a valuable essay 'De l'Influence des Milieux', &c. He lays much stress, in the case of plants, on the nature of the soil.

the innumerable structures are concerned, which are adapted for special ends. There can, however, be no doubt that changed conditions induce an almost indefinite amount of fluctuating variability, by which the whole organisation is rendered in some degree plastic.

In the United States, above 1,000,000 soldiers, who served in the late war, were measured, and the States in which they were born and reared were recorded.[17] From this astonishing number of observations it is proved that local influences of some kind act directly on stature; and we further learn that 'the State where the physical growth has in great measure taken place, and the State of birth, which indicates the ancestry, seem to exert a marked influence on the stature'. For instance, it is established, 'that residence in the Western States, during the years of growth, tends to produce increase of stature'. On the other hand, it is certain that with sailors, their life delays growth, as shewn 'by the great difference between the statures of soldiers and sailors at the ages of seventeen and eighteen years'. Mr B. A. Gould endeavoured to ascertain the nature of the influences which thus act on stature; but he arrived only at negative results, namely, that they did not relate to climate, the elevation of the land, soil, nor even 'in any controlling degree' to the abundance or the need of the comforts of life. This latter conclusion is directly opposed to that arrived at by Villermé, from the statistics of the height of the conscripts in different parts of France. When we compare the differences in stature between the Polynesian chiefs and the lower orders within the same islands, or between the inhabitants of the fertile volcanic and low barren coral islands of the same ocean,[18] or

17. 'Investigations in Military and Anthrop. Statistics', &c. 1869, by B. A. Gould, pp. 93, 107, 126, 131, 134.
18. For the Polynesians, see Prichard's 'Physical Hist. of Mankind', vol. v. 1847, pp. 145, 283. Also Godron, 'De l'Espèce', tom. ii. p. 289. There is also a remarkable difference in appearance between the closely-allied Hindoos inhabiting the Upper Ganges and Bengal; see Elphinstone's 'History of India', vol. i. p. 324.

again between the Fuegians on the eastern and western shores
of their country, where the means of subsistence are very different, it is scarcely possible to avoid the conclusion that better
food and greater comfort do influence stature. But the preced-
ing statements shew how difficult it is to arrive at any precise
result. Dr Beddoe has lately proved that, with the inhabitants of
Britain, residence in towns and certain occupations have a deteriorating influence on height; and he infers that the result is to
a certain extent inherited, as is likewise the case in the United
States. Dr Beddoe further believes that wherever a 'race attains
its maximum of physical development, it rises highest in energy
and moral vigour'.[19]

Whether external conditions produce any other direct effect on man is not known. It might have been expected that
differences of climate would have had a marked influence, in as
much as the lungs and kidneys are brought into activity under
a low temperature, and the liver and skin under a high one.[20]
It was formerly thought that the colour of the skin and the
character of the hair were determined by light or heat; and
although it can hardly be denied that some effect is thus produced, almost all observers now agree that the effect has been
very small, even after exposure during many ages. But this subject will be more properly discussed when we treat of the different races of mankind. With our domestic animals there are
grounds for believing that cold and damp directly affect the
growth of the hair; but I have not met with any evidence on
this head in the case of man.

Effects of the increased Use and Disuse of Parts—It is well known
that use strengthens the muscles in the individual, and com-

19. 'Memoirs, Anthropolog. Soc.', vol. iii. 1867–69, pp. 561, 565, 567.
20. Dr Brakenridge, 'Theory of Diathesis', 'Medical Times', June 19 and July 17,
1869.

plete disuse, or the destruction of the proper nerve, weakens them. When the eye is destroyed, the optic nerve often becomes atrophied. When an artery is tied, the lateral channels increase not only in diameter, but in the thickness and strength of their coats. When one kidney ceases to act from disease, the other increases in size, and does double work. Bones increase not only in thickness, but in length, from carrying a greater weight.[21] Different occupations, habitually followed, lead to changed proportions in various parts of the body. Thus it was ascertained by the United States Commission[22] that the legs of the sailors employed in the late war were longer by 0.217 of an inch than those of the soldiers, though the sailors were on an average shorter men; whilst their arms were shorter by 1.09 of an inch, and therefore, out of proportion, shorter in relation to their lesser height. This shortness of the arms is apparently due to their greater use, and is an unexpected result: but sailors chiefly use their arms in pulling, and not in supporting weights. With sailors, the girth of the neck and the depth of the instep are greater, whilst the circumference of the chest, waist, and hips is less, than in soldiers.

Whether the several foregoing modifications would become hereditary, if the same habits of life were followed during many generations, is not known, but it is probable. Rengger[23] attributes the thin legs and thick arms of the Payaguas Indians to successive generations having passed nearly their whole lives in canoes, with their lower extremities motionless. Other writers have come to a similar conclusion in analogous cases. According to Cranz,[24] who lived for a long time with the Esquimaux, 'the na-

---

21. I have given authorities for these several statements in my 'Variation of Animals under Domestication', vol. ii. pp. 297–300. Dr Jaeger, 'Ueber das Längenwachsthum der Knochen', 'Jenaischen Zeitschrift', B. v. Heft I.

22. 'Investigations', &c. By B. A. Gould, 1869, p. 288.

23. 'Säugethiere von Paraguay, 1830, s. 4.

24. 'History of Greenland', Eng. translat. 1767, vol. i. p. 230.

tives believe that ingenuity and dexterity in seal-catching (their highest art and virtue) is hereditary; there is really something in it, for the son of a celebrated seal-catcher will distinguish himself, though he lost his father in childhood'. But in this case it is mental aptitude, quite as much as bodily structure, which appears to be inherited. It is asserted that the hands of English labourers are at birth larger than those of the gentry.[25] From the correlation which exists, at least in some cases,[26] between the development of the extremities and of the jaws, it is possible that in those classes which do not labour much with their hands and feet, the jaws would be reduced in size from this cause. That they are generally smaller in refined and civilised men than in hard-working men or savages, is certain. But with savages, as Mr Herbert Spencer[27] has remarked, the greater use of the jaws in chewing coarse, uncooked food, would act in a direct manner on the masticatory muscles, and on the bones to which they are attached. In infants, long before birth, the skin on the soles of the feet is thicker than on any other part of the body;[28] and it can hardly be doubted that this is due to the inherited effects of pressure during a long series of generations.

It is familiar to every one that watchmakers and engravers are liable to be short-sighted, whilst men living much out of doors, and especially savages, are generally long-sighted.[29] Short-sight and long-sight certainly tend to be inherited.[30] The inferiority of Europeans, in comparison with savages, in eyesight and in

25. 'Intermarriage'. By Alex. Walker, 1838, p. 377.
26. 'The Variation of Animals under Domestication', vol. i. p. 173.
27. 'Principles of Biology', vol. i. p. 455.
28. Paget, 'Lectures on Surgical Pathology', vol. ii. 1853, p. 209.
29. It is a singular and unexpected fact that sailors are inferior to landsmen in their mean distance of distinct vision. Dr B. A. Gould ('Sanitary Memoirs of the War of the Rebellion', 1869, p. 530), has proved this to be the case; and he accounts for it by the ordinary range of vision in sailors being 'restricted to the length of the vessel and the height of the masts'.
30. 'The Variation of Animals under Domestication', vol. i. p. 8.

the other senses, is no doubt the accumulated and transmitted effect of lessened use during many generations; for Rengger[31] states that he has repeatedly observed Europeans, who had been brought up and spent their whole lives with the wild Indians, who nevertheless did not equal them in the sharpness of their senses. The same naturalist observes that the cavities in the skull for the reception of the several sense-organs are larger in the American aborigines than in Europeans; and this probably indicates a corresponding difference in the dimensions of the organs themselves. Blumenbach has also remarked on the large size of the nasal cavities in the skulls of the American aborigines, and connects this fact with their remarkably acute power of smell. The Mongolians of the plains of Northern Asia, according to Pallas, have wonderfully perfect senses; and Prichard believes that the great breadth of their skulls across the zygomas follows from their highly-developed sense-organs.[32]

The Quechua Indians inhabit the lofty plateaux of Peru; and Alcide d'Orbigny states[33] that, from continually breathing a highly rarefied atmosphere, they have acquired chests and lungs of extraordinary dimensions. The cells, also, of the lungs are larger and more numerous than in Europeans. These observations have been doubted; but Mr D. Forbes carefully measured many Aymaras, an allied race, living at the height of between 10,000 and 15,000 feet; and he informs me[34] that they differ

31. 'Säugethiere von Paraguay', s. 8, 10. I have had good opportunities for observing the extraordinary power of eyesight in the Fuegians. See also Lawrence ('Lectures on Physiology,' &c., 1822, p. 404) on this same subject. M. Giraud-Teulon has recently collected ('Revue des Cours Scientifiques', 1870, p. 625) a large and valuable body of evidence proving that the cause of short-sight, 'C'est le travail assidu, de près'.

32. Prichard, 'Phys. Hist. of Mankind', on the authority of Blumenbach, vol. i. 1851, p. 311; for the statement by Pallas, vol. iv. 1844, p. 407.

33. Quoted by Prichard, 'Researches into the Phys. Hist. of Mankind', vol. v. p. 463.

34. Mr Forbes' valuable paper is now published in the 'Journal of the Ethnological Soc. of London', new series, vol. ii. 1870, p. 193.

conspicuously from the men of all other races seen by him in the circumference and length of their bodies. In his table of measurements, the stature of each man is taken at 1000, and the other measurements are reduced to this standard. It is here seen that the extended arms of the Aymaras are shorter than those of Europeans, and much shorter than those of Negroes. The legs are likewise shorter; and they present this remarkable peculiarity, that in every Aymara measured, the femur is actually shorter than the tibia. On an average, the length of the femur to that of the tibia is as 211 to 252; whilst in two Europeans, measured at the same time, the femora to the tibiae were as 244 to 230; and in three Negroes as 258 to 241. The humerus is likewise shorter relatively to the forearm. This shortening of that part of the limb which is nearest to the body, appears to be, as suggested to me by Mr Forbes, a case of compensation in relation with the greatly increased length of the trunk. The Aymaras present some other singular points of structure, for instance, the very small projection of the heel.

These men are so thoroughly acclimatised to their cold and lofty abode, that when formerly carried down by the Spaniards to the low eastern plains, and when now tempted down by high wages to the gold-washings, they suffer a frightful rate of mortality. Nevertheless Mr Forbes found a few pure families which had survived during two generations: and he observed that they still inherited their characteristic peculiarities. But it was manifest, even without measurement, that these peculiarities had all decreased; and on measurement, their bodies were found not to be so much elongated as those of the men on the high plateau; whilst their femora had become somewhat lengthened, as had their tibiae, although in a less degree. The actual measurements may be seen by consulting Mr Forbes's memoir. From these observations, there can, I think, be no doubt that residence during many generations at a great elevation tends, both directly and indirectly, to induce inherited modifications in the proportions of the body.[35]

Although man may not have been much modified during the latter stages of his existence through the increased or decreased use of parts, the facts now given shew that his liability in this respect has not been lost; and we positively know that the same law holds good with the lower animals. Consequently we may infer that when at a remote epoch the progenitors of man were in a transitional state, and were changing from quadrupeds into bipeds, natural selection would probably have been greatly aided by the inherited effects of the increased or diminished use of the different parts of the body.

Arrests of Development—There is a difference between arrested development and arrested growth, for parts in the former state continue to grow whilst still retaining their early condition. Various monstrosities come under this head; and some, as a cleft-palate, are known to be occasionally inherited. It will suffice for our purpose to refer to the arrested brain-development of microcephalous idiots, as described in Vogt's memoir.[36] Their skulls are smaller, and the convolutions of the brain are less complex than in normal men. The frontal sinus, or the projection over the eye-brows, is largely developed, and the jaws are prognathous to an '*effrayant*' degree; so that these idiots somewhat resemble the lower types of mankind. Their intelligence, and most of their mental faculties, are extremely feeble. They cannot acquire the power of speech, and are wholly incapable of prolonged attention, but are much given to imitation. They are strong and remarkably active, continually gamboling and jumping about, and making

---

35. Dr Wilckens ('Landwirthschaft. Wochenblatt', No. 10, 1869) has lately published an interesting Essay shewing how domestic animals, which live in mountainous regions, have their frames modified.

36. 'Mémoire sur les Microcéphales', 1867, pp. 50, 125, 169, 171, 184–198.

grimaces. They often ascend stairs on all-fours; and are curiously fond of climbing up furniture or trees. We are thus reminded of the delight shewn by almost all boys in climbing trees; and this again reminds us how lambs and kids, originally alpine animals, delight to frisk on any hillock, however small. Idiots also resemble the lower animals in some other respects; thus several cases are recorded of their carefully smelling every mouthful of food before eating it. One idiot is described as often using his mouth in aid of his hands, whilst hunting for lice. They are often filthy in their habits, and have no sense of decency; and several cases have been published of their bodies being remarkably hairy.[37]

Reversion—Many of the cases to be here given, might have been introduced under the last heading. When a structure is arrested in its development, but still continues growing, until it closely resembles a corresponding structure in some lower and adult member of the same group, it may in one sense be considered as a case of reversion. The lower members in a group give us some idea how the common progenitor was probably constructed; and it is hardly credible that a complex part, arrested at an early phase of embryonic development, should go on growing so as ultimately to perform its proper function, unless it had acquired such power during some earlier state of existence, when the present exceptional or arrested structure was normal. The simple brain of a microcephalous idiot, in as far as it resembles that of an ape, may in this sense be said to offer a case of reversion.[38] There

37. Prof. Laycock sums up the character of brute-like idiots by calling them *theroid*; 'Journal of Mental Science', July 1863. Dr Scott ('The Deaf and Dumb', 2nd edit., 1870, p. 10) has often observed the imbecile smelling their food. See, on this same subject, and on the hairiness of idiots, Dr Maudsley, 'Body and Mind', 1870, pp. 46–51. Pinel has also given a striking case of hairiness in an idiot.

38. In my 'Variation of Animals under Domestication' (vol. ii, p. 57), I attributed the not very rare cases of supernumerary mammae in women to reversion. I was led to this as a probable conclusion, by the additional mammae being generally placed symmetrically on the breast; and more especially from one case, in which a single

are other cases which come more strictly under our present head of reversion. Certain structures, regularly occurring in the lower members of the group to which man belongs, occasionally make their appearance in him, though not found in the normal human

efficient mamma occurred in the inguinal region of a woman, the daughter of another woman with supernumerary mammae. But I now find (see, for instance, Prof. Preyer, 'Der Kampf um das Dasein', 1869, s. 45) that *mammae erraticae* occur in other situations, as on the back, in the armpit, and on the thigh; the mammae in this latter instance having given so much milk that the child was thus nourished. The probability that the additional mammae are due to reversion is thus much weakened; nevertheless, it still seems to me probable, because two pairs are often found symmetrically on the breast; and of this I myself have received information in several cases. It is well known that some Lemurs normally have two pairs of mammae on the breast. Five cases have been recorded of the presence of more than a pair of mammae (of course rudimentary) in the male sex of mankind; see 'Journal of Anat. and Physiology', 1872, p. 56, for a case given by Dr Handyside, in which two brothers exhibited this peculiarity; see also a paper by Dr Bartels, in 'Reichert's and du Bois Reymond's Archiv', 1872, p. 304. In one of the cases alluded to by Dr Bartels, a man bore five mammae, one being medial and placed above the navel; Meckel von Hamsbach thinks that this latter case is illustrated by a medial mamma occurring in certain Cheiroptera. On the whole, we may well doubt if additional mammae would ever have been developed in both sexes of mankind, had not his early progenitors been provided with more than a single pair.

In the above work (vol. ii. p. 12), I also attributed, though with much hesitation, the frequent cases of polydactylism in men and various animals to reversion. I was partly led to this through Prof. Owen's statement, that some of the Ichthyopterygia possess more than five digits, and therefore, as I supposed, had retained a primordial condition; but Prof. Gegenbaur ('Jenaischen Zeitschrift', B. v. Heft 3, s. 341), disputes Owen's conclusion. On the other hand, according to the opinion lately advanced by Dr Gunther, on the paddle of Ceratodus, which is provided with articulated bony rays on both sides of a central chain of bones, there seems no great difficulty in admitting that six or more digits on one side, or on both sides, might reappear through reversion. I am informed by Dr Zouteveen that there is a case on record of a man having twenty-four fingers and twenty-four toes! I was chiefly led to the conclusion that the presence of supernumerary digits might be due to reversion from the fact that such digits, not only are strongly inherited, but, as I then believed, had the power of regrowth after amputation, like the normal digits of the lower vertebrata. But I have explained in the Second Edition of my Variation under Domestication why I now place little reliance on the recorded cases of such regrowth. Nevertheless it deserves notice, in as much as arrested development and reversion are intimately related processes; that various structures in an embryonic or arrested condition, such as a cleft palate, bifid uterus, &c., are frequently accompanied by polydactylism. This has been strongly insisted on by Meckel and Isidore Geoffroy St-Hilaire. But at present it is the safest course to give up altogether the idea that there is any relation between the development of supernumerary digits and reversion to some lowly organised progenitor of man.

embryo; or, if normally present in the human embryo, they become abnormally developed, although in a manner which is normal in the lower members of the group. These remarks will be rendered clearer by the following illustrations.

In various mammals the uterus graduates from a double organ with two distinct orifices and two passages, as in the marsupials, into a single organ, which is in no way double except from having a slight internal fold, as in the higher apes and man. The rodents exhibit a perfect series of gradations between these two extreme states. In all mammals the uterus is developed from two simple primitive tubes, the inferior portions of which form the cornua; and it is in the words of Dr Farre, 'by the coalescence of the two cornua at their lower extremities that the body of the uterus is formed in man; while in those animals in which no middle portion or body exists, the cornua remain ununited. As the development of the uterus proceeds, the two cornua become gradually shorter, until at length they are lost, or, as it were, absorbed into the body of the uterus.' The angles of the uterus are still produced into cornua, even in animals as high up in the scale as the lower apes and lemurs.

Now in women, anomalous cases are not very infrequent, in which the mature uterus is furnished with cornua, or is partially divided into two organs; and such cases, according to Owen, repeat 'the grade of concentrative development', attained by certain rodents. Here perhaps we have an instance of a simple arrest of embryonic development, with subsequent growth and perfect functional development; for either side of the partially double uterus is capable of performing the proper office of gestation. In other and rarer cases, two distinct uterine cavities are formed, each having its proper orifice and passage.[39] No such stage is

39. See Dr A. Farre's well-known article in the 'Cyclopaedia of Anatomy and Physiology', vol. v. 1859, p. 642. Owen, 'Anatomy of Vertebrates,' vol. iii., 1868, p. 687. Professor Turner in 'Edinburgh Medical Journal', February 1865.

passed through during the ordinary development of the embryo; and it is difficult to believe, though perhaps not impossible, that the two simple, minute, primitive tubes should know how (if such an expression may be used) to grow into two distinct uteri, each with a well-constructed orifice and passage, and each furnished with numerous muscles, nerves, glands and vessels, if they had not formerly passed through a similar course of development, as in the case of existing marsupials. No one will pretend that so perfect a structure as the abnormal double uterus in woman could be the result of mere chance. But the principle of reversion, by which a long-lost structure is called back into existence, might serve as the guide for its full development, even after the lapse of an enormous interval of time.

Professor Canestrini, after discussing the foregoing and various analogous cases, arrives at the same conclusion as that just given. He adduces another instance, in the case of the malar bone,[40] which, in some of the Quadrumana and other mammals, normally consists of two portions. This is its condition in the human foetus when two months old; and through arrested development, it sometimes remains thus in man when adult, more especially in the lower prognathous races. Hence Canestrini concludes that some ancient progenitor of man must have had this bone normally divided into two portions, which afterwards

40. 'Annuario della Soc. dei Naturalisti in Modena', 1867, p. 83. Prof. Canestrini gives extracts on this subject from various authorities. Laurillard remarks, that as he has found a complete similarity in the form, proportions, and connection of the two malar bones in several human subjects and in certain apes, he cannot consider this disposition of the parts as simply accidental. Another paper on this same anomaly has been published by Dr Saviotti in the 'Gazzetta delle Cliniche', Turin, 1871, where he says that traces of the division may be detected in about two per cent of adult skulls; he also remarks that it more frequently occurs in prognathous skulls, not of the Aryan race, than in others. See also G. Delorenzi on the same subject; 'Tre nuovi casi d'anomalia dell'osso, malare', Torino, 1872. Also, E. Morselli, 'Sopra una rara anomalia dell' osso malare', Modena, 1872. Still more recently Gruber has written a pamphlet on the division of this bone. I give these references because a reviewer, without any grounds or scruples, has thrown doubts on my statements.

became fused together. In man the frontal bone consists of a single piece, but in the embryo, and in children, and in almost all the lower mammals, it consists of two pieces separated by a distinct suture. This suture occasionally persists more or less distinctly in man after maturity; and more frequently in ancient than in recent crania, especially, as Canestrini has observed, in those exhumed from the Drift, and belonging to the brachycephalic type. Here again he comes to the same conclusion as in the analogous case of the malar bones. In this, and other instances presently to be given, the cause of ancient races approaching the lower animals in certain characters more frequently than do the modern races, appears to be, that the latter stand at a somewhat greater distance in the long line of descent from their early semi-human progenitors.

Various other anomalies in man, more or less analogous to the foregoing, have been advanced by different authors, as cases of reversion; but these seem not a little doubtful, for we have to descend extremely low in the mammalian series, before we find such structures normally present.[41]

In man, the canine teeth are perfectly efficient instruments for mastication. But their true canine character, as Owen[42] remarks, 'is indicated by the conical form of the crown, which terminates in an obtuse point, is convex outward and flat or

41. A whole series of cases is given by Isid. Geoffroy St-Hilaire, 'Hist. des Anomalies', tom. iii. p. 437. A reviewer ('Journal of Anat. and Physiology', 1871, p. 366) blames me much for not having discussed the numerous cases, which have been recorded, of various parts arrested in their development. He says that, according to my theory, 'every transient condition of an organ, during its development, is not only a means to an end, but once was an end in itself'. This does not seem to me necessarily to hold good. Why should not variations occur during an early period of development, having no relation to reversion; yet such variations might be preserved and accumulated, if in any way serviceable, for instance, in shortening and simplifying the course of development? And again, why should not injurious abnormalities, such as atrophied or hypertrophied parts, which have no relation to a former state of existence, occur at an early period, as well as during maturity?

42. 'Anatomy of Vertebrates', vol. iii. 1868, p. 323.

sub-concave within, at the base of which surface there is a feeble prominence. The conical form is best expressed in the Melanian races, especially the Australian. The canine is more deeply implanted, and by a stronger fang than the incisors.' Nevertheless, this tooth no longer serves man as a special weapon for tearing his enemies or prey; it may, therefore, as far as its proper function is concerned, be considered as rudimentary. In every large collection of human skulls some may be found, as Häckel[43] observes, with the canine teeth projecting considerably beyond the others in the same manner as in the anthropomorphous apes, but in a less degree. In these cases, open spaces between the teeth in the one jaw are left for the reception of the canines of the opposite jaw. An interspace of this kind in a Kaffir skull, figured by Wagner, is surprisingly wide.[44] Considering how few are the ancient skulls which have been examined, compared to recent skulls, it is an interesting fact that in at least three cases the canines project largely; and in the Naulette jaw they are spoken of as enormous.[45]

Of the anthropomorphous apes the males alone have their canines fully developed; but in the female gorilla, and in a less degree in the female orang, these teeth project considerably beyond the others; therefore the fact, of which I have been assured, that women sometimes have considerably projecting canines, is no serious objection to the belief that their occasional great development in man is a case of reversion to an ape-like progenitor. He who rejects with scorn the belief that the shape of his own canines, and their occasional great development in other men, are due to our early forefathers having been provided with these formidable weapons, will probably reveal, by sneering, the line of

43. 'Generelle Morphologie', 1866, B. ii. s. clv.
44. Carl Vogt's 'Lectures on Man', Eng. translat. 1864, p. 151.
45. C. Carter Blake, on a jaw from La Naulette, 'Anthropolog. Review', 1867, p. 295. Schaaffhausen, ibid. 1868, p. 426.

his descent. For though he no longer intends, nor has the power, to use these teeth as weapons, he will unconsciously retract his 'snarling muscles' (thus named by Sir C. Bell),[46] so as to expose them ready for action, like a dog prepared to fight.

Many muscles are occasionally developed in man, which are proper to the Quadrumana or other mammals. Professor Vlacovich[47] examined forty male subjects, and found a muscle, called by him the ischio-pubic, in nineteen of them; in three others there was a ligament which represented this muscle; and in the remaining eighteen no trace of it. In only two out of thirty female subjects was this muscle developed on both sides, but in three others the rudimentary ligament was present. This muscle, therefore, appears to be much more common in the male than in the female sex; and on the belief in the descent of man from some lower form, the fact is intelligible; for it has been detected in several of the lower animals, and in all of these it serves exclusively to aid the male in the act of reproduction.

Mr J. Wood, in his valuable series of papers,[48] has minutely described a vast number of muscular variations in man, which resemble normal structures in the lower animals. The muscles which closely resemble those regularly present in our nearest allies, the Quadrumana, are too numerous to be here even specified. In a single male subject, having a strong bodily frame, and well-formed skull, no less than seven muscular variations were

46. 'The Anatomy of Expression', 1844, pp. 110, 131.
47. Quoted by Prof. Canestrini in the 'Annuario', &c., 1867, p. 90.
48. These papers deserve careful study by any one who desires to learn how frequently our muscles vary, and in varying come to resemble those of the Quadrumana. The following references relate to the few points touched on in my text: 'Proc. Royal Soc.', vol. xiv. 1865, pp. 379–384; vol. xv. 1866, pp. 241, 242; vol. xv. 1867, p. 544; vol. xvi. 1868, p. 524. I may here add that Dr Murie and Mr St George Mivart have shewn in their Memoir on the Lemuroidea ('Transact. Zoolog. Soc', vol. vii. 1869, p. 96), how extraordinarily variable some of the muscles are in these animals, the lowest members of the Primates. Gradations, also, in the muscles leading to structures found in animals still lower in the scale, are numerous in the Lemuroidea.

observed, all of which plainly represented muscles proper to various kinds of apes. This man, for instance, had on both sides of his neck a true and powerful '*levator claviculae*', such as is found in all kinds of apes, and which is said to occur in about one out of sixty human subjects.[49] Again, this man had 'a special abductor of the metatarsal bone of the fifth digit, such as Professor Huxley and Mr Flower have shewn to exist uniformly in the higher and lower apes'. I will give only two additional cases; the *acromio-basilar* muscle is found in all mammals below man, and seems to be correlated with a quadrupedal gait,[50] and it occurs in about one out of sixty human subjects. In the lower extremities Mr Bradley[51] found an *abductor ossis metatarsi quinti* in both feet of man; this muscle had not up to that time been recorded in mankind, but is always present in the anthropomorphous apes. The muscles of the hands and arms—parts which are so eminently characteristic of man—are extremely liable to vary, so as to resemble the corresponding muscles in the lower animals.[52] Such resemblances are either perfect or imperfect; yet in the latter case they are manifestly of a transitional nature. Certain variations are more common in man, and others in woman, without our being able to assign any reason. Mr Wood, after describing numerous variations, makes the following pregnant remark: 'Notable departures from the ordinary type of the muscular structures run in grooves or directions, which must be taken to indicate some unknown factor, of much importance to a comprehensive knowledge of general and scientific anatomy.'[53]

49. See also Prof. Macalister in 'Proc. R. Irish Academy', vol. x. 1868, p. 124.
50. Mr Champneys in 'Journal of Anat. and Phys.', Nov., 1871, p. 178.
51. 'Journal of Anat. and Phys.', May, 1872, p. 421.
52. Prof. Macalister (ibid. p. 121) has tabulated his observations, and finds that muscular abnormalities are most frequent in the fore-arms, secondly, in the face, thirdly, in the foot, &c.
53. The Rev. Dr Haughton, after giving ('Proc. R. Irish Academy', June 27, 1864, p. 715) a remarkable case of variation in the human *flexor pollicis longus*, adds, 'This remarkable example shews that man may sometimes possess the arrangement of

That this unknown factor is reversion to a former state of existence may be admitted as in the highest degree probable.[54] It is quite incredible that a man should through mere accident abnormally resemble certain apes in no less than seven of his muscles, if there had been no genetic connection between them. On the other hand, if man is descended from some ape-like creature, no valid reason can be assigned why certain muscles should not suddenly reappear after an interval of many thousand generations, in the same manner as with horses, asses, and mules, dark-coloured stripes suddenly reappear on the legs, and shoulders, after an interval of hundreds, or more probably of thousands of generations.

These various cases of reversion are so closely related to those of rudimentary organs given in the first chapter, that many of them might have been indifferently introduced either there or here. Thus a human uterus furnished with cornua may be said to represent, in a rudimentary condition, the same organ in its normal state in certain mammals. Some parts which are rudimentary in man, as the os coccyx in both sexes, and the mammae in the male sex, are always present; whilst others, such as the

tendons of thumb and fingers characteristic of the macaque; but whether such a case should be regarded as a macaque passing upwards into a man, or a man passing downwards into a macaque, or as a congenital freak of nature, I cannot undertake to say.' It is satisfactory to hear so capable an anatomist, and so embittered an opponent of evolutionism, admitting even the possibility of either of his first propositions. Prof. Macalister has also described ('Proc. R. Irish Acad.', vol. x. 1864, p. 138) variations in the *flexor pollicis longus*, remarkable from their relations to the same muscle in the Quadrumana.

54. Since the first edition of this book appeared, Mr Wood has published another memoir in the 'Phil. Transactions', 1870, p. 83, on the varieties of the muscles of the human neck, shoulder, and chest. He here shews how extremely variable these muscles are, and how often and how closely the variations resemble the normal muscles of the lower animals. He sums up by remarking, 'It will be enough for my purpose if I have succeeded in shewing the more important forms which, when occurring as varieties in the human subject, tend to exhibit in a sufficiently marked manner what may be considered as proofs and examples of the Darwinian principle of reversion, or law of inheritance, in this department of anatomical science.'

supracondyloid foramen, only occasionally appear, and therefore might have been introduced under the head of reversion. These several reversionary structures, as well as the strictly rudimentary ones, reveal the descent of man from some lower form in an unmistakable manner.

Correlated Variation—In man, as in the lower animals, many structures are so intimately related, that when one part varies so does another, without our being able, in most cases, to assign any reason. We cannot say whether the one part governs the other, or whether both are governed by some earlier developed part. Various monstrosities, as I. Geoffroy repeatedly insists, are thus intimately connected. Homologous structures are particularly liable to change together, as we see on the opposite sides of the body, and in the upper and lower extremities. Meckel long ago remarked, that when the muscles of the arm depart from their proper type, they almost always imitate those of the leg; and so, conversely, with the muscles of the legs. The organs of sight and hearing, the teeth and hair, the colour of the skin and of the hair, colour and constitution, are more or less correlated.[55] Professor Schaaffhausen first drew attention to the relation apparently existing between a muscular frame and the strongly-pronounced supra-orbital ridges, which are so characteristic of the lower races of man.

Besides the variations which can be grouped with more or less probability under the foregoing heads, there is a large class of variations which may be provisionally called spontaneous, for to our ignorance they appear to arise without any exciting cause. It can, however, be shewn that such variations, whether consisting of slight individual differences, or of strongly-marked and abrupt

55. The authorities for these several statements are given in my 'Variation of Animals under Domestication', vol. ii. pp. 320–335.

deviations of structure, depend much more on the constitution of the organism than on the nature of the conditions to which it has been subjected.[56]

Rate of Increase—Civilised populations have been known under favourable conditions, as in the United States, to double their numbers in twenty-five years; and, according to a calculation by Euler, this might occur in a little over twelve years.[57] At the former rate, the present population of the United States (thirty millions), would in 657 years cover the whole terraqueous globe so thickly, that four men would have to stand on each square yard of surface. The primary or fundamental check to the continued increase of man is the difficulty of gaining subsistence, and of living in comfort. We may infer that this is the case from what we see, for instance, in the United States, where subsistence is easy, and there is plenty of room. If such means were suddenly doubled in Great Britain, our number would be quickly doubled. With civilised nations this primary check acts chiefly by restraining marriages. The greater death-rate of infants in the poorest classes is also very important; as well as the greater mortality, from various diseases, of the inhabitants of crowded and miserable houses, at all ages. The effects of severe epidemics and wars are soon counterbalanced, and more than counterbalanced, in nations placed under favourable conditions. Emigration also comes in aid as a temporary check, but, with the extremely poor classes, not to any great extent.

There is reason to suspect, as Malthus has remarked, that the reproductive power is actually less in barbarous, than in civilised races. We know nothing positively on this head, for with sav-

---

56. This whole subject has been discussed in chap. xxiii. vol. ii. of my 'Variation of Animals and Plants under Domestication'.

57. See the ever memorable 'Essay on the Principle of Population', by the Rev. T. Malthas, vol. i. 1826, pp. 6, 517.

ages no census has been taken; but from the concurrent testimony of missionaries, and of others who have long resided with such people, it appears that their families are usually small, and large ones rare. This may be partly accounted for, as it is believed, by the women suckling their infants during a long time; but it is highly probable that savages, who often suffer much hardship, and who do not obtain so much nutritious food as civilised men, would be actually less prolific. I have shewn in a former work,[58] that all our domesticated quadrupeds and birds, and all our cultivated plants, are more fertile than the corresponding species in a state of nature. It is no valid objection to this conclusion that animals suddenly supplied with an excess of food, or when grown very fat; and that most plants on sudden removal from very poor to very rich soil, are rendered more or less sterile. We might, therefore, expect that civilised men, who in one sense are highly domesticated, would be more prolific than wild men. It is also probable that the increased fertility of civilised nations would become, as with our domestic animals, an inherited character: it is at least known that with mankind a tendency to produce twins runs in families.[59]

Notwithstanding that savages appear to be less prolific than civilised people, they would no doubt rapidly increase if their numbers were not by some means rigidly kept down. The Santali, or hill-tribes of India, have recently afforded a good illustration of this fact; for, as shewn by Mr Hunter,[60] they have increased at an extraordinary rate since vaccination has been introduced, other pestilences mitigated, and war sternly repressed. This increase, however, would not have been possible had not these rude people spread into the adjoining districts, and worked for hire. Savages almost always marry; yet there is some prudential restraint, for they do not commonly marry at the earliest pos-

58. 'Variation of Animals and Plants under Domestication', vol. ii. pp. 111–113, 163.
59. Mr Sedgwick, 'British and Foreign Medico-Chirurg. Review', July, 1863, p. 170.
60. 'The Annals of Rural Bengal', by W. W. Hunter, 1868, p. 259.

sible age. The young men are often required to shew that they can support a wife; and they generally have first to earn the price with which to purchase her from her parents. With savages the difficulty of obtaining subsistence occasionally limits their number in a much more direct manner than with civilised people, for all tribes periodically suffer from severe famines. At such times savages are forced to devour much bad food, and their health can hardly fail to be injured. Many accounts have been published of their protruding stomachs and emaciated limbs after and during famines. They are then, also, compelled to wander much, and, as I was assured in Australia, their infants perish in large numbers. As famines are periodical, depending chiefly on extreme seasons, all tribes must fluctuate in number. They cannot steadily and regularly increase, as there is no artificial increase in the supply of food. Savages, when hard pressed, encroach on each other's territories, and war is the result; but they are indeed almost always at war with their neighbours. They are liable to many accidents on land and water in their search for food; and in some countries they suffer much from the larger beasts of prey. Even in India, districts have been depopulated by the ravages of tigers.

Malthus has discussed these several checks, but he does not lay stress enough on what is probably the most important of all, namely infanticide, especially of female infants, and the habit of procuring abortion. These practices now prevail in many quarters of the world; and infanticide seems formerly to have prevailed, as Mr M'Lennan[61] has shewn, on a still more extensive scale. These practices appear to have originated in savages recognising the difficulty, or rather the impossibility of supporting all the infants that are born. Licentiousness may also be added to the foregoing checks; but this does not follow from failing means of subsistence; though there is reason to believe that in some cases

61. 'Primitive Marriage', 1865.

(as in Japan) it has been intentionally encouraged as a means of keeping down the population.

If we look back to an extremely remote epoch, before man had arrived at the dignity of manhood, he would have been guided more by instinct and less by reason than are the lowest savages at the present time. Our early semi-human progenitors would not have practised infanticide or polyandry; for the instincts of the lower animals are never so perverted[62] as to lead them regularly to destroy their own offspring, or to be quite devoid of jealousy. There would have been no prudential restraint from marriage, and the sexes would have freely united at an early age. Hence the progenitors of man would have tended to increase rapidly; but checks of some kind, either periodical or constant, must have kept down their numbers, even more severely than with existing savages. What the precise nature of these checks were, we cannot say, any more than with most other animals. We know that horses and cattle, which are not extremely prolific animals, when first turned loose in South America, increased at an enormous rate. The elephant, the slowest breeder of all known animals, would in a few thousand years stock the whole world. The increase of every species of monkey must be checked by some means; but not, as Brehm remarks, by the attacks of beasts of prey. No one will assume that the actual power of reproduction in the wild horses and cattle of America, was at first in any sensible degree increased; or that, as each district became fully stocked, this same

62. A writer in the 'Spectator' (March 12th, 1871, p. 320) comments as follows on this passage:—'Mr Darwin finds himself compelled to reintroduce a new doctrine of the fall of man. He shews that the instincts of the higher animals are far nobler than the habits of savage races of men, and he finds himself, therefore, compelled to re-introduce—in a form of the substantial orthodoxy of which he appears to be quite unconscious—and to introduce as a scientific hypothesis the doctrine that man's gain of *knowledge* was the cause of a temporary but long-enduring moral deterioration, as indicated by the many foul customs, especially as to marriage, of savage tribes. What does the Jewish tradition of the moral degeneration of man through his snatching at a knowledge forbidden him by his highest instinct assert beyond this?'

power was diminished. No doubt in this case, and in all others, many checks concur, and different checks under different circumstances; periodical dearths, depending on unfavourable seasons, being probably the most important of all. So it will have been with the early progenitors of man.

Natural Selection—We have now seen that man is variable in body and mind; and that the variations are induced, either directly or indirectly, by the same general causes, and obey the same general laws, as with the lower animals. Man has spread widely over the face of the earth, and must have been exposed, during his incessant migrations,[63] to the most diversified conditions. The inhabitants of Tierra del Fuego, the Cape of Good Hope, and Tasmania in the one hemisphere, and of the Arctic regions in the other, must have passed through many climates, and changed their habits many times, before they reached their present homes.[64] The early progenitors of man must also have tended, like all other animals, to have increased beyond their means of subsistence; they must, therefore, occasionally have been exposed to a struggle for existence, and consequently to the rigid law of natural selection. Beneficial variations of all kinds will thus, either occasionally or habitually, have been preserved, and injurious ones eliminated. I do not refer to strongly-marked deviations of structure, which occur only at long intervals of time, but to mere individual differences. We know, for instance, that the muscles of our hands and feet, which determine our powers of movement, are liable, like those of the lower animals,[65] to incessant variability. If then

---

63. See some good remarks to this effect by W. Stanley Jevons, 'A Deduction from Darwin's Theory', 'Nature', 1869, p. 231.
64. Latham, 'Man and his Migrations', 1851, p. 135.
65. Messrs, Murie and Mivart in their 'Anatomy of the Lemuroidea' ('Transact. Zoolog. Soc.' vol. vii. 1869, pp. 96–98) say, 'some muscles are so irregular in their distribution that they cannot be well classed in any of the above groups.' These muscles differ even on the opposite sides of the same individual.

the progenitors of man inhabiting any district, especially one undergoing some change in its conditions, were divided into two equal bodies, the one half which included all the individuals best adapted by their powers of movement for gaining subsistence, or for defending themselves, would on an average survive in greater numbers, and procreate more offspring than the other and less well endowed half.

Man in the rudest state in which he now exists is the most dominant animal that has ever appeared on this earth. He has spread more widely that any other highly organised form: and all others have yielded before him. He manifestly owes this immense superiority to his intellectual faculties, to his social habits, which lead him to aid and defend his fellows, and to his corporeal structure. The supreme importance of these characters has been proved by the final arbitrament of the battle for life. Through his powers of intellect, articulate language has been evolved; and on this his wonderful advancement has mainly depended. As Mr Chauncey Wright remarks:[66] 'a psychological analysis of the faculty of language shews, that even the smallest proficiency in it might require more brain power than the greatest proficiency in any other direction.' He has invented and is able to use various weapons, tools, traps, &c., with which he defends himself, kills or catches prey, and otherwise obtains food. He has made rafts or canoes for fishing or crossing over to neighbouring fertile islands. He has discovered the art of making fire, by which hard and stringy roots can be rendered digestible, and poisonous roots or herbs innocuous. This discovery of fire, probably the greatest ever made by man, excepting language, dates from before the dawn of history. These several inventions, by which man in the rudest state has become so pre-eminent, are the direct results of the development of his powers of observation, memory, curios-

66. Limits of Natural Selection, 'North American Review', Oct. 1870, p. 295.

ity, imagination, and reason. I cannot, therefore, understand how it is that Mr Wallace[67] maintains, that 'natural selection could only have endowed the savage with a brain a little superior to that of an ape'.

Although the intellectual powers and social habits of man are of paramount importance to him, we must not underrate the importance of his bodily structure, to which subject the remainder of this chapter will be devoted; the development of the intellectual and social or moral faculties being discussed in a later chapter.

Even to hammer with precision is no easy matter, as every one who has tried to learn carpentry will admit. To throw a stone with as true an aim as a Fuegian in defending himself, or in killing birds, requires the most consummate perfection in the correlated action of the muscles of the hand, arm, and shoulder, and, further, a fine sense of touch. In throwing a stone or spear, and in many other actions, a man must stand firmly on his feet; and this again demands the perfect co-adaptation of numerous muscles. To chip a flint into the rudest tool, or to form a barbed spear or hook from a bone, demands the use of a perfect hand; for, as a most capable judge, Mr Schoolcraft,[68]

67. 'Quarterly Review', April 1869, p. 392. This subject is more fully discussed in Mr Wallace's 'Contributions to the Theory of Natural Selection', 1870, in which all the essays referred to in this work are republished. The 'Essay on Man' has been ably criticised by Prof. Claparède, one of the most distinguished zoologists in Europe, in an article published in the 'Bibliothèque Universelle', June 1870. The remark quoted in my text will surprise every one who has read Mr Wallace's celebrated paper on 'The Origin of Human Races deduced from the Theory of Natural Selection', originally published in the 'Anthropological Review', May 1864, p. clviii. I cannot here resist quoting a most just remark by Sir J. Lubbock ('Prehistoric Times', 1865, p. 479) in reference to this paper, namely, that Mr Wallace, 'with characteristic unselfishness, ascribes it (i.e. the idea of natural selection) unreservedly to Mr Darwin, although, as is well known, he struck out the idea independently, and published it, though not with the same elaboration, at the same time'.

68. Quoted by Mr Lawson Tait in his 'Law of Natural Selection'—'Dublin Quarterly Journal of Medical Science', Feb. 1869. Dr Keller is likewise quoted to the same effect.

remarks, the shaping fragments of stone into knives, lances, or arrow-heads, shews 'extraordinary ability and long practice'. This is to a great extent proved by the fact that primeval men practised a division of labour; each man did not manufacture his own flint tools or rude pottery, but certain individuals appear to have devoted themselves to such work, no doubt receiving in exchange the produce of the chase. Archaeologists are convinced that an enormous interval of time elapsed before our ancestors thought of grinding chipped flints into smooth tools. One can hardly doubt, that a man-like animal who possessed a hand and arm sufficiently perfect to throw a stone with precision, or to form a flint into a rude tool, could, with sufficient practice, as far as mechanical skill alone is concerned, make almost anything which a civilised man can make. The structure of the hand in this respect may be compared with that of the vocal organs, which in the apes are used for uttering various signal-cries, or, as in one genus, musical cadences; but in man the closely similar vocal organs have become adapted through the inherited effects of use for the utterance of articulate language.

Turning now to the nearest allies of men, and therefore to the best representatives of our early progenitors, we find that the hands of the Quadrumana are constructed on the same general pattern as our own, but are far less perfectly adapted for diversified uses. Their hands do not serve for locomotion so well as the feet of a dog; as may be seen in such monkeys as the chimpanzee and orang, which walk on the outer margins of the palms, or on the knuckles.[69] Their hands, however, are admirably adapted for climbing trees. Monkeys seize thin branches or ropes, with the thumb on one side and the fingers and palm on the other, in the same manner as we do. They can thus also lift rather large ob-

69. Owen, 'Anatomy of Vertebrates', vol. iii, p. 71.

jects, such as the neck of a bottle, to their mouths. Baboons turn over stones, and scratch up roots with their hands. They seize nuts, insects, or other small objects with the thumb in opposition to the fingers, and no doubt they thus extract eggs and the young from the nests of birds. American monkeys beat the wild oranges on the branches until the rind is cracked, and then tear it off with the fingers of the two hands. In a wild state they break open hard fruits with stones. Other monkeys open mussel-shells with the two thumbs. With their fingers they pull out thorns and burs, and hunt for each other's parasites. They roll down stones, or throw them at their enemies: nevertheless, they are clumsy in these various actions, and, as I have myself seen, are quite unable to throw a stone with precision.

It seems to me far from true that because 'objects are grasped clumsily' by monkeys, 'a much less specialised organ of prehension' would have served them[70] equally well with their present hands. On the contrary, I see no reason to doubt that more perfectly constructed hands would have been an advantage to them, provided that they were not thus rendered less fitted for climbing trees. We may suspect that a hand as perfect as that of man would have been disadvantageous for climbing; for the most arboreal monkeys in the world, namely, Ateles in America, Colobus in Africa, and Hylobates in Asia, are either thumbless, or their toes partially cohere, so that their limbs are converted into mere grasping hooks.[71]

As soon as some ancient member in the great series of the Primates came to be less arboreal, owing to a change in its manner

70. 'Quarterly Review', April 1869, p. 392.
71. In *Hylobates syndactylus*, as the name expresses, two of the toes regularly cohere; and this, as Mr Blyth informs me, is occasionally the case with 'the toes of *H. agilis*, *lar*, and *leuciscus*. Colobus is strictly arboreal and extraordinarily active (Brehm, 'Thierleben', B. i. s. 50), but whether a better climber than the species of the allied genera, I do not know. It deserves notice that the feet of the sloths, the most arboreal animals in the world, are wonderfully hook-like.

of procuring subsistence, or to some change in the surrounding conditions, its habitual manner of progression would have been modified: and thus it would have been rendered more strictly quadrupedal or bipedal. Baboons frequent hilly and rocky districts, and only from necessity climb high trees;[72] and they have acquired almost the gait of a dog. Man alone has become a biped; and we can, I think, partly see how he has come to assume his erect attitude, which forms one of his most conspicuous characters. Man could not have attained his present dominant position in the world without the use of his hands, which are so admirably adapted to act in obedience to his will. Sir C. Bell[73] insists that 'the hand supplies all instruments, and by its correspondence with the intellect gives him universal dominion'. But the hands and arms could hardly have become perfect enough to have manufactured weapons, or to have hurled stones and spears with a true aim, as long as they were habitually used for locomotion and for supporting the whole weight of the body, or, as before remarked, so long as they were especially fitted for climbing trees. Such rough treatment would also have blunted the sense of touch, on which their delicate use largely depends. From these causes alone it would have been an advantage to man to become a biped; but for many actions it is indispensable that the arms and whole upper part of the body should be free; and he must for this end stand firmly on his feet. To gain this great advantage, the feet have been rendered flat; and the great toe has been peculiarly modified, though this has entailed the almost complete loss of its power of prehension. It accords with the principle of the division of physiological labour, prevailing throughout the animal kingdom, that as the hands became perfected for prehension, the feet should have become perfected for support and locomo-

72. Brehm, 'Thierleben', B. i. s. 80.
73. 'The Hand', &c. 'Bridgewater Treatise', 1833, p. 38.

tion. With some savages, however, the foot has not altogether lost its prehensile power, as shewn by their manner of climbing trees, and of using them in other ways.[74]

If it be an advantage to man to stand firmly on his feet and to have his hands and arms free, of which, from his pre-eminent success in the battle of life, there can be no doubt, then I can see no reason why it should not have been advantageous to the progenitors of man to have become more and more erect or bipedal. They would thus have been better able to defend themselves with stones or clubs, to attack their prey, or otherwise to obtain food. The best built individuals would in the long run have succeeded best, and have survived in larger numbers. If the gorilla and a few allied forms had become extinct, it might have been argued, with great force and apparent truth, that an animal could not have been gradually converted from a quadruped into a biped, as all the individuals in an intermediate condition would have been miserably ill-fitted for progression. But we know (and this is well worthy of reflection) that the anthropomorphous apes are now actually in an intermediate condition; and no one doubts that they are on the whole well adapted for their conditions of life. Thus the gorilla runs with a sidelong shambling gait, but more commonly progresses by resting on its bent hands. The long-armed apes occasionally use their arms like crutches, swinging their bodies forward between them, and some kinds of Hylobates, without having been taught, can walk or run upright with tolerable quickness; yet they move awkwardly, and much less securely than man. We see, in short, in existing monkeys a manner of progression intermediate between that of a quadru-

---

74. Häckel has an excellent discussion on the steps by which man became a biped: 'Natürliche Schöpfungsgeschichte', 1868, s. 507. Dr Büchner ('Conférences sur la Théorie Darwinienne', 1869, p. 135) has given good cases of the use of the foot as a prehensile organ by man; and has also written on the manner of progression of the higher apes, to which I allude in the following paragraph: see also Owen ('Anatomy of Vertebrates', vol. iii. p. 71) on this latter subject.

ped and a biped; but, as an unprejudiced judge[75] insists, the anthropomorphous apes approach in structure more nearly to the bipedal than to the quadrupedal type.

As the progenitors of man became more and more erect, with their hands and arms more and more modified for prehension and other purposes, with their feet and legs at the same time transformed for firm support and progression, endless other changes of structure would have become necessary. The pelvis would have to be broadened, the spine peculiarly curved, and the head fixed in an altered position, all which changes have been attained by man. Prof. Schaaffhausen[76] maintains that 'the powerful mastoid processes of the human skull are the result of his erect position', and these processes are absent in the orang, chimpanzee, &c., and are smaller in the gorilla than in man. Various other structures, which appear connected with man's erect position, might here have been added. It is very difficult to decide how far these correlated modifications are the result of natural selection, and how far of the inherited effects of the increased use of certain parts, or of the action of one part on another. No doubt these means of change often co-operate: thus when certain muscles, and the crests of bone to which they are attached, become enlarged by habitual use, this shews that certain actions are habitually performed and must be serviceable. Hence the individuals which performed them best, would tend to survive in greater numbers.

The free use of the arms and hands, partly the cause and partly the result of man's erect position, appears to have led in an indirect manner to other modifications of structure. The early male forefathers of man were, as previously stated, probably fur-

---

75. Prof. Broca, La Constitution des Vertèbres caudales; 'La Revue d'Anthropologie', 1872, p. 26, (separate copy).
76. 'On the Primitive Form of the Skull', translated in 'Anthropological Review', Oct. 1868, p. 428. Owen ('Anatomy of Vertebrates', vol. ii, 1866, p. 551) on the mastoid processes in the higher apes.

nished with great canine teeth; but as they gradually acquired the habit of using stones, clubs, or other weapons, for fighting with their enemies or rivals, they would use their jaws and teeth less and less. In this case, the jaws, together with the teeth, would become reduced in size, as we may feel almost sure from innumerable analogous cases. In a future chapter we shall meet with a closely parallel case, in the reduction or complete disappearance of the canine teeth in male ruminants, apparently in relation with the development of their horns; and in horses, in relation to their habit of fighting with their incisor teeth and hoofs.

In the adult male anthropomorphous apes, as Rütimeyer,[77] and others, have insisted, it is the effect on the skull of the great development of the jaw-muscles that causes it to differ so greatly in many respects from that of man, and has given to these animals 'a truly frightful physiognomy'. Therefore, as the jaws and teeth in man's progenitors gradually became reduced in size, the adult skull would have come to resemble more and more that of existing man. As we shall hereafter see, a great reduction of the canine teeth in the males would almost certainly affect the teeth of the females through inheritance.

As the various mental faculties gradually developed themselves the brain would almost certainly become larger. No one, I presume, doubts that the large proportion which the size of man's brain bears to his body, compared to the same proportion in the gorilla or orang, is closely connected with his higher mental powers. We meet with closely analogous facts with insects, for in ants the cerebral ganglia are of extraordinary dimensions, and in all the Hymenoptera these ganglia are many times larger than in the less intelligent orders, such as beetles.[78] On the other

---

77. 'Die Grenzen der Thierwelt, eine Betrachtung zu Darwin's Lehre', 1868, s. 51.
78. Dujardin, 'Annales des Sc. Nat', 3rd series Zoolog. tom. xiv. 1850, p. 203. See also Mr Lowne, 'Anatomy and Phys. of the *Musca vomitoria*', 1870, p. 14. My son, Mr F. Darwin, dissected for me the cerebral ganglia of the *Formica rufa*.

hand, no one supposes that the intellect of any two animals or of any two men can be accurately gauged by the cubic contents of their skulls. It is certain that there may be extraordinary mental activity with an extremely small absolute mass of nervous matter: thus the wonderfully diversified instincts, mental powers, and affections of ants are notorious, yet their cerebral ganglia are not so large as the quarter of a small pin's head. Under this point of view, the brain of an ant is one of the most marvellous atoms of matter in the world, perhaps more so than the brain of a man.

The belief that there exists in man some close relation between the size of the brain and the development of the intellectual faculties is supported by the comparison of the skulls of savage and civilised races, of ancient and modern people, and by the analogy of the whole vertebrate series. Dr J. Barnard Davis has proved,[79] by many careful measurements, that the mean internal capacity of the skull in Europeans is 92.3 cubic inches; in Americans 87.5; in Asiatics 87.1; and in Australians only 81.9 cubic inches. Professor Broca[80] found that the nineteenth century skulls from graves in Paris were larger than those from vaults of the twelfth century, in the proportion of 1484 to 1426; and that the increased size, as ascertained by measurements, was exclusively in the frontal part of the skull—the seat of the intellectual faculties. Prichard is persuaded that the present inhabitants of Britain have 'much more capacious brain-cases' than the ancient inhabitants. Nevertheless, it must be admitted that some skulls of very high antiquity, such as the famous one of Neanderthal, are well developed and capacious.[81] With respect to the lower

79. 'Philosophical Transactions', 1869, p. 513.
80. 'Les Sélections', M. P. Broca, 'Revue d'Anthropologies', 1873; see also, as quoted in C. Vogt's 'Lectures on Man', Eng. translat. 1864, pp. 88, 90. Prichard, 'Phys. Hist. of Mankind', vol. i. 1838, p. 305.
81. In the interesting article just referred to, Prof. Broca has well remarked, that in civilised nations, the average capacity of the skull must be lowered by the preservation of a considerable number of individuals, weak in mind and body, who would have been promptly eliminated in the savage state. On the other hand,

animals, M. E. Lartet,[82] by comparing the crania of tertiary and recent mammals belonging to the same groups, has come to the remarkable conclusion that the brain is generally larger and the convolutions are more complex in the more recent forms. On the other hand, I have shewn[83] that the brains of domestic rabbits are considerably reduced in bulk, in comparison with those of the wild rabbit or hare; and this may be attributed to their having been closely confined during many generations, so that they have exerted their intellect, instincts, senses and voluntary movements but little.

The gradually increasing weight of the brain and skull in man must have influenced the development of the supporting spinal column, more especially whilst he was becoming erect. As this change of position was being brought about, the internal pressure of the brain will also have influenced the form of the skull; for many facts show how easily the skull is thus affected. Ethnologists believe that it is modified by the kind of cradle in which infants sleep. Habitual spasms of the muscles, and a cicatrix from a severe burn, have permanently modified the facial bones. In young persons whose heads have become fixed either sideways or backwards, owing to disease, one of the two eyes has changed its position, and the shape of the skull has been altered apparently by the pressure of the brain in a new direction.[84] I have shewn that with long-eared rabbits even so trifling a cause as the lopping forward of one

with savages, the average includes only the more capable individuals, who have been able to survive under extremely hard conditions of life. Broca thus explains the otherwise inexplicable fact, that the mean capacity of the skull of the ancient Troglodytes of Lozère is greater than that of modern Frenchmen.

82. 'Comptes-rendus des Sciences', &c. June 1, 1868.

83. 'The Variation of Animals and Plants under Domestication', vol. i. pp. 124–129.

84. Schaaffhausen gives from Blumenbach and Busch, the cases of the spasms and cicatrix, in 'Anthropolog. Review', Oct. 1868, p. 420. Dr Jarrold ('Anthropologia', 1808, pp. 115, 116) adduces from Camper and from his own observations, cases of the modification of the skull from the head being fixed in an unnatural position. He believes that in certain trades, such as that of a shoemaker, where the head is habitually held forward, the forehead becomes more rounded and prominent.

ear drags forward almost every bone of the skull on that side; so that the bones on the opposite side no longer strictly correspond. Lastly, if any animal were to increase or diminish much in general size, without any change in its mental powers, or if the mental powers were to be much increased or diminished, without any great change in the size of the body, the shape of the skull would almost certainly be altered. I infer this from my observations on domestic rabbits, some kinds of which have become very much larger than the wild animal, whilst others have retained nearly the same size, but in both cases the brain has been much reduced relatively to the size of the body. Now I was at first much surprised on finding that in all these rabbits the skull had become elongated or dolichocephalic; for instance, of two skulls of nearly equal breadth, the one from a wild rabbit and the other from a large domestic kind, the former was 3.15 and the latter 4.3 inches in length.[85] One of the most marked distinctions in different races of men is that the skull in some is elongated, and in others rounded; and here the explanation suggested by the case of the rabbits may hold good; for Welcker finds that short 'men incline more to brachycephaly, and tall men to dolichocephaly';[86] and tall men may be compared with the larger and longer-bodied rabbits, all of which have elongated skulls, or are dolichocephalic.

From these several facts we can understand, to a certain extent, the means by which the great size and more or less rounded form of the skull have been acquired by man; and these are characters eminently distinctive of him in comparison with the lower animals.

Another most conspicuous difference between man and the lower animals is the nakedness of his skin. Whales and porpoises (Cetacea), dugongs (Sirenia) and the hippopotamus are naked; and this may be advantageous to them for gliding through the water;

85. 'Variation of Animals', &c., vol. i. p. 117, on the elongation of the skull; p. 119, on the effect of the lopping of one ear.
86. Quoted by Schaaffhausen, in 'Anthropolog. Review', Oct. 1868, p. 419.

nor would it be injurious to them from the loss of warmth, as the species, which inhabit the colder regions, are protected by a thick layer of blubber, serving the same purpose as the fur of seals and otters. Elephants and rhinoceroses are almost hairless; and as certain extinct species, which formerly lived under an Arctic climate, were covered with long wool or hair, it would almost appear as if the existing species of both genera had lost their hairy covering from exposure to heat. This appears the more probable, as the elephants in India which live on elevated and cool districts are more hairy[87] than those on the lowlands. May we then infer that man became divested of hair from having aboriginally inhabited some tropical land? That the hair is chiefly retained in the male sex on the chest and face, and in both sexes at the junction of all four limbs with the trunk, favours this inference—on the assumption that the hair was lost before man became erect; for the parts which now retain most hair would then have been most protected from the heat of the sun. The crown of the head, however, offers a curious exception, for at all times it must have been one of the most exposed parts, yet it is thickly clothed with hair. The fact, however, that the other members of the order of Primates, to which man belongs, although inhabiting various hot regions, are well clothed with hair, generally thickest on the upper surface,[88] is opposed to the supposition that man became naked through the action of the sun. Mr Belt believes[89]

87. Owen, 'Anatomy of Vertebrates', vol. iii. p. 619.
88. Isidore Geoffroy St-Hilaire remarks ('Hist. Nat. Générale', tom. ii. 1859, pp. 215–217) on the head of man being covered with long hair; also on the upper surfaces of monkeys and of other mammals being more thickly clothed than the lower surfaces. This has likewise been observed by various authors. Prof. P. Gervais ('Hist. Nat. des Mammifères', tom. i. 1854, p. 28), however, states that in the Gorilla the hair is thinner on the back, where it is partly rubbed off, than on the lower surface.
89. The 'Naturalist in Nicaragua', 1874, p. 209. As some confirmation of Mr Belt's view, I may quote the following passage from Sir W. Denison ('Varieties of Vice-Regal Life', vol. i. 1870, p. 440); 'It is said to be a practice with the Australians, when the vermin get troublesome, to singe themselves.'

that within the tropics it is an advantage to man to be destitute of hair, as he is thus enabled to free himself of the multitude of ticks (acari) and other parasites, with which he is often infested, and which sometimes cause ulceration. But whether this evil is of sufficient magnitude to have led to the denudation of his body through natural selection, may be doubted, since none of the many quadrupeds inhabiting the tropics have, as far as I know, acquired any specialised means of relief. The view which seems to me the most probable is that man, or rather primarily woman, became divested of hair for ornamental purposes, as we shall see under Sexual Selection; and, according to this belief, it is not surprising that man should differ so greatly in hairiness from all other Primates, for characters, gained through sexual selection, often differ to an extraordinary degree in closely-related forms.

According to a popular impression, the absence of a tail is eminently distinctive of man; but as those apes which come nearest to him are destitute of this organ, its disappearance does not relate exclusively to man. The tail often differs remarkably in length within the same genus: thus in some species of Macacus it is longer than the whole body, and is formed of twenty-four vertebrae; in others it consists of a scarcely visible stump, containing only three or four vertebrae. In some kinds of baboons there are twenty-five, whilst in the mandrill there are ten very small stunted caudal vertebrae, or, according to Cuvier,[90] sometimes only five. The tail, whether it be long or short, almost always tapers towards the end; and this, I presume, results from the atrophy of the terminal muscles, together with their arteries and nerves, through disuse, leading to the atrophy of the terminal bones. But no explanation can at present be given

90. Mr St George Mivart, 'Proc. Zoolog. Soc.', 1865, pp. 562, 583. Dr J. E. Gray, 'Cat. Brit. Mus: Skeletons'. Owen, 'Anatomy of Vertebrates', vol. ii. p. 517. Isidore Geoffroy, 'Hist. Nat. Gén.', tom. ii, p. 244.

of the great diversity which often occurs in its length. Here, however, we are more specially concerned with the complete external disappearance of the tail. Professor Broca has recently shewn[91] that the tail in all quadrupeds consists of two portions, generally separated abruptly from each other; the basal portion consists of vertebrae, more or less perfectly channelled and furnished with apophyses like ordinary vertebrae; whereas those of the terminal portion are not channelled, are almost smooth, and scarcely resemble true vertebrae. A tail, though not externally visible, is really present in man and the anthropomorphous apes, and is constructed on exactly the same pattern in both. In the terminal portion the vertebrae, constituting the *os coccyx*, are quite rudimentary, being much reduced in size and number. In the basal portion, the vertebrae are likewise few, are united firmly together, and are arrested in development; but they have been rendered much broader and flatter than the corresponding vertebrae in the tails of other animals: they constitute what Broca calls the accessory sacral vertebrae. These are of functional importance by supporting certain internal parts and in other ways; and their modification is directly connected with the erect or semi-erect attitude of man and the anthropomorphous apes. This conclusion is the more trustworthy, as Broca formerly held a different view, which he has now abandoned. The modification, therefore, of the basal caudal vertebrae in man and the higher apes may have been effected, directly or indirectly, through natural selection.

But what are we to say about the rudimentary and variable vertebrae of the terminal portion of the tail, forming the *os coccyx*? A notion which has often been, and will no doubt again be ridiculed, namely, that friction has had something to do with the disappearance of the external portion of the tail, is not so ridicu-

91. 'Revue d'Anthropologie', 1872; 'La Constitution des Vertèbres caudales'.

lous as it at first appears. Dr Anderson[92] states that the extremely short tail of *Macacus brunneus* is formed of eleven vertebrae, including the imbedded basal ones. The extremity is tendinous and contains no vertebrae; this is succeeded by five rudimentary ones, so minute that together they are only one line and a half in length, and these are permanently bent to one side in the shape of a hook. The free part of the tail, only a little above an inch in length, includes only four more small vertebrae. This short tail is carried erect; but about a quarter of its total length is doubled on to itself to the left; and this terminal part, which includes the hook-like portion, serves 'to fill up the interspace between the upper divergent portion of the callosities'; so that the animal sits on it, and thus renders it rough and callous. Dr Anderson thus sums up his observations: 'These facts seem to me to have only one explanation; this tail, from its short size, is in the monkey's way when it sits down, and frequently becomes placed under the animal while it is in this attitude; and from the circumstance that it does not extend beyond the extremity of the ischial tuberosities it seems as if the tail originally had been bent round, by the will of the animal, into the interspace between the callosities, to escape being pressed between them and the ground, and that in time the curvature became permanent, fitting in of itself when the organ happens to be sat upon.' Under these circumstances it is not surprising that the surface of the tail should have been roughened and rendered callous; and Dr Murie,[93] who carefully observed this species in the Zoological Gardens, as well as three other closely allied forms with slightly longer tails, says that when the animal sits down, the tail 'is necessarily thrust to one side of the buttocks; and whether long or short its root is consequently liable to be rubbed or chafed'. As we now have

92. 'Proc. Zoolog. Soc.', 1872, p. 210.
93. 'Proc. Zoolog. Soc.', 1872, p. 786.

evidence that mutilations occasionally produce an inherited effect,[94] it is not very improbable that in short-tailed monkeys, the projecting part of the tail, being functionally useless, should after many generations have become rudimentary and distorted, from being continually rubbed and chafed. We see the projecting part in this condition in the *Macacus brunneus*, and absolutely aborted in the *M. ecaudatus* and in several of the higher apes. Finally, then, as far as we can judge, the tail has disappeared in man and the anthropomorphous apes, owing to the terminal portion having been injured by friction during a long lapse of time; the basal and embedded portion having been reduced and modified, so as to become suitable to the erect or semi-erect position.

I have now endeavoured to shew that some of the most distinctive characters of man have in all probability been acquired, either directly, or more commonly indirectly, through natural selection. We should bear in mind that modifications in structure or constitution, which do not serve to adapt an organism to its habits of life, to the food which it consumes, or passively to the surrounding conditions, cannot have been thus acquired. We must not, however, be too confident in deciding what modifications are of service to each being: we should remember how little we know about the use of many parts, or what changes in the blood or tissues may serve to fit an organism for a new climate or new kinds of food. Nor must we forget the principle of correlation, by which, as Isidore Geoffroy has shewn in the case of man, many strange deviations of structure are tied together.

94. I allude to Dr Brown-Séquard's observations on the transmitted effect of an operation causing epilepsy in guinea-pigs, and likewise more recently on the analogous effects of cutting the sympathetic nerve in the neck. I shall hereafter have occasion to refer to Mr Salvin's interesting case of the apparently inherited effects of mot-mots biting off the barbs of their own tail-feathers. See also on the general subject 'Variation of Animals and Plants under Domestication', vol. ii. pp. 22–24.

Independently of correlation, a change in one part often leads, through the increased or decreased use of other parts, to other changes of a quite unexpected nature. It is also well to reflect on such facts, as the wonderful growth of galls on plants caused by the poison of an insect, and on the remarkable changes of colour in the plumage of parrots when fed on certain fishes, or inoculated with the poison of toads;[95] for we can thus see that the fluids of the system, if altered for some special purpose, might induce other changes. We should especially bear in mind that modifications acquired and continually used during past ages for some useful purpose, would probably become firmly fixed, and might be long inherited.

Thus a large yet undefined extension may safely be given to the direct and indirect results of natural selection; but I now admit, after reading the essay by Nägeli on plants, and the remarks by various authors with respect to animals, more especially those recently made by Professor Broca, that in the earlier editions of my 'Origin of Species' I perhaps attributed too much to the action of natural selection or the survival of the fittest. I have altered the fifth edition of the 'Origin' so as to confine my remarks to adaptive changes of structure; but I am convinced, from the light gained during even the last few years that very many structures which now appear to us useless, will hereafter be proved to be useful, and will therefore come within the range of natural selection. Nevertheless, I did not formerly consider sufficiently the existence of structures, which, as far as we can at present judge, are neither beneficial nor injurious and this I believe to be one of the greatest oversights as yet detected in my work. I may be permitted to say, as some excuse, that I had two distinct objects in view; firstly, to shew that species had not been separately created, and secondly, that natural selection had been

---

95. 'The Variation of Animals and Plants under Domestication', vol. ii. pp. 280, 282.

the chief agent of change, though largely aided by the inherited effects of habit, and slightly by the direct action of the surrounding conditions. I was not, however, able to annul the influence of my former belief, then almost universal, that each species had been purposely created; and this led to my tacit assumption that every detail of structure, excepting rudiments, was of some special, though unrecognised, service. Any one with this assumption in his mind would naturally extend too far the action of natural selection, either during past or present times. Some of those who admit the principle of evolution, but reject natural selection, seem to forget, when criticising my book, that I had the above two objects in view; hence if I have erred in giving to natural selection great power, which I am very far from admitting, or in having exaggerated its power, which is in itself probable, I have at least, as I hope, done good service in aiding to overthrow the dogma of separate creations.

It is, as I can now see, probable that all organic beings, including man, possess peculiarities of structure, which neither are now, nor were formerly of any service to them, and which, therefore, are of no physiological importance. We know not what produces the numberless slight differences between the individuals of each species, for reversion only carries the problem a few steps backwards; but each peculiarity must have had its efficient cause. If these causes, whatever they may be, were to act more uniformly and energetically during a lengthened period (and against this no reason can be assigned), the result would probably be not a mere slight individual difference, but a well-marked and constant modification, though one of no physiological importance. Changed structures, which are in no way beneficial, cannot be kept uniform through natural selection, though the injurious will be thus eliminated. Uniformity of character would, however, naturally follow from the assumed uniformity of the exciting causes, and likewise from the free intercrossing of many individuals. During successive periods, the same organism might in this manner acquire successive modifications, which would be transmitted in a nearly uniform state as long as the exciting causes remained the same and there was free intercrossing. With respect to the exciting causes we can only say, as when speaking of so-called spontaneous variations, that they relate much more closely to the constitution of the varying organism, then to the nature of the conditions to which it has been subjected.

Conclusion—In this chapter we have seen that as man at the present day is liable, like every other animal, to multiform individual differences or slight variations, so no doubt were the early progenitors of man; the variations being formerly induced by the same general causes, and governed by the same general and complex laws as at present. As all animals tend to multiply beyond their means of subsistence, so it must have been with the progenitors of man; and this would inevitably lead to a struggle for existence and to natural selection. The latter process would be greatly aided by the inherited effects of the increased use of parts, and these two processes would incessantly react on each other. It appears, also, as we shall hereafter see, that various unimportant characters have been acquired by man through sexual selection. An unexplained residuum of change must be left to the assumed uniform action of those unknown agencies, which occasionally induce strongly marked and abrupt deviations of structure in our domestic productions.

Judging from the habits of savages and of the greater number of the Quadrumana, primeval men, and even their ape-like progenitors, probably lived in society. With strictly social animals, natural selection sometimes acts on the individual, through the preservation of variations which are beneficial to the community. A community which includes a large number of well-endowed individuals increases in number, and is victorious over other less favoured ones; even although each separate member gains no advantage over the others of the same community. Associated insects have thus acquired many remarkable structures, which are of little or no service to the individual, such as the pollen-collecting apparatus, or the sting of the worker-bee, or the great jaws of soldier-ants. With the higher social animals, I am not aware that any structure has been modified solely for the good of the community, though some are of secondary service to it. For instance, the horns of ruminants and the great canine teeth of baboons appear to have been acquired by the males as weapons for sexual strife, but they are used in defence of the herd or troop. In regard to certain mental powers the case, as we shall see in the fifth chapter, is wholly different; for these faculties have been chiefly, or even exclusively, gained for the benefit of the community, and the individuals thereof, have at the same time gained an advantage indirectly.

It has often been objected to such views as the foregoing, that man is one of the most helpless and defenceless creatures in the world; and that during his early and less well-developed condition he would have been still more helpless. The Duke of Argyll, for instance, insists[96] that 'the human frame has diverged from the structure of brutes, in the direction of greater physical helplessness and weakness. That is to say, it is a divergence which of all others it is most impossible to ascribe to mere natural selection.' He adduces the naked and unprotected state of the body, the absence of great teeth or claws for defence, the small strength and speed of man, and his slight power of discovering food or of avoiding danger by smell. To these deficiencies there might be added one still more serious, namely, that he cannot climb quickly, and so escape from enemies. The loss of hair would not

96. 'Primeval Man', 1869, p. 66.

have been a great injury to the inhabitants of a warm country. For we know that the unclothed Fuegians can exist under a wretched climate. When we compare the defenceless state of man with that of apes, we must remember that the great canine teeth with which the latter are provided, are possessed in their full development by the males alone, and are chiefly used by them for fighting with their rivals; yet the females, which are not thus provided, manage to survive.

In regard to bodily size or strength, we do not know whether man is descended from some small species, like the chimpanzee, or from one as powerful as the gorilla; and, therefore, we cannot say whether man has become larger and stronger, or smaller and weaker, than his ancestors. We should, however, bear in mind that an animal possessing great size, strength, and ferocity, and which, like the gorilla, could defend itself from all enemies, would not perhaps have become social; and this would most effectually have checked the acquirement of the higher mental qualities, such as sympathy and the love of his fellows. Hence it might have been an immense advantage to man to have sprung from some comparatively weak creature.

The small strength and speed of man, his want of natural weapons, &c., are more than counterbalanced, firstly, by his intellectual powers, through which he has formed for himself weapons, tools, &c., though still remaining in a barbarous state, and, secondly, by his social qualities which lead him to give and receive aid from his fellow-men. No country in the world abounds in a greater degree with dangerous beasts than Southern Africa; no country presents more fearful physical hardships than the Arctic regions; yet one of the puniest of races, that of the Bushmen, maintains itself in Southern Africa, as do the dwarfed Esquimaux in the Arctic regions. The ancestors of man were, no doubt, inferior in intellect, and probably in social disposition, to the lowest existing savages; but it is quite conceivable that they might have existed, or even flourished, if they had advanced in intellect, whilst gradually losing their brute-like powers, such as that of climbing trees, &c. But these ancestors would not have been exposed to any special danger, even if far more helpless and defenceless than any existing savages, had they inhabited some warm continent or large island, such as Australia, New Guinea, or Borneo, which is now the home of the orang. And natural selection arising from the competition of tribe with tribe, in some such large area as one of these, together with the inherited effects of habit, would, under favourable conditions, have sufficed to raise man to his present high position in the organic scale.

4

Mental Powers

In the third chapter of *The Descent of Man*, Darwin moves from the physical to the mental. He seeks now to establish homologies of instinct, behavior, and thought, just as he established homologies in flesh and bone.

For physical homologies, Darwin relied on the evidence gathered by the anatomists of his day. But Darwin could not turn to an equivalent body of knowledge about animal behavior. It would take many decades before a rigorous science of animal behavior would emerge—one in which scientists ran carefully controlled experiments and amassed statistically powerful observations in the wild. Darwin had to content himself with the anecdotes of naturalists.

Even in that meager supply of data, Darwin found enough evidence to mount a powerful attack on the belief that humans had a mental life utterly distinct from that of animals. Darwin granted that the differences were real, but argued they were differences of degree, not of kind. Humans used tools; chimpanzees broke nuts with rocks. Humans could carry out abstract reasoning; foxes could learn simple rules. Humans used language; monkeys used a range of distinct calls.

Darwin's arguments about mental powers are radical, but he does not break entirely from his Victorian peers. When Darwin discusses the faculties of animals and humans, he places them along a continuum of perfection, from lower to higher. He even arranges different populations of humans along the same scale, from the lowly "savages" beset by superstitions to the lofty powers of reason in the highest reaches of civilization.

Darwin may have wanted to use this scale to further his argument for the continuity between man and beast. But it does not fit comfortably with the other image that dominates his theory of evolution: the tree of life. Darwin had published a tree in *The Origin of Species* to illustrate the branching process of evolution. As species split apart into new ones, life becomes more diverse and the tree becomes broader. They do not climb a ladder of perfection. Darwin leaves this contradiction unresolved in this chapter.

Another difficulty arises when Darwin discusses how these different mental powers arose. He argues that the capacities found in all humans, such as language, evolved by natural selection from older behaviors. Darwin also claims that different peoples have different mental powers. But he is not clear about whether those differences emerged by natural selection. Did Europeans invent calculus because their mathematicians had more offspring than those who were bad with numbers? Darwin gives us little grounds to know how he'd answer that question in this chapter. But as with many topics in *The Descent of Man*, he will return to it later.

Despite these inconsistencies, much of Darwin's argument in this chapter stands strong today. Scientists have found a vast number of similarities in the mental faculties of humans and other animals—similarities that can only be explained through a shared evolutionary history. Along with experiments on and observations of wild animals, scientists are also starting to dissect the genetic changes underlying the rise of human mental powers.

Consider human language. Scientists suspect that a number of genes are essential to the development of language, but they've only identified one strong candidate, known as FOXP2. Mutations to FOXP2 can cause devastating difficulties in speaking and understanding grammar. Scientists have discovered other versions of FOXP2 in other mammals. To learn its function in other species, researchers have experimentally shut down FOXP2 in mouse embryos. Normally mouse pups squeak to their mothers to ensure they are properly cared for. Without FOXP2, however, mouse pups cannot squeak properly. These results suggest that FOXP2 took on a role in animal communication starting at least one hundred million years ago. In most mammals scientists have surveyed, FOXP2's sequence has barely changed. The FOXP2 gene carried by a chimpanzee is practically identical to that of a mouse. However, humans—and humans alone—have undergone a drastic evolution in the FOXP2 gene, and probably in the past two hundred thousand years.

The FOXP2 story illustrates precisely the distinction Darwin was making between humans and animals. In some ways we truly are unique, but underneath that uniqueness lies a common bond with other species.

CHAPTER 3:
Comparison of the Mental Powers of Man and the Lower Animals

WE HAVE SEEN in the last two chapters that man bears in his bodily structure clear traces of his descent from some lower form; but it may be urged that, as man differs so greatly in his mental power from all other animals, there must be some error in this conclusion. No doubt the difference in this respect is enormous, even if we compare the mind of one of the lowest savages, who has no words to express any number higher than four, and who uses hardly any abstract terms for common objects or for the affections,[1] with that of the most highly organised ape. The difference would, no doubt, still remain immense, even if one of the higher apes had been improved or civilised as much as a dog has been in comparison with its parent-form, the wolf or jackal. The Fuegians rank amongst the lowest barbarians; but I was continually struck with surprise how closely the three natives on board HMS 'Beagle', who had lived some years in England, and could talk a little English, resembled us in disposition and in most of our mental faculties. If no organic being excepting man had possessed any mental power, or if his powers had been of a wholly different nature from those of the lower animals, then we should

---

1. See the evidence on those points, as given by Lubbock, 'Prehistoric Times', p. 354, &c.

never have been able to convince ourselves that our high faculties had been gradually developed. But it can be shewn that there is no fundamental difference of this kind. We must also admit that there is a much wider interval in mental power between one of the lowest fishes, as a lamprey or lancelet, and one of the higher apes, than between an ape and man; yet this interval is filled up by numberless gradations.

Nor is the difference slight in moral disposition between a barbarian, such as the man described by the old navigator Byron, who dashed his child on the rocks for dropping a basket of sea-urchins, and a Howard or Clarkson; and in intellect, between a savage who uses hardly any abstract terms, and a Newton or Shakespeare. Differences of this kind between the highest men of the highest races and the lowest savages, are connected by the finest gradations. Therefore it is possible that they might pass and be developed into each other.

My object in this chapter is to shew that there is no fundamental difference between man and the higher mammals in their mental faculties. Each division of the subject might have been extended into a separate essay, but must here be treated briefly. As no classification of the mental powers has been universally accepted, I shall arrange my remarks in the order most convenient for my purpose; and will select those facts which have struck me most, with the hope that they may produce some effect on the reader.

With respect to animals very low in the scale, I shall give some additional facts under Sexual Selection, shewing that their mental powers are much higher than might have been expected. The variability of the faculties in the individuals of the same species is an important point for us, and some few illustrations will here be given. But it would be superfluous to enter into many details on this head, for I have found on frequent enquiry, that it is the unanimous opinion of all those who have long attended to animals of many kinds, including birds, that the individuals differ greatly in every mental characteristic. In what manner the mental powers were first developed in the lowest organisms, is as hopeless an enquiry as how life itself first originated. These are problems for the distant future, if they are ever to be solved by man.

As man possesses the same senses as the lower animals, his fundamental intuitions must be the same. Man has also some few instincts in common, as that of self-preservation, sexual love, the love of the mother for her new-born offspring, the desire possessed by the latter to suck, and so forth. But man, perhaps, has somewhat fewer instincts than those possessed by the animals which come next to him in the series. The orang in the Eastern islands, and the chimpanzee in Africa, build platforms on which they sleep; and, as both species follow the same habit, it might be argued that this was due to instinct, but we cannot feel sure that it is not the result of both animals having similar wants, and possessing similar powers of reasoning. These apes, as we may assume, avoid the many poisonous fruits of the tropics, and man has no such knowledge: but as our domestic animals, when taken to foreign lands, and when first turned out in the spring, often eat poisonous herbs, which they afterwards avoid, we cannot feel sure that the apes do not learn from their own experience or from that of their parents what fruits to select. It is, however, certain, as we shall presently see, that apes have an instinctive dread of serpents, and probably of other dangerous animals.

The fewness and the comparative simplicity of the instincts in the higher animals are remarkable in contrast with those of the lower animals. Cuvier maintained that instinct and intelligence stand in an inverse ratio to each other; and some have thought that the intellectual faculties of the higher animals have been gradually developed from their instincts. But Pouchet, in an interesting essay,[2] has shewn that no such inverse ratio really

2. 'L'Instinct chez les Insectes', 'Revue des Deux Mondes', Feb. 1870, p. 690.

exists. Those insects which possess the most wonderful instincts are certainly the most intelligent. In the vertebrate series, the least intelligent members, namely fishes and amphibians, do not possess complex instincts; and amongst mammals the animal most remarkable for its instincts, namely the beaver, is highly intelligent, as will be admitted by every one who has read Mr Morgan's excellent work.[3]

Although the first dawnings of intelligence, according to Mr Herbert Spencer,[4] have been developed through the multiplication and co-ordination of reflex actions, and although many of the simpler instincts graduate into reflex actions, and can hardly be distinguished from them, as in the case of young animals sucking, yet the more complex instincts seem to have originated independently of intelligence. I am, however, very far from wishing to deny that instinctive actions may lose their fixed and untaught character, and be replaced by others performed by the aid of the free will. On the other hand, some intelligent actions, after being performed during several generations, become converted into instincts and are inherited, as when birds on oceanic islands learn to avoid man. These actions may then be said to be degraded in character, for they are no longer performed through reason or from experience. But the greater number of the more complex instincts appear to have been gained in a wholly different manner, through the natural selection of variations of simpler instinctive actions. Such variations appear to arise from the same unknown causes acting on the cerebral organisation, which induce slight variations or individual differences in other parts of the body; and these variations, owing to our ignorance, are often said to arise spontaneously. We can, I think, come to no other conclusion

3. 'The American Beaver and his Works', 1868.
4. 'The Principles of Psychology', 2nd edit. 1870, pp. 418–443.

with respect to the origin of the more complex instincts, when we reflect on the marvellous instincts of sterile worker-ants and bees, which leave no offspring to inherit the effects of experience and of modified habits.

Although, as we learn from the above-mentioned insects and the beaver, a high degree of intelligence is certainly compatible with complex instincts, and although actions, at first learnt voluntarily can soon through habit be performed with the quickness and certainty of a reflex action, yet it is not improbable that there is a certain amount of interference between the development of free intelligence and of instinct—which latter implies some inherited modification of the brain. Little is known about the functions of the brain, but we can perceive that as the intellectual powers become highly developed, the various parts of the brain must be connected by very intricate channels of the freest intercommunication; and as a consequence, each separate part would perhaps tend to be less well fitted to answer to particular sensations or associations in a definite and inherited—that is instinctive—manner. There seems even to exist some relation between a low degree of intelligence and a strong tendency to the formation of fixed, though not inherited habits; for as a sagacious physician remarked to me, persons who are slightly imbecile tend to act in everything by routine or habit; and they are rendered much happier if this is encouraged.

I have thought this digression worth giving, because we may easily underrate the mental powers of the higher animals, and especially of man, when we compare their actions founded on the memory of past events, on foresight, reason, and imagination, with exactly similar actions instinctively performed by the lower animals; in this latter case the capacity of performing such actions has been gained, step by step, through the variability of the mental organs and natural selection, without any conscious intelligence on the part of the animal during each successive generation. No doubt, as Mr Wallace has argued,[5] much of the intelligent work done by man is due to imitation and not to reason; but there is this great difference between his actions and many of those performed by the lower animals, namely, that man cannot, on his first trial, make, for instance, a stone hatchet or a canoe, through his power of imitation. He has to learn his work by practice; a beaver, on the other hand, can make its dam or canal, and a bird its nest, as well, or nearly as well, and a spider its wonderful web, quite as well,[6] the first time it tries, as when old and experienced.

To return to our immediate subject: the lower animals, like man, manifestly feel pleasure and pain, happiness and misery. Happiness is never better exhibited than by young animals, such as puppies, kittens, lambs, &c., when playing together, like our own children. Even insects play together, as has been described by that excellent observer, P. Huber,[7] who saw ants chasing and pretending to bite each other, like so many puppies.

The fact that the lower animals are excited by the same emotions as ourselves is so well established, that it will not be necessary to weary the reader by many details. Terror acts in the same manner on them as on us, causing the muscles to tremble, the heart to palpitate, the sphincters to be relaxed, and the hair to stand on end. Suspicion, the offspring of fear, is eminently characteristic of most wild animals. It is, I think, impossible to read the account given by Sir E. Tennent, of the behaviour of the female elephants, used as decoys, without admitting that they intentionally practise deceit, and well know what they are about. Courage and timidity are extremely variable qualities in the individuals of the same species, as is plainly seen in our dogs. Some

5. 'Contributions to the Theory of Natural Selection', 1870, p. 212.
6. For the evidence on this head, see Mr J. Traherne Moggridge's most interesting work, 'Harvesting Ants and Trap-door Spiders', 1873, pp. 126, 128.
7. 'Recherches sur les Moeurs des Fourmis', 1810, p. 173.

dogs and horses are ill-tempered, and easily turn sulky; others are good-tempered; and these qualities are certainly inherited. Every one knows how liable animals are to furious rage, and how plainly they show it. Many, and probably true, anecdotes have been published on the long-delayed and artful revenge of various animals. The accurate Rengger, and Brehm[8] state that the American and African monkeys which they kept tame, certainly revenged themselves. Sir Andrew Smith, a zoologist whose scrupulous accuracy was known to many persons, told me the following story of which he was himself an eye-witness; at the Cape of Good Hope an officer had often plagued a certain baboon, and the animal, seeing him approaching one Sunday for parade, poured water into a hole and hastily made some thick mud, which he skilfully dashed over the officer as he passed by, to the amusement of many bystanders. For long afterwards the baboon rejoiced and triumphed whenever he saw his victim.

The love of a dog for his master is notorious; as an old writer quaintly says,[9] 'A dog is the only thing on this earth that luvs you more than he luvs himself.'

In the agony of death a dog has been known to caress his master, and every one has heard of the dog suffering under vivisection, who licked the hand of the operator; this man, unless the operation was fully justified by an increase of our knowledge, or unless he had a heart of stone, must have felt remorse to the last hour of his life.

As Whewell[10] has well asked, 'who that reads the touching instances of maternal affection, related so often of the women of all nations, and of the females of all animals, can doubt that the

---

8. All the following statements, given on the authority of these two naturalists, are taken from Rengger's 'Naturgesch, der Säugethiere von Paraguay', 1830, s. 41–57, and from Brehm's 'Thierleben', B. i. s. 10–87.
9. Quoted by Dr Lauder Lindsay, in his 'Physiology of Mind in the Lower Animals'; 'Journal of Mental Science', April 1871, p. 38.
10. 'Bridgewater Treatise', p. 263.

principle of action is the same in the two cases?' We see maternal affection exhibited in the most trifling details; thus Rengger
observed an American monkey (a Cebus) carefully driving away
the flies which plagued her infant; and Duvaucel saw a Hylobates washing the faces of her young ones in a stream. So intense
is the grief of female monkeys for the loss of their young, that
it invariably caused the death of certain kinds kept under confinement by Brehm in N. Africa. Orphan monkeys were always
adopted and carefully guarded by the other monkeys, both males
and females. One female baboon had so capacious a heart that
she not only adopted young monkeys of other species, but stole
young dogs and cats, which she continually carried about. Her
kindness, however, did not go so far as to share her food with her
adopted offspring, at which Brehm was surprised, as his monkeys always divided everything quite fairly with their own young
ones. An adopted kitten scratched this affectionate baboon, who
certainly had a fine intellect, for she was much astonished at
being scratched, and immediately examined the kitten's feet, and
without more ado bit off the claws.[11] In the Zoological Gardens,
I heard from the keeper that an old baboon (*C. chacma*) had adopted a Rhesus monkey; but when a young drill and mandrill
were placed in the cage, she seemed to perceive that these monkeys, though distinct species, were her nearer relatives, for she at
once rejected the Rhesus and adopted both of them. The young
Rhesus, as I saw, was greatly discontented at being thus rejected,
and it would, like a naughty child, annoy and attack the young
drill and mandrill whenever it could do so with safety; this conduct exciting great indignation in the old baboon. Monkeys will
also, according to Brehm, defend their master when attacked by

11. A critic, without any grounds ('Quarterly Review', July 1871, p. 72), disputes
the possibility of this act as described by Brehm, for the sake of discrediting my
work. Therefore I tried, and found that I could readily seize with my own teeth
the sharp little claws of a kitten nearly five weeks old.

any one, as well as dogs to whom they are attached, from the attacks of other dogs. But we here trench on the subjects of sympathy and fidelity, to which I shall recur. Some of Brehm's monkeys took much delight in teasing a certain old dog whom they disliked, as well as other animals, in various ingenious ways.

Most of the more complex emotions are common to the higher animals and ourselves. Every one has seen how jealous a dog is of his master's affection, if lavished on any other creature; and I have observed the same fact with monkeys. This shews that animals not only love, but have desire to be loved. Animals manifestly feel emulation. They love approbation or praise; and a dog carrying a basket for his master exhibits in a high degree self-complacency or pride. There can, I think, be no doubt that a dog feels shame, as distinct from fear, and something very like modesty when begging too often for food. A great dog scorns the snarling of a little dog, and this may be called magnanimity. Several observers have stated that monkeys certainly dislike being laughed at; and they sometimes invent imaginary offences. In the Zoological Gardens I saw a baboon who always got into a furious rage when his keeper took out a letter or book and read it aloud to him; and his rage was so violent that, as I witnessed on one occasion, he bit his own leg till the blood flowed. Dogs show what may be fairly called a sense of humour, as distinct from mere play; if a bit of stick or other such object be thrown to one, he will often carry it away for a short distance; and then squatting down with it on the ground close before him, will wait until his master comes quite close to take it away. The dog will then seize it and rush away in triumph, repeating the same manoeuvre, and evidently enjoying the practical joke.

We will now turn to the more intellectual emotions and faculties, which are very important, as forming the basis for the development of the higher mental powers. Animals manifestly enjoy excitement, and suffer from ennui, as may be seen with dogs, and, according to Rengger, with monkeys. All animals feel *Wonder*, and many exhibit *Curiosity*. They sometimes suffer from this latter quality, as when the hunter plays antics and thus attracts them; I have witnessed this with deer, and so it is with the wary chamois, and with some kinds of wild-ducks. Brehm gives a curious account of the instinctive dread, which his monkeys exhibited, for snakes; but their curiosity was so great that they could not desist from occasionally satiating their horror in a most human fashion, by lifting up the lid of the box in which the snakes were kept. I was so much surprised at his account, that I took a stuffed and coiled-up snake into the monkey-house at the Zoological Gardens, and the excitement thus caused was one of the most curious spectacles which I ever beheld. Three species of Cercopithecus were the most alarmed; they dashed about their cages, and uttered sharp signal cries of danger, which were understood by the other monkeys. A few young monkeys and one old Anubis baboon alone took no notice of the snake. I then placed the stuffed specimen on the ground in one of the larger compartments. After a time all the monkeys collected round it in a large circle, and staring intently, presented a most ludicrous appearance. They became extremely nervous; so that when a wooden ball, with which they were familiar as a plaything, was accidentally moved in the straw, under which it was partly hidden, they all instantly started away. These monkeys behaved very differently when a dead fish, a mouse,[12] a living turtle, and other new objects were placed in their cages; for though at first frightened, they soon approached, handled and examined them. I then placed a live snake in a paper bag, with the mouth loosely closed, in one of the larger compartments. One of the monkeys im-

---

12. I have given a short account of their behaviour on this occasion in my 'Expression of the Emotions', p. 43.

mediately approached, cautiously opened the bag a little, peeped in, and instantly dashed away. Then I witnessed what Brehm has described, for monkey after monkey, with head raised high and turned on one side, could not resist taking a momentary peep into the upright bag, at the dreadful object lying quietly at the bottom. It would almost appear as if monkeys had some notion of zoological affinities, for those kept by Brehm exhibited a strange, though mistaken, instinctive dread of innocent lizards and frogs. An orang, also, has been known to be much alarmed at the first sight of a turtle.[13]

The principle of *Imitation* is strong in man, and especially, as I have myself observed, with savages. In certain morbid states of the brain this tendency is exaggerated to an extraordinary degree; some hemiplegic patients and others, at the commencement of inflammatory softening of the brain, unconsciously imitate every word which is uttered, whether in their own or in a foreign language, and every gesture or action which is performed near them.[14] Desor[15] has remarked that no animal voluntarily imitates an action performed by man, until in the ascending scale we come to monkeys, which are well known to be ridiculous mockers. Animals, however, sometimes imitate each other's actions: thus two species of wolves, which had been reared by dogs, learned to bark, as does sometimes the jackal,[16] but whether this can be called voluntary imitation is another question. Birds imitate the songs of their parents, and sometimes of other birds; and parrots are notorious imitators of any sound which they often hear. Dureau de la Malle gives an account[17] of a dog reared by a cat, who learnt to imitate the well-known action of a cat licking

---

13. W. C. L. Martin, 'Nat. Hist. of Mammalia', 1841, p. 405.
14. Dr Bateman 'On Aphasia', 1870, p. 110.
15. Quoted by Vogt, 'Mémoire sur les Microcéphales', 1867, p. 168.
16. 'The Variation of Animals and Plants under Domestication', vol. i. p. 27.
17. 'Annales des Sc. Nat.' (1st Series), tom. xxii. p. 397.

her paws, and thus washing her ears and face; this was also witnessed by the celebrated naturalist Audouin. I have received several confirmatory accounts; in one of these, a dog had not been suckled by a cat, but had been brought up with one, together with kittens, and had thus acquired the above habit, which he ever afterwards practised during his life of thirteen years. Durean de la Malle's dog likewise learnt from the kittens to play with a ball by rolling it about with his fore paws, and springing on it. A correspondent assures me that a cat in his house used to put her paws into jugs of milk having too narrow a mouth for her head. A kitten of this cat soon learned the same trick, and practised it ever afterwards, whenever there was an opportunity.

The parents of many animals, trusting to the principle of imitation in their young, and more especially to their instinctive or inherited tendencies, may be said to educate them. We see this when a cat brings a live mouse to her kittens; and Dureau de la Malle has given a curious account (in the paper above quoted) of his observations on hawks which taught their young dexterity, as well as judgment of distances, by first dropping through the air dead mice and sparrows, which the young generally failed to catch, and then bringing them live birds and letting them loose.

Hardly any faculty is more important for the intellectual progress of man than *Attention*. Animals clearly manifest this power, as when a cat watches by a hole and prepares to spring on its prey. Wild animals sometimes become so absorbed when thus engaged, that they may be easily approached. Mr Bartlett has given me a curious proof how variable this faculty is in monkeys. A man who trains monkeys to act in plays, used to purchase common kinds from the Zoological Society at the price of five pounds for each; but he offered to give double the price, if he might keep three or four of them for a few days, in order to select one. When asked how he could possibly learn so soon, whether a particular monkey would turn out a good actor, he answered that it all depended on their power of attention. If, when he was talking and explaining anything to a monkey, its attention was easily distracted, as by a fly on the wall or other trifling object, the case was hopeless. If he tried by punishment to make an inattentive monkey act, it turned sulky. On the other hand, a monkey which carefully attended to him could always be trained.

It is almost superfluous to state that animals have excellent *Memories* for persons and places. A baboon at the Cape of Good Hope, as I have been informed by Sir Andrew Smith, recognised him with joy after an absence of nine months. I had a dog who was savage and averse to all strangers, and I purposely tried his memory after an absence of five years and two days. I went near the stable where he lived, and shouted to him in my old manner; he shewed no joy, but instantly followed me out walking, and obeyed me, exactly as if I had parted with him only half an hour before. A train of old associations, dormant during five years, had thus been instantaneously awakened in his mind. Even ants, as P. Huber[18] has clearly shewn, recognised their fellow-ants belonging to the same community after a separation of four months. Animals can certainly by some means judge of the intervals of time between recurrent events.

The *Imagination* is one of the highest prerogatives of man. By this faculty he unites former images and ideas, independently of the will, and thus creates brilliant and novel results. A poet, as Jean Paul Richter remarks,[19] 'who must reflect whether he shall make a character say yes or no—to the devil with him; he is only a stupid corpse'. Dreaming gives us the best notion of this power; as Jean Paul again says, 'The dream is an involuntary art of poetry'. The value of the products of our imagination de-

18. 'Les Moeurs des Fourmis', 1810, p. 150.
19. Quoted in Dr Maudsley's 'Physiology and Pathology of Mind', 1868, pp. 19, 220.

pends of course on the number, accuracy, and clearness of our impressions, on our judgment and taste in selecting or rejecting the involuntary combinations, and to a certain extent on our power of voluntarily combining them. As dogs, cats, horses, and probably all the higher animals, even birds[20] have vivid dreams, and this is shewn by their movements and the sounds uttered, we must admit that they possess some power of imagination. There must be something special, which causes dogs to howl in the night, and especially during moonlight, in that remarkable and melancholy manner called baying. All dogs do not do so; and, according to Houzeau,[21] they do not then look at the moon, but at some fixed point near the horizon. Houzeau thinks that their imaginations are disturbed by the vague outlines of the surrounding objects, and conjure up before them fantastic images: if this be so, their feelings may almost be called superstitious.

Of all the faculties of the human mind, it will, I presume, be admitted that *Reason* stands at the summit. Only a few persons now dispute that animals possess some power of reasoning. Animals may constantly be seen to pause, deliberate, and resolve. It is a significant fact, that the more the habits of any particular animal are studied by a naturalist, the more he attributes to reason and the less to unlearnt instincts.[22] In future chapters we shall see that some animals extremely low in the scale apparently display a certain amount of reason. No doubt it is often difficult to distinguish between the power of reason and that of instinct. For instance, Dr Hayes, in his work on 'The Open Polar Sea', repeatedly remarks that his dogs, instead of continuing to draw the sledges in a compact body, diverged and separated when they

20. Dr Jerdon, 'Birds of India', vol. i. 1862, p. xxi. Houzeau says that his parakeets and canary-birds dreamt: 'Facultés Mentales', tom. ii. p. 136.
21. 'Facultés Mentales des Animaux', 1872, tom. ii. p. 181.
22. Mr L. H. Morgan's work on 'The American Beaver', 1868, offers a good illustration of this remark. I cannot help thinking, however, that he goes too far in underrating the power of Instinct.

came to thin ice, so that their weight might be more evenly distributed. This was often the first warning which the travellers received that the ice was becoming thin and dangerous. Now, did the dogs act thus from the experience of each individual, or from the example of the older and wiser dogs, or from an inherited habit, that is from instinct? This instinct, may possibly have arisen since the time, long ago, when dogs were first employed by the natives in drawing their sledges; or the Arctic wolves, the parent-stock of the Esquimaux dog, may have acquired an instinct, impelling them not to attack their prey in a close pack, when on thin ice.

We can only judge by the circumstances under which actions are performed, whether they are due to instinct, or to reason, or to the mere association of ideas: this latter principle, however, is intimately connected with reason. A curious case has been given by Prof. Möbius,[23] of a pike, separated by a plate of glass from an adjoining aquarium stocked with fish, and who often dashed himself with such violence against the glass in trying to catch the other fishes, that he was sometimes completely stunned. The pike went on thus for three months, but at last learnt caution, and ceased to do so. The plate of glass was then removed, but the pike would not attack these particular fishes, though he would devour others which were afterwards introduced; so strongly was the idea of a violent shock associated in his feeble mind with the attempt on his former neighbours. If a savage, who had never seen a large plate-glass window, were to dash himself even once against it, he would for a long time afterwards associate a shock with a window-frame; but very differently from the pike, he would probably reflect on the nature of the impediment, and be cautious under analogous circumstances. Now with monkeys, as we shall presently see, a painful or merely a disagreeable impres-

23. 'Die Bewegungen der Thiere', &c., 1873, p. 11.

sion, from an action once performed, is sometimes sufficient to prevent the animal from repeating it. If we attribute this difference between the monkey and the pike solely to the association of ideas being so much stronger and more persistent in the one than the other, though the pike often received much the more severe injury, can we maintain in the case of man that a similar difference implies the possession of a fundamentally different mind?

Houzeau relates[24] that, whilst crossing a wide and arid plain in Texas, his two dogs suffered greatly from thirst, and that between thirty and forty times they rushed down the hollows to search for water. These hollows were not valleys, and there were no trees in them, or any other difference in the vegetation, and as they were absolutely dry there could have been no smell of damp earth. The dogs behaved as if they knew that a dip in the ground offered them the best chance of finding water, and Houzeau has often witnessed the same behaviour in other animals.

I have seen, as I daresay have others, that when a small object is thrown on the ground beyond the reach of one of the elephants in the Zoological Gardens, he blows through his trunk on the ground beyond the object, so that the current reflected on all sides may drive the object within his reach. Again a well-known ethnologist, Mr Westropp, informs me that he observed in Vienna a bear deliberately making with his paw a current in some water, which was close to the bars of his cage, so as to draw a piece of floating bread within his reach. These actions of the elephant and bear can hardly be attributed to instinct or inherited habit, as they would be of little use to an animal in a state of nature. Now, what is the difference between such actions, when performed by an uncultivated man, and by one of the higher animals?

24. 'Facultés Mentales des Animaux', 1872, tom. ii. p. 265.

The savage and the dog have often found water at a low level, and the coincidence under such circumstances has become associated in their minds. A cultivated man would perhaps make some general proposition on the subject; but from all that we know of savages it is extremely doubtful whether they would do so, and a dog certainly would not. But a savage, as well as a dog, would search in the same way, though frequently disappointed; and in both it seems to be equally an act of reason, whether or not any general proposition on the subject is consciously placed before the mind.[25] The same would apply to the elephant and the bear making currents in the air or water. The savage would certainly neither know nor care by what law the desired movements were effected; yet his act would be guided by a rude process of reasoning, as surely as would a philosopher in his longest chain of deductions. There would no doubt be this difference between him and one of the higher animals, that he would take notice of much slighter circumstances and conditions, and would observe any connection between them after much less experience, and this would be of paramount importance. I kept a daily record of the actions of one of my infants, and when he was about eleven months old, and before he could speak a single word, I was continually struck with the greater quickness, with which all sorts of objects and sounds were associated together in his mind, compared with that of the most intelligent dogs I ever knew. But the higher animals differ in exactly the same way in this power of association from those low in the scale, such as the pike, as well as in that of drawing inferences and of observation.

The promptings of reason, after very short experience, are well shewn by the following actions of American monkeys,

---

25. Prof. Huxley has analysed with admirable clearness the mental steps by which a man, as well as a dog, arrives at a conclusion in a case analogous to that given in my text. See his article, 'Mr Darwin's Critics', in the 'Contemporary Review', Nov. 1871, p. 462, and in his 'Critiques and Essays', 1873, p. 279.

which stand low in their order. Rengger, a most careful observer, states that when he first gave eggs to his monkeys in Paraguay, they smashed them, and thus lost much of their contents; afterwards they gently hit one end against some hard body, and picked off the bits of shell with their fingers. After cutting themselves only *once* with any sharp tool, they would not touch it again, or would handle it with the greatest caution. Lumps of sugar were often given them wrapped up in paper; and Rengger sometimes put a live wasp in the paper, so that in hastily unfolding it they got stung; after this had *once* happened, they always first held the packet to their ears to detect any movement within.[26]

The following cases relate to dogs. Mr Colquhoun[27] winged two wild-ducks, which fell on the further side of a stream; his retriever tried to bring over both at once, but could not succeed; she then, though never before known to ruffle a feather, deliberately killed one, brought over the other, and returned for the dead bird. Col. Hutchinson relates that two partridges were shot at once, one being killed, the other wounded; the latter ran away, and was caught by the retriever, who on her return came across the dead bird; 'she stopped, evidently greatly puzzled, and after one or two trials, finding she could not take it up without permitting the escape of the winged bird, she considered a moment, then deliberately murdered it by giving it a severe crunch, and afterwards brought away both together. This was the only known instance of her ever having wilfully injured any game.' Here we have reason though not quite perfect, for the retriever might have brought the wounded bird first and then returned for the dead one, as in the case of the two wild-ducks. I give

26. Mr Belt, in his most interesting work, 'The Naturalist in Nicaragua', 1874 (p. 119), likewise describes various actions of a tamed Cebus, which, I think, clearly shew that this animal possessed some reasoning power.
27. 'The Moor and the Loch', p. 45. Col. Hutchinson on 'Dog Breaking', 1850, p. 46.

the above cases, as resting on the evidence of two independent witnesses, and because in both instances the retrievers, after deliberation, broke through a habit which is inherited by them (that of not killing the game retrieved), and because they shew how strong their reasoning faculty must have been to overcome a fixed habit.

I will conclude by quoting a remark by the illustrious Humboldt.[28] 'The muleteers in S. America say, "I will not give you the mule whose step is easiest, but *la mas racional*—the one that reasons best";' and as he adds, 'this popular expression, dictated by long experience, combats the system of animated machines, better perhaps than all the arguments of speculative philosophy.' Nevertheless some writers even yet deny that the higher animals possess a trace of reason; and they endeavour to explain away, by what appears to be mere verbiage,[29] all such facts as those above given.

It has, I think, now been shewn that man and the higher animals, especially the Primates, have some few instincts in common. All have the same senses, intuitions, and sensations—similar passions, affections, and emotions, even the more complex ones, such as jealousy, suspicion, emulation, gratitude, and magnanimity; they practise deceit and are revengeful; they are sometimes susceptible to ridicule, and even have a sense of humour; they feel wonder and curiosity; they possess the same faculties of imitation, attention, deliberation, choice, memory, imagination, the

---

28. 'Personal Narrative', Eng. translat., vol. iii. p. 106.
29. I am glad to find that so acute a reasoner as Mr Leslie Stephen ('Darwinism and Divinity, Essays on Free-thinking', 1873, p. 80), in speaking of the supposed impassable barrier between the minds of man and the lower animals, says, 'The distinctions, indeed, which have been drawn, seem to us to rest upon no better foundation than a great many other metaphysical distinctions; that is, the assumption that because you can give two things different names, they must therefore have different natures. It is difficult to understand how anybody who has ever kept a dog, or seen an elephant, can have any doubts as to an animal's power of performing the essential processes of reasoning.'

association of ideas, and reason, though in very different degrees. The individuals of the same species graduate in intellect from absolute imbecility to high excellence. They are also liable to insanity, though far less often than in the case of man.[30] Nevertheless, many authors have insisted that man is divided by an insuperable barrier from all the lower animals in his mental faculties. I formerly made a collection of above a score of such aphorisms, but they are almost worthless, as their wide difference and number prove the difficulty, if not the impossibility, of the attempt. It has been asserted that man alone is capable of progressive improvement; that he alone makes use of tools or fire, domesticates other animals, or possesses property; that no animal has the power of abstraction, or of forming general concepts, is self-conscious and comprehends itself; that no animal employs language; that man alone has a sense of beauty, is liable to caprice, has the feeling of gratitude, mystery, &c.; believes in God, or is endowed with a conscience. I will hazard a few remarks on the more important and interesting of these points.

Archbishop Sumner formerly maintained[31] that man alone is capable of progressive improvement. That he is capable of incomparably greater and more rapid improvement than is any other animal, admits of no dispute; and this is mainly due to his power of speaking and handing down his acquired knowledge. With animals, looking first to the individual, every one who has had any experience in setting traps, knows that young animals can be caught much more easily than old ones; and they can be much more easily approached by an enemy. Even with respect to old animals, it is impossible to catch many in the same place and in the same kind of trap, or to destroy them by the same kind of poison; yet it is improbable that all should have partaken of

30. See 'Madness in Animals', by Dr W. Lauder Lindsay, in 'Journal of Mental Science', July 1871.
31. Quoted by Sir C. Lyell, 'Antiquity of Man', p. 497.

the poison, and impossible that all should have been caught in a trap. They must learn caution by seeing their brethren caught or poisoned. In North America, where the fur-bearing animals have long been pursued, they exhibit, according to the unanimous testimony of all observers, an almost incredible amount of sagacity, caution and cunning; but trapping has been there so long carried on, that inheritance may possibly have come into play. I have received several accounts that when telegraphs are first set up in any district, many birds kill themselves by flying against the wires, but that in the course of a very few years they learn to avoid this danger, by seeing, as it would appear, their comrades killed.[32]

If we look to successive generations, or to the race, there is no doubt that birds and other animals gradually both acquire and lose caution in relation to man or other enemies;[33] and this caution is certainly in chief part an inherited habit or instinct, but in part the result of individual experience. A good observer, Leroy,[34] states, that in districts where foxes are much hunted, the young, on first leaving their burrows, are incontestably much more wary than the old ones in districts where they are not much disturbed.

Our domestic dogs are descended from wolves and jackals,[35] and though they may not have gained in cunning, and may have lost in wariness and suspicion, yet they have progressed in certain moral qualities, such as in affection, trust-worthiness, temper, and probably in general intelligence. The common rat has conquered and beaten several other species throughout Europe, in parts of North America, New Zealand, and recently in Formosa, as well

32. For additional evidence, with details, see M. Houzeau, 'Les Facultés Mentales', tom. ii. 1872, p. 147.
33. See, with respect to birds on oceanic islands, my 'Journal of Researches during the voyage of the "Beagle" ', 1845, p. 398. 'Origin of Species', 5th edit. p. 260.
34. 'Lettres Phil. sur l'Intelligence des Animaux', nouvelle édit. 1802, p. 86.
35. See the evidence on this head in chap. i. vol. i. 'On the Variation of Animals and Plants under Domestication'.

as on the mainland of China. Mr Swinhoe,[36] who describes these two latter cases, attributes the victory of the common rat over the large *Mus coninga* to its superior cunning; and this latter quality may probably be attributed to the habitual exercise of all its faculties in avoiding extirpation by man, as well as to nearly all the less cunning or weak-minded rats having been continuously destroyed by him. It is, however, possible that the success of the common rat may be due to its having possessed greater cunning than its fellow-species, before it became associated with man. To maintain, independently of any direct evidence, that no animal during the course of ages has progressed in intellect or other mental faculties, is to beg the question of the evolution of species. We have seen that, according to Lartet, existing mammals belonging to several orders have larger brains than their ancient tertiary prototypes.

It has often been said that no animal uses any tool; but the chimpanzee in a state of nature cracks a native fruit, somewhat like a walnut, with a stone.[37] Rengger[38] easily taught an American monkey thus to break open hard palm-nuts; and afterwards of its own accord, it used stones to open other kinds of nuts, as well as boxes. It thus also removed the soft rind of fruit that had a disagreeable flavour. Another monkey was taught to open the lid of a large box with a stick, and afterwards it used the stick as a lever to move heavy bodies; and I have myself seen a young orang put a stick into a crevice, slip his hand to the other end, and use it in the proper manner as a lever. The tamed elephants in India are well known to break off branches of trees and use them to drive away the flies; and this same act has been observed in an elephant in a state of nature.[39] I have seen a young orang, when she thought she was going to be whipped, cover and protect herself with a blanket

36. 'Proc. Zoolog. Soc.', 1864, p. 186.
37. Savage and Wyman in 'Boston Journal of Nat. Hist.', vol. iv. 1843–44, p. 383.
38. 'Saugethiere von Paraguay', 1830, s. 51–56.
39. The 'Indian Field', March 4, 1871.

or straw. In these several cases stones and sticks were employed as implements; but they are likewise used as weapons. Brehm[40] states, on the authority of the well-known traveller Schimper, that in Abyssinia when the baboons belonging to one species (*C. gelada*) descend in troops from the mountains to plunder the fields, they sometimes encounter troops of another species (*C. hamadryas*), and then a fight ensues. The Geladas roll down great stones, which the Hamadryas try to avoid, and then both species, making a great uproar, rush furiously against each other. Brehm, when, accompanying the Duke of Coburg-Gotha, aided in an attack with fire-arms on a troop of baboons in the pass of Mensa in Abyssinia. The baboons in return rolled so many stones down the mountain, some as large as a man's head, that the attackers had to beat a hasty retreat; and the pass was actually closed for a time against the caravan. It deserves notice that these baboons thus acted in concert. Mr Wallace[41] on three occasions saw female orangs, accompanied by their young, 'breaking off branches and the great spiny fruit of the Durian tree, with every appearance of rage; causing such a shower of missiles as effectually kept us from approaching too near the tree'. As I have repeatedly seen, a chimpanzee will throw any object at hand at a person who offends him; and the before mentioned baboon at the Cape of Good Hope prepared mud for the purpose.

In the Zoological Gardens, a monkey, which had weak teeth, used to break open nuts with a stone; and I was assured by the keepers that after using the stone, he hid it in the straw, and would not let any other monkey touch it. Here, then, we have the idea of property; but this idea is common to every dog with a bone, and to most or all birds with their nests.

The Duke of Argyll[42] remarks, that the fashioning of an im-

40. 'Thierleben', B. i. s. 79, 82.
41. 'The Malay Archipelago', vol. i. 1869, p. 87.
42. 'Primeval Man', 1869. pp. 145, 147.

plement for a special purpose is absolutely peculiar to man; and he considers that this forms an immeasurable gulf between him and the brutes. This is no doubt a very important distinction; but there appears to me much truth in Sir J. Lubbock's suggestion,[43] that when primeval man first used flint-stones for any purpose, he would have accidentally splintered them, and would then have used the sharp fragments. From this step it would be a small one to break the flints on purpose, and not a very wide step to fashion them rudely. This latter advance, however, may have taken long ages, if we may judge by the immense interval of time which elapsed before the men of the neolithic period took to grinding and polishing their stone tools. In breaking the flints, as Sir J. Lubbock likewise remarks, sparks would have been emitted, and in grinding them heat would have been evolved: thus the two usual methods of 'obtaining fire may have originated'. The nature of fire would have been known in the many volcanic regions where lava occasionally flows through forests. The anthropomorphous apes, guided probably by instinct, build for themselves temporary platforms; but as many instincts are largely controlled by reason, the simpler ones, such as this of building a platform, might readily pass into a voluntary and conscious act. The orang is known to cover itself at night with the leaves of the Pandanus; and Brehm states that one of his baboons used to protect itself from the heat of the sun by throwing a straw-mat over its head. In these several habits, we probably see the first steps towards some of the simpler arts, such as rude architecture and dress, as they arose amongst the early progenitors of man.

Abstraction, General Conceptions, Self-consciousness, Mental Individuality—It would be very difficult for any one with even much more knowledge than I possess, to determine how far ani-

43. 'Prehistoric Times', 1865, p. 473, &c.

mals exhibit any traces of these high mental powers. This difficulty arises from the impossibility of judging what passes through the mind of an animal; and again, the fact that writers differ to a great extent in the meaning which they attribute to the above terms, causes a further difficulty. If one may judge from various articles which have been published lately, the greatest stress seems to be laid on the supposed entire absence in animals of the power of abstraction, or of forming general concepts. But when a dog sees another dog at a distance, it is often clear that he perceives that it is a dog in the abstract; for when he gets nearer his whole manner suddenly changes, if the other dog be a friend. A recent writer remarks, that in all such cases it is a pure assumption to assert that the mental act is not essentially of the same nature in the animal as in man. If either refers what he perceives with his senses to a mental concept, then so do both.[44] When I say to my terrier, in an eager voice (and I have made the trial many times), 'Hi, hi, where is it?' she at once takes it as a sign that something is to be hunted, and generally first looks quickly all around, and then rushes into the nearest thicket, to scent for any game, but finding nothing, she looks up into any neighbouring tree for a squirrel. Now do not these actions clearly shew that she had in her mind a general idea or concept that some animal is to be discovered and hunted?

It may be freely admitted that no animal is self-conscious, if by this term it is implied, that he reflects on such points, as whence he comes or whither he will go, or what is life and death, and so forth. But how can we feel sure that an old dog with an excellent memory and some power of imagination, as shewn by his dreams, never reflects on his past pleasures or pains in the chase? And this would be a form of self-consciousness. On the other

---

44. Mr Hookham, in a letter to Prof. Max Müller, in the 'Birmingham News', May 1873.

hand, as Büchner[45] has remarked, how little can the hard-worked wife of a degraded Australian savage, who uses very few abstract words, and cannot count above four, exert her self-consciousness, or reflect on the nature of her own existence. It is generally admitted, that the higher animals possess memory, attention, association, and even some imagination and reason. If these powers, which differ much in different animals, are capable of improvement, there seems no great improbability in more complex faculties, such as the higher forms of abstraction, and self-consciousness, &c., having been evolved through the development and combination of the simpler ones. It has been urged against the views here maintained, that it is impossible to say at what point in the ascending scale animals become capable of abstraction, &c.; but who can say at what age this occurs in our young children? We see at least that such powers are developed in children by imperceptible degrees.

That animals retain their mental individuality is unquestionable. When my voice awakened a train of old associations in the mind of the before-mentioned dog, he must have retained his mental individuality, although every atom of his brain had probably undergone change more than once during the interval of five years. This dog might have brought forward the argument lately advanced to crush all evolutionists, and said, 'I abide amid all mental moods and all material changes . . . The teaching that atoms leave their impressions as legacies to other atoms falling into the places they have vacated is contradictory of the utterance of consciousness, and is therefore false; but it is the teaching necessitated by evolutionism, consequently the hypothesis is a false one.'[46]

Language—This faculty has justly been considered as one of the chief distinctions between man and the lower animals. But man,

45. 'Conférences sur la Théorie Darwinienne', French translat. 1869, p. 132.
46. The Rev. Dr J. M'Cann, 'Anti-Darwinism', 1869, p. 13.

as a highly competent judge, Archbishop Whately remarks, 'is not the only animal that can make use of language to express what is passing in his mind, and can understand, more or less, what is so expressed by another'.[47] In Paraguay the *Cebus azarae* when excited utters at least six distinct sounds, which excite in other monkeys similar emotions.[48] The movements of the features and gestures of monkeys are understood by us, and they partly understand ours, as Rengger and others declare. It is a more remarkable fact that the dog, since being domesticated, has learnt to bark[49] in at least four or five distinct tones. Although barking is a new art, no doubt the wild parent-species of the dog expressed their feelings by cries of various kinds. With the domesticated dog we have the bark of eagerness, as in the chase; that of anger, as well as growling; the yelp or howl of despair, as when shut up; the baying at night; the bark of joy, as when starting on a walk with his master; and the very distinct one of demand or supplication, as when wishing for a door or window to be opened. According to Houzeau, who paid particular attention to the subject, the domestic fowl utters at least a dozen significant sounds.[50]

The habitual use of articulate language is, however, peculiar to man; but he uses, in common with the lower animals, inarticulate cries to express his meaning, aided by gestures and the movements of the muscles of the face.[51] This especially holds good with the more simple and vivid feelings, which are but little connected with our higher intelligence. Our cries of pain, fear, surprise, anger, together with their appropriate actions, and the murmur of a mother to her beloved child, are more expressive than any words. That which distinguishes man from

47. Quoted in 'Anthropological Review', 1864, p. 158.
48. Rengger, ibid. s. 45.
49. See my 'Variation of Animals and Plants under Domestication', vol. i. p. 27.
50. 'Facultés Mentales des Animaux', tom. ii, 1872, pp. 346–349.
51. See a discussion on this subject in Mr E. B. Tylor's very interesting work, 'Researches into the Early History of Mankind', 1865, chaps. ii. to iv.

the lower animals is not the understanding of articulate sounds, for, as every one knows, dogs understand many words and sentences. In this respect they are at the same stage of development as infants, between the ages of ten and twelve months, who understand many words and short sentences, but cannot yet utter a single word. It is not the mere articulation which is our distinguishing character, for parrots and other birds possess this power. Nor is it the mere capacity of connecting definite sounds with definite ideas; for it is certain that some parrots, which have been taught to speak, connect unerringly words with things, and persons with events.[52] The lower animals differ from man solely in his almost infinitely larger power of associating together the most diversified sounds and ideas; and this obviously depends on the high development of his mental powers.

As Horne Took, one of the founders of the noble science of philology, observes, language is an art, like brewing or baking; but writing would have been a better simile. It certainly is not a true instinct, for every language has to be learnt. It differs, however, widely from all ordinary arts, for man has an instinctive tendency to speak, as we see in the babble of our young children; whilst no child has an instinctive tendency to brew, bake, or write. Moreover, no philologist now supposes that any language has been deliberately invented; it has been slowly and unconsciously developed by

52. I have received several detailed accounts to this effect. Admiral Sir J. Sulivan, whom I know to be a careful observer, assures me that an African parrot, long kept in his father's house, invariably, called certain persons of the household, as well as visitors, by their names. He said 'good morning' to every one at breakfast, and 'good night' to each as they left the room at night, and never reversed these salutations. To Sir J. Sulivan's father, he used to add to the 'good morning' a short sentence, which was never once repeated after his father's death. He scolded violently a strange dog which came into the room through the open window; and he scolded another parrot (saying 'you naughty polly') which had got out of its cage, and was eating apples on the kitchen table. See also, to the same effect, Houzeau on parrots, 'Facultés Mentales', tom. ii. p. 309. Dr A. Moschkau informs me that he knew a starling which never made a mistake in saying in German 'good morning' to persons arriving, and 'goodbye, old fellow', to those departing. I could add several other such cases.

many steps.[53] The sounds uttered by birds offer in several respects
the nearest analogy to language, for all the members of the same
species utter the same instinctive cries expressive of their emotions; and all the kinds which sing, exert their power instinctively;
but the actual song, and even the call notes, are learnt from their
parents or foster-parents. These sounds, as Daines Barrington[54] has
proved, 'are no more innate than language is in man'. The first attempts to sing 'may be compared to the imperfect endeavour in
a child to babble'. The young males continue practising, or as the
bird-catchers say, 'recording', for ten or eleven months. Their first
essays show hardly a rudiment of the future song; but as they grow
older we can perceive what they are aiming at; and at last they
are said 'to sing their song round'. Nestlings which have learnt
the song of a distinct species, as with the canary-birds educated
in the Tyrol, teach and transmit their new song to their offspring.
The slight natural differences of song in the same species inhabiting different districts may be appositely compared, as Barrington
remarks, 'to provincial dialects'; and the songs of allied, though
distinct species may be compared with the languages of distinct
races of man. I have given the foregoing details to shew that an
instinctive tendency to acquire an art is not peculiar to man.

With respect to the origin of articulate language, after having read on the one side the highly interesting works of Mr
Hensleigh Wedgwood, the Rev. F. Farrar, and Prof. Schleicher,[55]

53. See some good remarks on this head by Prof. Whitney, in his 'Oriental and
Linguistic Studies', 1873, p. 354. He observes that the desire of communication
between man is the living force, which, in the development of language, 'works
both consciously and unconsciously; consciously as regards the immediate end
to be attained; unconsciously as regards the further consequences of the act'.
54. Hon. Daines Barrington in 'Philosoph. Transactions', 1778, p. 262. See also Dureau de la Malle in 'Ann. des. Sc. Nat.', 3rd series, Zoolog. tom. x. p. 119.
55. 'On the Origin of Language', by H. Wedgwood, 1866. 'Chapters on Language',
by the Rev. F. W. Farrar, 1865. These works are most interesting. See also 'De la
Phys. et de Parole', par Albert Lemoine, 1865, p. 190. The work on this subject,
by the late Prof. Aug. Schleicher, has been translated by Dr Bikkers into English,
under the title of 'Darwinism tested by the Science of Language', 1869.

and the celebrated lectures of Prof. Max Müller on the other side, I cannot doubt that language owes its origin to the imitation and modification of various natural sounds, the voices of other animals, and man's own instinctive cries, aided by signs and gestures. When we treat of sexual selection we shall see that primeval man, or rather some early progenitor of man, probably first used his voice in producing true musical cadences, that is in singing, as do some of the gibbon-apes at the present day; and we may conclude from a widely-spread analogy, that this power would have been especially exerted during the courtship of the sexes—would have expressed various emotions, such as love, jealousy, triumph—and would have served as a challenge to rivals. It is, therefore, probable that the imitation of musical cries by articulate sounds may have given rise to words expressive of various complex emotions. The strong tendency in our nearest allies, the monkeys, in microcephalous idiots,[56] and in the barbarous races of mankind, to imitate whatever they hear deserves notice, as bearing on the subject of imitation. Since monkeys certainly understand much that is said to them by man, and when wild, utter signal-cries of danger to their fellows;[57] and since fowls give distinct warnings for danger on the ground, or in the sky from hawks (both, as well as a third cry, intelligible to dogs),[58] may not some unusually wise ape-like animal have imitated the growl of a beast of prey, and thus told his fellow-monkeys the nature of the expected danger? This would have been a first step in the formation of a language.

As the voice was used more and more, the vocal organs would have been strengthened and perfected through the principle of the inherited effects of use; and this would have reacted on the

56. Vogt, 'Mémoire sur les Microcéphales', 1867, p. 169. With respect to savages, I have given some facts in my 'Journal of Researches', &c., 1845, p. 206.
57. See clear evidence on this head in the two works so often quoted, by Brehm and Rengger.
58. Houzeau gives a very curious account of his observations on this subject in his 'Facultés Mentales des Animaux', tom. ii., p. 348.

power of speech. But the relation between the continued use of language and the development of the brain, has no doubt been far more important. The mental powers in some early progenitor of man must have been more highly developed than in any existing ape, before even the most imperfect form of speech could have come into use; but we may confidently believe that the continued use and advancement of this power would have reacted on the mind itself, by enabling and encouraging it to carry on long trains of thought. A complex train of thought can no more be carried on without the aid of words, whether spoken or silent, than a long calculation without the use of figures or algebra. It appears, also, that even an ordinary train of thought almost requires, or is greatly facilitated by some form of language, for the dumb, deaf, and blind girl, Laura Bridgman, was observed to use her fingers whilst dreaming.[59] Nevertheless, a long succession of vivid and connected ideas may pass through the mind without the aid of any form of language, as we may infer from the movements of dogs during their dreams. We have, also, seen that animals are able to reason to a certain extent, manifestly without the aid of language. The intimate connection between the brain, as it is now developed in us, and the faculty of speech, is well shewn by those curious cases of brain-disease in which speech is specially affected, as when the power to remember substantives is lost, whilst other words can be correctly used, or where substantives of a certain class, or all except the initial letters of substantives and proper names are forgotten.[60] There is no more improbability in the continued use of the mental and vocal organs leading to inherited changes in their structure and func-

---

59. See remarks on this head by Dr Maudsley, 'The Physiology and Pathology of Mind', 2nd edit. 1868, p. 199.
60. Many curious cases have been recorded. See, for instance, Dr Bateman, 'On Aphasia', 1870, pp. 27, 31, 53, 100, &c. Also, 'Inquiries Concerning the Intellectual Powers', by Dr Abercrombie, 1838, p. 150.

tions, than in the case of handwriting, which depends partly on the form of the hand and partly on the disposition of the mind; and hand-writing is certainly inherited.[61]

Several writers, more especially Prof. Max Müller,[62] have lately insisted that the use of language implies the power of forming general concepts; and that as no animals are supposed to possess this power, an impossible barrier is formed between them and man.[63] With respect to animals, I have already endeavoured to show that they have this power, at least in a rude and incipient degree. As far as concerns infants of from ten to eleven months old, and deaf-mutes, it seems to me incredible, that they should be able to connect certain sounds with certain general ideas as quickly as they do, unless such ideas were already formed in their minds. The same remark may be extended to the more intelligent animals; as Mr Leslie Stephen observes,[64] 'A dog frames a general concept of cats or sheep, and knows the corresponding words as well as a philosopher. And the capacity to understand is as good a proof of vocal intelligence, though in an inferior degree, as the capacity to speak.'

61. 'The Variation of Animals and Plants under Domestication', vol. ii. p. 6.
62. Lectures on 'Mr Darwin's Philosophy of Language', 1873.
63. The judgment of a distinguished philologist, such as Prof. Whitney, will have far more weight on this point than anything that I can say. He remarks ('Oriental and Linguistic Studies', 1873, p. 297), in speaking of Bleek's views: 'Because on the grand scale language is the necessary auxiliary of thought, indispensable to the development of the power of thinking, to the distinctness and variety and complexity of cognitions to the full mastery of consciousness; therefore he would fain make thought absolutely impossible without speech, identifying the faculty with its instrument. He might just as reasonably assert that the human hand cannot act without a tool. With such a doctrine to start from, he cannot stop short of Müller's worst paradoxes, that an infant (*in fans*, not speaking) is not a human being, and that deaf-mutes do not become possessed of reason until they learn to twist their fingers into imitation of spoken words.' Max Müller gives in italics ('Lectures on Mr Darwin's Philosophy of Language', 1873, third lecture) the following aphorism: 'There is no thought without words, as little as there are words without thought.' What a strange definition must here be given to the word thought!
64. 'Essays on Free-thinking', &c., 1873, p. 82.

Why the organs now used for speech should have been originally perfected for this purpose, rather than any other organs,
it is not difficult to see. Ants have considerable powers of intercommunication by means of their antennae, as shewn by Huber,
who devotes a whole chapter to their language. We might have
used our fingers as efficient instruments, for a person with practice can report to a deaf man every word of a speech rapidly
delivered at a public meeting; but the loss of our hands, whilst
thus employed, would have been a serious inconvenience. As all
the higher mammals possess vocal organs, constructed on the
same general plan as ours, and used as a means of communication, it was obviously probable that these same organs would be
still further developed if the power of communication had to
be improved; and this has been effected by the aid of adjoining
and well adapted parts, namely the tongue and lips.[65] The fact
of the higher apes not using their vocal organs for speech, no
doubt depends on their intelligence not having been sufficiently
advanced. The possession by them of organs, which with long-
continued practice might have been used for speech, although
not thus used, is paralleled by the case of many birds which possess organs fitted for singing, though they never sing. Thus, the
nightingale and crow have vocal organs similarly constructed,
these being used by the former for diversified song, and by the
latter only for croaking.[66] If it be asked why apes have not had
their intellects developed to the same degree as that of man, general causes only can be assigned in answer, and it is unreasonable
to expect anything more definite, considering our ignorance

65. See some good remarks to this effect by Dr Maudsley, 'The Physiology and
    Pathology of Mind', 1868, p. 199.
66. Macgillivray, 'Hist. of British Birds', vol. ii. 1839, p. 29. An excellent observer, Mr
    Blackwall, remarks that the magpie learns to pronounce single words, and even
    short sentences, more readily than almost any other British bird; yet, as he adds,
    after long and closely investigating its habits, he has never known it, in a state
    of nature, display any unusual capacity for imitation. 'Researches in Zoology',
    1834, p. 158.

with respect to the successive stages of development through which each creature has passed.

The formation of different languages and of distinct species, and the proofs that both have been developed through a gradual process, are curiously parallel.[67] But we can trace the formation of many words further back than that of species, for we can perceive how they actually arose from the imitation of various sounds. We find in distinct languages striking homologies due to community of descent, and analogies due to a similar process of formation. The manner in which certain letters or sounds change when others change is very like correlated growth. We have in both cases the reduplication of parts, the effects of long-continued use, and so forth. The frequent presence of rudiments, both in languages and in species, is still more remarkable. The letter *m* in the word *am*, means *I*; so that in the expression *I am*, a superfluous and useless rudiment has been retained. In the spelling also of words, letters often remain as the rudiments of ancient forms of pronunciation. Languages, like organic beings, can be classed in groups under groups; and they can be classed either naturally according to descent, or artificially by other characters. Dominant languages and dialects spread widely, and lead to the gradual extinction of other tongues. A language, like a species, when once extinct, never, as Sir C. Lyell remarks, reappears. The same language never has two birth-places. Distinct languages may be crossed or blended together.[68] We see variability in every tongue, and new words are continually cropping up; but as there is a limit to the powers of the memory, single words, like whole languages, gradually become extinct. As Max Müller[69] has well remarked:—'A struggle

---

67. See the very interesting parallelism between the development of species and languages, given by Sir C. Lyell in 'The Geolog. Evidences of the Antiquity of Man', 1863, chap. xxiii.

68. See remarks to this effect by the Rev. F. W. Farrar, in an interesting article, entitled 'Philology and Darwinism' in 'Nature', March 24th, 1870, p. 528.

69. 'Nature', Jan. 6th, 1870, p. 257.

for life is constantly going on amongst the words and grammatical forms in each language. The better, the shorter, the easier forms are constantly gaining the upper hand, and they owe their success to their own inherent virtue.' To these more important causes of the survival of certain words, mere novelty and fashion may be added; for there is in the mind of man a strong love for slight changes in all things. The survival or preservation of certain favoured words in the struggle for existence is natural selection.

The perfectly regular and wonderfully complex construction of the langauges of many barbarous nations has often been advanced as a proof, either of the divine origin of these languages, or of the high art and former civilisation of their founders. Thus F. von Schlegel writes: 'In those languages which appear to be at the lowest grade of intellectual culture, we frequently observe a very high and elaborate degree of art in their grammatical structure. This is especially the case with the Basque and the Lapponian, and many of the American languages'.[70] But it is assuredly an error to speak of any language as an art, in the sense of its having been elaborately and methodically formed. Philologists now admit that conjugations, declensions, &c., originally existed as distinct words, since joined together; and as such words express the most obvious relations between objects and persons, it is not surprising that they should have been used by the men of most races during the earliest ages. With respect to perfection, the following illustration will best shew how easily we may err: a Crinoid sometimes consists of no less than 150,000 pieces of shell,[71] all arranged with perfect symmetry in radiating lines; but a naturalist does not consider an animal of this kind as more perfect than a bilateral one with comparatively few parts, and with none of these parts alike, excepting on the opposite sides of the

70. Quoted by C. S. Wake, 'Chapters on Man', 1868, p. 101.
71. Buckland, 'Bridgewater Treatise', p. 411.

body. He justly considers the differentiation and specialisation of organs as the test of perfection. So with languages; the most symmetrical and complex ought not to be ranked above irregular, abbreviated, and bastardised languages, which have borrowed expressive words and useful forms of construction from various conquering, conquered, or immigrant races.

From these few and imperfect remarks I conclude that the extremely complex and regular construction of many barbarous languages, is no proof that they owe their origin to a special act of creation.[72] Nor, as we have seen, does the faculty of articulate speech in itself offer any insuperable objection to the belief that man has been developed from some lower form.

Sense of Beauty—This sense has been declared to be peculiar to man. I refer here only to the pleasure given by certain colours, forms, and sounds, and which may fairly be called a sense of the beautiful; with cultivated men such sensations are, however, intimately associated with complex ideas and trains of thought. When we behold a male bird elaborately displaying his graceful plumes or splendid colours before the female, whilst other birds, not thus decorated, make no such display, it is impossible to doubt that she admires the beauty of her male partner. As women everywhere deck themselves with these plumes, the beauty of such ornaments cannot be disputed. As we shall see later, the nests of humming-birds, and the playing passages of bower-birds are tastefully ornamented with gaily-coloured objects; and this shews that they must receive some kind of pleasure from the sight of such things. With the great majority of animals, however, the taste for the beautiful is confined, as far as we can judge, to the attractions of the opposite sex. The sweet

72. See some good remarks on the simplification of languages, by Sir J. Lubbock, 'Origin of Civilisation', 1870, p. 278.

strains poured forth by many male birds during the season of love, are certainly admired by the females, of which fact evidence will hereafter be given. If female birds had been incapable of appreciating the beautiful colours, the ornaments, and voices of their male partners, all the labour and anxiety exhibited by the latter in displaying their charms before the females would have been thrown away; and this it is impossible to admit. Why certain bright colours should excite pleasure cannot, I presume, be explained, any more than why certain flavours and scents are agreeable; but habit has something to do with the result, for that which is at first unpleasant to our senses, ultimately becomes pleasant, and habits are inherited. With respect to sounds, Helmholtz has explained to a certain extent on physiological principles, why harmonies and certain cadences are agreeable. But besides this, sounds frequently recurring at irregular intervals are highly disagreeable, as every one will admit who has listened at night to the irregular flapping of a rope on board ship. The same principle seems to come into play with vision, as the eye prefers symmetry or figures with some regular recurrence. Patterns of this kind are employed by even the lowest savages as ornaments; and they have been developed through sexual selection for the adornment of some male animals. Whether we can or not give any reason for the pleasure thus derived from vision and hearing, yet man and many of the lower animals are alike pleased by the same colours, graceful shading and forms, and the same sounds.

The taste for the beautiful, at least as far as female beauty is concerned, is not of a special nature in the human mind; for it differs widely in the different races of man, and is not quite the same even in the different nations of the same race. Judging from the hideous ornaments, and the equally hideous music admired by most savages, it might be argued that their aesthetic faculty was not so highly developed as in certain animals, for instance, as in birds. Obviously no animal would be capable of admiring such scenes as the heavens at night, a beautiful landscape, or refined music; but such high tastes are acquired through culture, and depend on complex associations; they are not enjoyed by barbarians or by uneducated persons.

Many of the faculties, which have been of inestimable service to man for his progressive advancement, such as the powers of the imagination, wonder, curiosity, an undefined sense of beauty, a tendency to imitation, and the love of excitement or novelty, could hardly fail to lead to capricious changes of customs and fashions. I have alluded to this point, because a recent writer[73] has oddly fixed on Caprice 'as one of the most remarkable and typical differences between savages and brutes'. But not only can we partially understand how it is that man is from various conflicting influences rendered capricious, but that the lower animals are, as we shall hereafter see, likewise capricious in their affections, aversions, and sense of beauty. There is also reason to suspect that they love novelty, for its own sake.

Belief in God—Religion—There is no evidence that man was aboriginally endowed with the ennobling belief in the existence of an Omnipotent God. On the contrary there is ample evidence, derived not from hasty travellers, but from men who have long resided with savages, that numerous races have existed, and still exist, who have no idea of one or more gods, and who have no words in their languages to express such an idea.[74] The question is of course wholly distinct from that higher one, whether there exists a Creator and Ruler of the universe; and this has been

73. 'The Spectator', Dec. 4th, 1869, p. 1430.
74. See an excellent article on this subject by the Rev. F.W. Farrar, in the 'Anthropological Review', Aug. 1864, p. ccxvii. For further facts see Sir J. Lubbock, 'Prehistoric Times', 2nd edit. 1869, p. 564; and especially the chapters on Religion in his 'Origin of Civilisation', 1870.

answered in the affirmative by some of the highest intellects that have ever existed.

If, however, we include under the term 'religion' the belief in unseen or spiritual agencies, the case is wholly different; for this belief seems to be universal with the less civilised races. Nor is it difficult to comprehend how it arose. As soon as the important faculties of the imagination, wonder, and curiosity, together with some power of reasoning, had become partially developed, man would naturally crave to understand what was passing around him, and would have vaguely speculated on his own existence. As Mr M'Lennan[75] has remarked, 'Some explanation of the phenomena of life, a man must feign for himself; and to judge from the universality of it, the simplest hypothesis and the first to occur to men, seems to have been that natural phenomena are ascribable to the presence in animals, plants, and things, and in the forces of nature, of such spirits prompting to action as men are conscious they themselves possess.' It is also probable, as Mr Tylor has shewn, that dreams may have first given rise to the notion of spirits; for savages do not readily distinguish between subjective and objective impressions. When a savage dreams, the figures which appear before him are believed to have come from a distance, and to stand over him; or 'the soul of the dreamer goes out on its travels, and comes home with a remembrance of what it has seen'.[76] But until the faculties of imagination, curiosity,

75. 'The Worship of Animals and Plants', in the 'Fortnightly Review', Oct. 1, 1869, p. 422.
76. Tylor, 'Early History of Mankind', 1865, p. 6. See also the three striking chapters on the Development of Religion, in Lubbock's 'Origin of Civilisation', 1870. In a like manner Mr Herbert Spencer, in his ingenious essay in the 'Fortnightly Review' (May 1st, 1870, p. 535), accounts for the earliest forms of religious belief throughout the world, by man being led through dreams, shadows, and other causes, to look at himself as a double essence, corporeal and spiritual. As the spiritual being is supposed to exist after death and to be powerful, it is propitiated by various gifts and ceremonies, and its aid invoked. He then further shews that names or nicknames given from some animal or other object, to the early progenitors or founders of a tribe, are supposed after a long interval to represent the

reason, &c., had been fairly well developed in the mind of man, his dreams would not have led him to believe in spirits, any more than in the case of a dog.

The tendency in savages to imagine that natural objects and agencies are animated by spiritual or living essences, is perhaps illustrated by a little fact which I once noticed: my dog, a full-grown and very sensible animal, was lying on the lawn during a hot and still day; but at a little distance a slight breeze occasionally moved an open parasol, which would have been wholly disregarded by the dog, had any one stood near it. As it was, every time that the parasol slightly moved, the dog growled fiercely and barked. He must, I think, have reasoned to himself in a rapid and unconscious manner, that movement without any apparent cause indicated the presence of some strange living agent, and that no stranger had a right to be on his territory.

The belief in spiritual agencies would easily pass into the belief in the existence of one or more gods. For savages would naturally attribute to spirits the same passions, the same love of vengeance or simplest form of justice, and the same affections which they themselves feel. The Fuegians appear to be in this respect in an intermediate condition, for when the surgeon on board the 'Beagle' shot some young ducklings as specimens, York Minster declared in the most solemn manner, 'Oh, Mr Bynoe, much rain, much snow, blow much'; and this was evidently a retributive punishment for wasting human food. So again he related how, when his brother killed a 'wild man', storms long raged, much rain and snow fell. Yet we could never discover that the Fuegians believed in what we should call a God, or practised

---

real progenitor of the tribe; and such animal or object is then naturally believed still to exist as a spirit, is held sacred, and worshipped as a god. Nevertheless I cannot but suspect that there is a still earlier and ruder stage, when anything which manifests power or movement is thought to be endowed with some form of life, and with mental faculties analogous to our own.

any religious rites; and Jemmy Button, with justifiable pride, stoutly maintained that there was no devil in his land. This latter assertion is the more remarkable, as with savages the belief in bad spirits is far more common than that in good ones.

The feeling of religious devotion is a highly complex one, consisting of love, complete submission to an exalted and mysterious superior, a strong sense of dependence,[77] fear, reverence, gratitude, hope for the future, and perhaps other elements. No being could experience so complex an emotion until advanced in his intellectual and moral faculties to at least a moderately high level. Nevertheless, we see some distant approach to this state of mind in the deep love of a dog for his master, associated with complete submission, some fear, and perhaps other feelings. The behaviour of a dog when returning to his master after an absence, and, as I may add, of a monkey to his beloved keeper, is widely different from that towards their fellows. In the latter case the transports of joy appear to be somewhat less, and the sense of equality is shewn in every action. Professor Braubach goes so far as to maintain that a dog looks on his master as on a god.[78]

The same high mental faculties which first led man to believe in unseen spiritual agencies, then in fetishism, polytheism, and ultimately in monotheism, would infallibly lead him, as long as his reasoning powers remained poorly developed, to various strange superstitions and customs. Many of these are terrible to think of—such as the sacrifice of human beings to a blood-loving god; the trial of innocent persons by the ordeal of poison or fire; witchcraft, &c.—yet it is well occasionally to reflect on these superstitions, for they shew us what an infinite debt of

77. See an able article on the 'Physical Elements of Religion', by Mr L. Owen Pike, in 'Anthropolog. Review', April, 1870, p. lxiii.
78. 'Religion, Moral, &c., der Darwin'schen Art-Lehre', 1869, s. 53. It is said (Dr W. Lauder Lindsay, 'Journal of Mental Science', 1871, p. 43), that Bacon long ago, and the poet Burns, held the same notion.

gratitude we owe to the improvement of our reason, to science, and to our accumulated knowledge. As Sir J. Lubbock[79] has well observed, 'it is not too much to say that the horrible dread of unknown evil hangs like a thick cloud over savage life, and embitters every pleasure'. These miserable and indirect consequences of our highest faculties may be compared with the incidental and occasional mistakes of the instincts of the lower animals.

79. 'Prehistoric Times', 2nd edit. p. 571. In this work (p. 571) there will be found an excellent account of the many strange and capricious customs of savages.

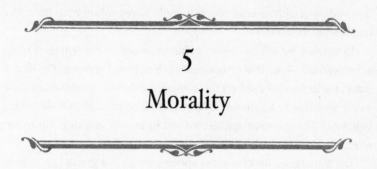

5

Morality

Long before *The Origin of Species*, evolution was considered a threat to morality. Adam Sedgwick predicted that if people accepted that they had evolved, they would abandon all the moral codes that allowed civilization to survive. Some critics leveled the same argument against *The Descent of Man*. A critic for the *Edinburgh Review* warned that if Darwin was right, "a revolution in thought is imminent, which will shake society to its very foundations, by destroying the sanctity of the conscience and the religious sense."

Readers today may come to *The Descent of Man* with the same assumption. For them, chapter four should come as quite a surprise. Darwin does not claim that morality is a delusion. In fact, he champions man as a moral being, capable of the noblest self-sacrifice, sympathy, and loyalty. But man's morality, he argues, did not spring up suddenly in humans. It is a quantitative difference, not a qualitative difference, from the social instincts of other animals.

In the first three chapters, Darwin surveyed physical and mental homologies shared by humans and animals. In chapter four Darwin does the same for morality. He argues that morality in humans is driven in large part by instinct and emotion rather than elaborate reasoning. He points out many animals that also appear to have strong social instincts. Just as language could have evolved from a simple communication system of calls, human morality could have evolved from the social bonds seen in animals.

In arguing for a deep history of human morality, Darwin runs into a serious puzzle—one that he discusses at length in *The Descent of Man*. If natural selection is based on the competition between individuals, how can it give rise to kindness, loyalty, and other sorts of self-sacrificing behavior? The question applies not just to human morality, but to any altruistic behavior in animals.

Darwin argues that for some species, living in a group can help an individual survive. Together, animals can find more food or defend against predators more effectively than they could alone. But they can only enjoy the advantages of living in a group if they behave nicely together. Darwin envisioned well-behaved groups of animals enjoying more reproductive success than groups of selfish individuals. Given enough time, an entire species would become altruists.

In the 1960s evolutionary biologists gave this puzzle another look, bringing with them knowledge about genes and how they spread from one generation to the next. Evolution favors mutations to genes that spread them through a population. In some cases, those mutations may contribute toward an individual having more healthy children. But an individual can also spread its own genes by increasing the odds that its relatives can reproduce.

Consider a sterile female worker bee. She has no hope of reproducing, it is true, but that does not mean that the genes she carries have no chance of getting into the next generation. Every female worker in a beehive is the offspring of the queen, as are the eggs she helps to raise. That means that she is helping rear bees that share some of the same genes she carries. In fact, thanks to a quirk in insect genetics, a worker bee shares more genes with the eggs of the queen than she would with her own offspring. If altruism is more likely to pass a set of genes to the next generation than reproducing oneself, it can be favored by natural selection. This process is known as kin selection, and experiments over the past thirty years have provided a great deal of support for it.

Kin selection can also explain a great deal of human behavior as well. Humans in cultures around the world tend to offer more help to closer relatives.

But humans also cooperate with others who are not closely related. Scientists have constructed mathematical models to analyze how this sort of cooperation can evolve. It's possible that groups of people who cooperate with nonrelatives may thrive better than groups that fight amongst themselves, as Darwin suggested. But scientists have recognized some serious threats to this harmony. If genes can promote altruism, they may also be able to promote cheating. Cheaters could take advantage of the altruism of others without reciprocating—by eating the meat that others are willing to share, for example, while hiding their own game. Cheating might translate into more reproductive success than altruism, which would lead to the rise of cheaters and the fall of altruism.

Evolution may not favor love so much as it favors tough love. Altruism may win only if it is backed up by punishment for cheaters. This rule appears to apply both to animals and humans. In many species of social insects, for example, sterile female workers can sometimes gain the ability to lay eggs of their own. They become cheaters, in other words, taking advantage of the selflessness of other females, which rear their eggs along with those of the queen. If every worker made this switch, the colony would collapse. Scientists have discovered that sterile workers are careful to distinguish between eggs from their queens and from imposters. They destroy the imposter eggs, and along with them the evolutionary advantage of cheating.

We humans also appear to be wired for punishment. Perhaps you have found yourself on a freeway, trying to get onto an exit ramp choked with traffic. Just as you are about to reach the exit, a driver races up on your left side and tries to cut ahead of you. Rage suddenly fills you toward this cheater, and you nudge your car to the left to keep him from barging in. But in your rage, you nearly drive your car into a guard rail. Your urge to punish has put you at risk.

Scientists are now looking for this urge to punish in the wiring of our brains. In one experiment published in 2004, Swiss economists and neuroscientists had a series of subjects play a simple game. Two people (call them Alice and Bill) play the game. They each got ten money units (let's call them dollars). Alice could choose whether to hand over her money to Bill or hold on to it. If she sent the money to Bill, the experimenters quadrupled her ten dollars to forty. Now Bill had fifty dollars and Alice had nothing. Bill could either send nothing back to Alice or send half of his money.

The game had three outcomes. If both Alice and Bill cooperated, they each ended up with twenty-five dollars—more than twice what they started with. But if Alice trusted Bill and he cheated her, he ended up with five times more, and Alice was broke. And if Alice didn't trust Bob, they both ended up with the ten dollars that they started with. If Bill cheated Alice, the scientists gave her a chance to punish him. She could inflict up to twenty "punishment points"—each point takes a dollar off of Bill's winnings. But she had to pay a dollar herself for every point.

The researchers monitored the brain activity of volunteers playing the role of Alice as they decided whether to punish or not. (They did so with PET scans, which can track radioactive tracers injected into the bloodstream.) The researchers then compared the brain activity in different versions of the game. In one version, for example, Alice could only inflict symbolic punishment points—Bill would be told she was angry with him, but her punishment didn't affect his winnings. In another version, Alice could take away some of Bill's money, but didn't have to give up any of her own.

The researchers found that forcing Bill to pay for his sins made Alice feel good. Human brains, like other mammal brains, have a circuit of neurons called the reward pathway, which produces pleasant feelings in response to rewards. In animals such as rats and monkeys, the reward pathway becomes active when they unexpectedly discover food. When humans think about romantic partners or win at gambling, it switches on as well. And the Swiss researchers found that punishing cheaters also switches it on.

Significantly, the symbolic punishment didn't create anything close to the response that the real punishment did. The volunteers were anticipating the satisfaction they'd have when their punishment was carried out. And just as tellingly, volunteers whose reward pathway responded more strongly were more willing to pay out of their own pocket to make the other player suffer for cheating.

Discovering a biological basis for morality does not diminish the importance of the differences between moral systems in different cultures. What is prized in one culture may be regarded with indifference in another. But many scientists now argue that moral systems are like languages: They may differ in the details, but they all develop from the same underlying foundation.

CHAPTER 4:
Comparison of the Mental Powers of Man and the Lower Animals—continued

I FULLY SUBSCRIBE to the judgment of those writers[1] who maintain that of all the differences between man and the lower animals, the moral sense or conscience is by far the most important. This sense, as Mackintosh[2] remarks, 'has a rightful supremacy over every other principle of human action', it is summed up in that short but imperious word *ought*, so full of high significance. It is the most noble of all the attributes of man, leading him without a moment's hesitation to risk his life for that of a fellow-creature; or after due deliberation, impelled simply by the deep feeling of right or duty, to sacrifice it in some great cause. Immanuel Kant exclaims, 'Duty! Wondrous thought, that workest neither by fond insinuation, flattery, nor by any threat, but merely by holding up thy naked law in the soul, and so extorting for thyself always reverence, if not always obedience; before whom all appetites are dumb, however secretly they rebel; whence thy original?'[3]

This great question has been discussed by many writers[4] of

1. See, for instance, on this subject, Quatrefages, 'Unité de l'Espèce Humaine', 1861, p. 21, &c.
2. 'Dissertation on Ethical Philosophy', 1837, p. 281, &c.
3. 'Metaphysics of Ethics', translated by J. W. Semple, Edinburgh, 1836, p. 136.
4. Mr Bain gives a list ('Mental and Moral Science', 1868, pp. 543–725) of twenty-six British authors who have written on this subject, and whose names are familiar to every reader; to these, Mr Bain's own name, and those of Mr Lecky, Mr Shadworth Hodgson, Sir J. Lubbock, and others, might be added.

consummate ability; and my sole excuse for touching on it, is the impossibility of here passing it over; and because, as far as I know, no one has approached it exclusively from the side of natural history. The investigation possesses, also, some independent interest, as an attempt to see how far the study of the lower animals throws light on one of the highest psychical faculties of man.

The following proposition seems to me in a high degree probably—namely, that any animal whatever, endowed with well-marked social instincts,[5] the parental and filial affections being here included, would inevitably acquire a moral sense or conscience, as soon as its intellectual powers had become as well, or nearly as well developed, as in man. For, *firstly*, the social instincts lead an animal to take pleasure in the society of its fellows, to feel a certain amount of sympathy with them, and to perform various services for them. The services may be of a definite and evidently instinctive nature; or there may be only a wish and readiness, as with most of the higher social animals, to aid their fellows in certain general ways. But these feelings and services are by no means extended to all the individuals of the same species, only to those of the same association. *Secondly*, as

5. Sir B. Brodie, after observing that man is a social animal ('Psychological Enquiries', 1854, p. 192), asks the pregnant question, 'ought not this to settle the disputed question as to the existence of a moral sense?' Similar ideas have probably occurred to many persons, as they did long ago to Marcus Aurelius. Mr J. S. Mill speaks, in his celebrated work, 'Utilitarianism' (1864, pp. 45, 46), of the social feelings as a 'powerful natural sentiment', and as 'the natural basis of sentiment for utilitarian morality'. Again he says, 'Like the other acquired capacities above referred to, the moral faculty, if not a part of our nature, is a natural out-growth from it; capable, like them, in a certain small degree of springing up spontaneously.' But in opposition to all this, he also remarks, 'if, as is my own belief, the moral feelings are not innate, but acquired, they are not for that reason less natural'. It is with hesitation that I venture to differ at all from so profound a thinker, but it can hardly be disputed that the social feelings are instinctive or innate in the lower animals; and why should they not be so in man? Mr Bain (see, for instance, 'The Emotions and the Will', 1865, p. 481) and others believe that the moral sense is acquired by each individual during his lifetime. On the general theory of evolution this is at least extremely improbable. The ignoring of all transmitted mental qualities will, as it seems to me, be hereafter judged as a most serious blemish in the works of Mr Mill.

soon as the mental faculties had become highly developed, images of all past actions and motives would be incessantly passing through the brain of each individual; and that feeling of dissatisfaction, or even misery, which invariably results, as we shall hereafter see, from any unsatisfied instinct, would arise, as often as it was perceived that the enduring and always present social instinct had yielded to some other instinct, at the time stronger, but neither enduring in its nature, nor leaving behind it a very vivid impression. It is clear that many instinctive desires, such as that of hunger, are in their nature of short duration; and after being satisfied, are not readily or vividly recalled. *Thirdly*, after the power of language had been acquired, and the wishes of the community could be expressed, the common opinion how each member ought to act for the public good, would naturally become in a paramount degree the guide to action. But it should be borne in mind that however great weight we may attribute to public opinion, our regard for the approbation and disapprobation of our fellows depends on sympathy, which, as we shall see, forms an essential part of the social instinct, and is indeed its foundation-stone. *Lastly*, habit in the individual would ultimately play a very important part in guiding the conduct of each member; for the social instinct, together with sympathy, is, like any other instinct, greatly strengthened by habit, and so consequently would be obedience to the wishes and judgment of the community. These several subordinate propositions must now be discussed, and some of them at considerable length.

It may be well first to premise that I do not wish to maintain that any strictly social animal, if its intellectual faculties were to become as active and as highly developed as in man, would acquire exactly the same moral sense as ours. In the same manner as various animals have some sense of beauty, though they admire widely different objects, so they might have a sense of right and wrong, though led by it to follow widely different lines of conduct. If, for instance, to take an extreme case, men were reared under precisely the same conditions as hive-bees, there can hardly be a doubt that our unmarried females would, like the worker-bees, think it a sacred duty to kill their brothers, and mothers would strive to kill their fertile daughters; and no one would think of interfering.[6] Nevertheless, the bee, or any other social animal, would gain in our supposed case, as it appears to me, some feeling of right or wrong, or a conscience. For each individual would have an inward sense of possessing certain stronger or more enduring instincts, and others less strong or enduring; so that there would often be a struggle as to which impulse should be followed; and satisfaction, dissatisfaction, or even misery would be felt, as past impressions were compared during their incessant passage through the mind. In this case an inward monitor would tell the animal that it would have been better to have followed the one impulse rather than the other. The one course ought to have been followed, and the other ought not; the one would have been right and the other wrong; but to these terms I shall recur.

Sociability—Animals of many kinds are social; we find even distinct species living together; for example, some American mon-

6. Mr H. Sidgwick remarks, in an able discussion on this subject [the 'Academy', June 15th, 1872, (p. 231)], 'a superior bee, we may feel sure, would aspire to a milder solution of the population question'. Judging, however, from the habits of many or most savages, man solves the problem by female infanticide, polyandry and promiscuous intercourse; therefore it may well be doubted whether it would be by a milder method. Miss Cobbe, in commenting ('Darwinism in Morals' 'Theological Review', April, 1872, pp. 188–191) on the same illustration, says, the *principles* of social duty would be thus reversed; and by this, I presume, she means that the fulfilment of a social duty would tend to the injury of individuals; but she overlooks the fact, which she would doubtless admit, that the instincts of the bee have been acquired for the good of the community. She goes so far as to say that if the theory of ethics advocated in this chapter were ever generally accepted, 'I cannot but believe that in the hour of their triumph would be sounded the knell of the virtue of mankind!' It is to be hoped that the belief in the permanence of virtue on this earth is not held by many persons on so weak a tenure.

keys; and united flocks of rooks, jackdaws, and starlings. Man shews the same feeling in his strong love for the dog, which the dog returns with interest. Every one must have noticed how miserable horses, dogs, sheep, &c., are when separated from their companions, and what strong mutual affection the two former kinds, at least, shew on their reunion. It is curious to speculate on the feelings of a dog, who will rest peacefully for hours in a room with his master or any of the family, without the least notice being taken of him; but if left for a short time by himself, barks or howls dismally. We will confine our attention to the higher social animals; and pass over insects, although some of these are social, and aid one another in many important ways. The most common mutual service in the higher animals is to warn one another of danger by means of the united senses of all. Every sportsman knows, as Dr Jaeger remarks,[7] how difficult it is to approach animals in a herd or troop. Wild horses and cattle do not, I believe, make any danger-signal; but the attitude of any one of them who first discovers an enemy, warns the others. Rabbits stamp loudly on the ground with their hind-feet, as a signal: sheep and chamois do the same with their forefeet, uttering likewise a whistle. Many birds, and some mammals, post sentinels, which in the case of seals are said[8] generally to be the females. The leader of a troop of monkeys acts as the sentinel, and utters cries expressive both of danger and of safety.[9] Social animals perform many little services for each other: horses nibble, and cows lick each other, on any spot which itches: monkeys search each other for external parasites; and Brehm states that after a troop of

7. 'Die Darwin'sche Theorie', s. 101.
8. Mr R. Brown in 'Proc. Zoolog. Soc.', 1868, p. 409.
9. Brehm, 'Thierleben', B. i. 1864, s. 52, 79. For the case of the monkeys extracting thorns from each other, see s. 54. With respect to the Hamadryas turning over stones, the fact is given (s. 76) on the evidence of Alvarez, whose observations Brehm thinks quite trustworthy. For the cases of the old male baboons attacking the dogs, see s. 79; and with respect to the eagle, s. 56.

the *Cercopithecus griseo-viridis* has rushed through a thorny brake, each monkey stretches itself on a branch, and another monkey sitting by, 'conscientiously' examines its fur, and extracts every thorn or burr.

Animals also render more important services to one another: thus wolves and some other beasts of prey hunt in packs, and aid one another in attacking their victims. Pelicans fish in concert. The Hamadryas baboons turn over stones to find insects, &c.; and when they come to a large one, as many as can stand round, turn it over together and share the booty. Social animals mutually defend each other. Bull bisons in N. America, when there is danger, drive the cows and calves into the middle of the herd, whilst they defend the outside. I shall also in a future chapter give an account of two young wild bulls at Chillingham attacking an old one in concert, and of two stallions together trying to drive away a third stallion from a troop of mares. In Abyssinia, Brehm encountered a great troop of baboons, who were crossing a valley: some had already ascended the opposite mountain, and some were still in the valley: the latter were attacked by the dogs, but the old males immediately hurried down from the rocks, and with mouths widely opened, roared so fearfully, that the dogs quickly drew back. They were again encouraged to the attack; but by this time all the baboons had reascended the heights, excepting a young one, about six months old, who, loudly calling for aid, climbed on a block of rock, and was surrounded. Now one of the largest males, a true hero, came down again from the mountain, slowly went to the young one, coaxed him, and triumphantly led him away—the dogs being too much astonished to make an attack. I cannot resist giving another scene which was witnessed by this same naturalist; an eagle seized a young Cercopithecus, which, by clinging to a branch, was not at once carried off; it cried loudly for assistance, upon which the other members of the troop, with much uproar, rushed to the rescue, surrounded the eagle, and pulled out so many feathers, that he no longer thought of his prey, but only how to escape. This eagle, as Brehm remarks, assuredly would never again attack a single monkey of a troop.[10]

It is certain that associated animals have a feeling of love for each other, which is not felt by non-social adult animals. How far in most cases they actually sympathise in the pains and pleasures of others, is more doubtful, especially with respect to pleasures. Mr Buxton, however, who had excellent means of observation,[11] states that his macaws, which lived free in Norfolk, took 'an extravagant interest' in a pair with a nest; and whenever the female left it, she was surrounded by a troop 'screaming horrible acclamations in her honour'. It is often difficult to judge whether animals have any feeling for the sufferings of others of their kind. Who can say what cows feel, when they surround and stare intently on a dying or dead companion; apparently, however, as Houzeau remarks, they feel no pity. That animals sometimes are far from feeling any sympathy is too certain; for they will expel a wounded animal from the herd, or gore or worry it to death. This is almost the blackest fact in natural history, unless, indeed, the explanation which has been suggested is true, that their instinct or reason leads them to expel an injured companion, lest beasts of prey, including man, should be tempted to follow the troop. In this case their conduct is not much worse than that of the North American Indians, who leave their feeble comrades to perish on the plains; or the Fijians, who, when their parents get old, or fall ill, bury them alive.[12]

---

10. Mr Belt gives the case of a spider-monkey (Ateles) in Nicaragua, which was heard screaming for nearly two hours in the forest, and was found with an eagle perched close by it. The bird apparently feared to attack as long as it remained face to face; and Mr Belt believes, from what he has seen of the habits of these monkeys, that they protect themselves from eagles by keeping two or three together. 'The Naturalist in Nicaragua', 1874, p. 118.
11. 'Annals and Mag. of Nat. Hist.', November 1868, p. 382.
12. Sir J. Lubbock, 'Prehistoric Times', 2nd edit., p. 446.

Many animals, however, certainly sympathise with each other's distress or danger. This is the case even with birds. Capt. Stansbury[13] found on a salt lake in Utah an old and completely blind pelican, which was very fat, and must have been well fed for a long time by his companions. Mr Blyth, as he informs me, saw Indian crows feeding two or three of their companions which were blind; and I have heard of an analogous case with the domestic cock. We may, if we choose, call these actions instinctive; but such cases are much too rare for the development of any special instinct.[14] I have myself seen a dog, who never passed a cat who lay sick in a basket, and was a great friend of his, without giving her a few licks with his tongue, the surest sign of kind feeling in a dog.

It must be called sympathy that leads a courageous dog to fly at any one who strikes his master, as he certainly will. I saw a person pretending to beat a lady, who had a very timid little dog on her lap, and the trial had never been made before; the little creature instantly jumped away, but after the pretended beating was over, it was really pathetic to see how perseveringly he tried to lick his mistress's face, and comfort her. Brehm[15] states that when a baboon in confinement was pursued to be punished, the others tried to protect him. It must have been sympathy in the cases above given which led the baboons and Cercopitheci to defend their young comrades from the dogs and the eagle. I will give only one other instance of sympathetic and heroic conduct, in the case of a little American monkey. Several years ago a keeper at the Zoological Gardens shewed me some deep and scarcely healed wounds on the nape of his

13. As quoted by Mr L. H. Morgan 'The American Beaver', 1868, p. 272. Capt. Stansbury also gives an interesting account of the manner in which a very young pelican, carried away by a strong stream, was guided and encouraged in its attempts to reach the shore by half a dozen old birds.
14. As Mr Bain states, 'effective aid to a sufferer springs from sympathy proper': 'Mental and Moral Science', 1868, p. 245.
15. 'Thierleben', B. i. s. 85.

own neck, inflicted on him, whilst kneeling on the floor, by a fierce baboon. The little American monkey, who was a warm friend of this keeper, lived in the same large compartment, and was dreadfully afraid of the great baboon. Nevertheless, as soon as he saw his friend in peril, he rushed to the rescue, and by screams and bites so distracted the baboon that the man was able to escape, after, as the surgeon thought, running great risk of his life.

Besides love and sympathy, animals exhibit other qualities connected with the social instincts, which in us would be called moral; and I agree with Agassiz[16] that dogs possess something very like a conscience.

Dogs possess some power of self-command, and this does not appear to be wholly the result of fear. As Braubach[17] remarks, they will refrain from stealing food in the absence of their master. They have long been accepted as the very type of fidelity and obedience. But the elephant is likewise very faithful to his driver or keeper, and probably considers him as the leader of the herd. Dr Hooker informs me that an elephant, which he was riding in India, became so deeply bogged that he remained stuck fast until the next day, when he was extricated by men with ropes. Under such circumstances elephants will seize with their trunks any object, dead or alive, to place under their knees, to prevent their sinking deeper in the mud; and the driver was dreadfully afraid lest the animal should have seized Dr Hooker and crushed him to death. But the driver himself, as Dr Hooker was assured, ran no risk. This forbearance under an emergency so dreadful for a heavy animal, is a wonderful proof of noble fidelity.[18]

All animals living in a body, which defend themselves or attack

16. 'De l'Espèce et de la Classe', 1869, p. 97.
17. 'Die Darwin'sche Art-Lehre', 1869, s. 54.
18. See also Hooker's 'Himalayan Journals', vol. ii, 1854, p. 333.

their enemies in concert, must indeed be in some degree faithful to one another; and those that follow a leader must be in some degree obedient. When the baboons in Abyssinia[19] plunder a garden, they silently follow their leader; and if an imprudent young animal makes a noise, he receives a slap from the others to teach him silence and obedience. Mr Galton, who has had excellent opportunities for observing the half-wild cattle in S. Africa, says,[20] that they cannot endure even a momentary separation from the herd. They are essentially slavish, and accept the common determination, seeking no better lot than to be led by any one ox who has enough self-reliance to accept the position. The men who break in these animals for harness, watch assiduously for those who, by grazing apart, shew a self-reliant disposition, and these they train as fore-oxen. Mr Galton adds that such animals are rare and valuable; and if many were born they would soon be eliminated, as lions are always on the lookout for the individuals which wander from the herd.

With respect to the impulse which leads certain animals to associate together, and to aid one another in many ways, we may infer that in most cases they are impelled by the same sense of satisfaction or pleasure which they experience in performing other instinctive actions; or by the same sense of dissatisfaction as when other instinctive actions are checked. We see this in innumerable instances, and it is illustrated in a striking manner by the acquired instincts of our domesticated animals; thus a young shepherd-dog delights in driving and running round a flock of sheep, but not in worrying them; a young fox-hound delights in hunting a fox, whilst some other kinds of dogs, as I have witnessed, utterly disregard foxes. What a strong feeling of inward satisfaction must impel a bird, so full of activity, to brood day after day over her

19. Brehm, 'Thierleben', B. i. s. 76.
20. See his extremely interesting paper on 'Gregariousness in Cattle, and in Man', 'Macmillan's Mag.', Feb. 1871, p. 353.

eggs. Migratory birds are quite miserable if stopped from migrating; perhaps they enjoy starting on their long flight; but it is hard to believe that the poor pinioned goose, described by Audubon, which started on foot at the proper time for its journey of probably more than a thousand miles, could have felt any joy in doing so. Some instincts are determined solely by painful feelings, as by fear, which leads to self-preservation, and is in some cases directed towards special enemies. No one, I presume, can analyse the sensations of pleasure or pain. In many instances, however, it is probable that instincts are persistently followed from the mere force of inheritance, without the stimulus of either pleasure or pain. A young pointer, when it first scents game, apparently cannot help pointing. A squirrel in a cage who pats the nuts which it cannot eat, as if to bury them in the ground, can hardly be thought to act thus, either from pleasure or pain. Hence the common assumption that men must be impelled to every action by experiencing some pleasure or pain may be erroneous. Although a habit may be blindly and implicitly followed, independently of any pleasure or pain felt at the moment, yet if it be forcibly and abruptly checked, a vague sense of dissatisfaction is generally experienced.

It has often been assumed that animals were in the first place rendered social, and that they feel as a consequence uncomfortable when separated from each other, and comfortable whilst together; but it is a more probable view that these sensations were first developed, in order that those animals which would profit by living in society, should be induced to live together, in the same manner as the sense of hunger and the pleasure of eating were, no doubt, first acquired in order to induce animals to eat. The feeling of pleasure from society is probably an extension of the parental or filial affections, since the social instinct seems to be developed by the young remaining for a long time with their parents; and this extension may be attributed in part to habit, but chiefly to natural selection. With those animals which were benefited by living in close association, the individuals which took the greatest pleasure in society would best escape various dangers; whilst those that cared least for their comrades, and lived solitary, would perish in greater numbers. With respect to the origin of the parental and filial affections, which apparently lie at the base of the social instincts, we know not the steps by which they have been gained; but we may infer that it has been to a large extent through natural selection. So it has almost certainly been with the unusual and opposite feeling of hatred between the nearest relations, as with the worker-bees which kill their brother-drones, and with the queen-bees which kill their daughter-queens; the desire to destroy their nearest relations having been in this case of service to the community. Parental affection, or some feeling which replaces it, has been developed in certain animals extremely low in the scale, for example, in star-fishes and spiders. It is also occasionally present in a few members alone in a whole group of animals, as in the genus Forficula, or earwigs.

The all-important emotion of sympathy is distinct from that of love. A mother may passionately love her sleeping and passive infant, but she can hardly at such times be said to feel sympathy for it. The love of a man for his dog is distinct from sympathy, and so is that of a dog for his master. Adam Smith formerly argued, as has Mr Bain recently, that the basis of sympathy lies in our strong retentiveness of former states of pain or pleasure. Hence, 'the sight of another person enduring hunger, cold, fatigue, revives in us some recollection of these states, which are painful even in idea'. We are thus impelled to relieve the sufferings of another, in order that our own painful feelings may be at the same time relieved. In like manner we are led to participate in the pleasures of others.[21] But I cannot see how this view explains the fact that sympathy is excited, in an immeasurably stronger degree, by a beloved, than by an indifferent person. The mere sight of suffering, independently of love, would suffice to call up in us vivid recollections and associations. The explanation may lie in the fact that, with all animals, sympathy is directed solely towards the members of the same community, and therefore towards known, and more or less beloved members, but not to all the individuals of the same species. This fact is not more surprising than that the fears of many animals should be directed against special enemies. Species which are not social, such as lions and tigers, no doubt feel sympathy for the suffering of their own young, but not for that of any other animal. With mankind, selfishness, experience, and imitation, probably add, as Mr Bain has shewn, to the power of sympathy; for we are led by the hope of receiving good in return to perform acts of sympathetic kindness to others; and sympathy is much strengthened by habit. In however complex a manner this feeling may have originated, as it is one of high importance to all those animals which aid and defend one another, it will have been increased through natural selection; for those communities, which included the greatest number of the most sympathetic members, would flourish best, and rear the greatest number of offspring.

It is, however, impossible to decide in many cases whether certain social instincts have been acquired through natural selection, or are the indirect result of other instincts and faculties, such as sympathy, reason, experience, and a tendency to imitation;

21. See the first and striking chapter in Adam Smith's 'Theory of Moral Sentiments'. Also Mr Bain's 'Mental and Moral Science', 1888, pp. 244, and 275–282. Mr Bain states, that 'sympathy is, indirectly, a source of pleasure to the sympathiser'; and he accounts for this through reciprocity. He remarks that 'the person benefited, or others in his stead, may make up, by sympathy and good offices returned, for all the sacrifice'. But if, as appears to be the case, sympathy is strictly an instinct, its exercise would give direct pleasure, in the same manner as the exercise, as before remarked, of almost every other instinct.

or again, whether they are simply the result of long-continued habit. So remarkable an instinct as the placing sentinels to warn the community of danger, can hardly have been the indirect result of any of these faculties; it must, therefore, have been directly acquired. On the other hand, the habit followed by the males of some social animals of defending the community, and of attacking their enemies or their prey in concert, may perhaps have originated from mutual sympathy; but courage, and in most cases strength, must have been previously acquired, probably through natural selection.

Of the various instincts and habits, some are much stronger than others; that is, some either give more pleasure in their performance, and more distress in their prevention, than others; or, which is probably quite as important, they are, through inheritance, more persistently followed, without exciting any special feeling of pleasure or pain. We are ourselves conscious that some habits are much more difficult to cure or change than others. Hence a struggle may often be observed in animals between different instincts, or between an instinct and some habitual disposition; as when a dog rushes after a hare, is rebuked, pauses, hesitates, pursues again, or returns ashamed to his master; or as between the love of a female dog for her young puppies and for her master—for she may be seen to slink away to them, as if half ashamed of not accompanying her master. But the most curious instance known to me of one instinct getting the better of another, is the migratory instinct conquering the maternal instinct. The former is wonderfully strong; a confined bird will at the proper season beat her breast against the wires of her cage, until it is bare and bloody. It causes young salmon to leap out of the fresh water, in which they could continue to exist, and thus unintentionally to commit suicide. Every one knows how strong the maternal instinct is, leading even timid birds to face great danger, though with hesitation, and in opposition to the instinct of self-preservation. Nevertheless, the migratory instinct is so powerful, that late in the autumn swallows, house-martins, and swifts frequently desert their tender young, leaving them to perish miserably in their nests.[22]

We can perceive that an instinctive impulse, if it be in any way more beneficial to a species than some other or opposed instinct, would be rendered the more potent of the two through natural selection; for the individuals which had it most strongly developed would survive in larger numbers. Whether this is the case with the migratory in comparison with the maternal instinct, may be doubted. The great persistence, or steady action of the former at certain seasons of the year during the whole day, may give it for a time paramount force.

Man a social animal—Every one will admit that man is a social being. We see this in his dislike of solitude, and in his wish for society beyond that of his own family. Solitary confinement is one of the severest punishments which can be inflicted. Some authors suppose that man primevally lived in single families; but at the present day, though single families, or only two or three together, roam the solitudes of some savage lands, they always, as far as I can discover, hold friendly relations with other families inhabiting the same district. Such families occasionally meet in council, and unite for their common defence. It

22. This fact, the Rev. L. Jenyns states (see his edition of 'White's Nat. Hist. of Selborne', 1853, p. 204) was first recorded by the illustrious Jenner, in 'Phil. Transact.', 1824, and has since been confirmed by several observers, especially by Mr Blackwall. This latter careful observer examined, late in the autumn, during two years, thirty-six nests; he found that twelve contained young dead birds, five contained eggs on the point of being hatched, and three, eggs not nearly hatched. Many birds, not yet old enough for a prolonged flight, are likewise deserted and left behind. See Blackwall, 'Researches in Zoology', 1834, pp. 108, 118. For some additional evidence, although this is not wanted, see Leroy, 'Lettres Phil.', 1802, p. 217. For Swifts, Gould's 'Introduction to the Birds of Great Britain', 1823, p. 5. Similar cases have been observed in Canada by Mr Adams; 'Pop. Science Review', July 1873, p. 283.

is no argument against savage man being a social animal, that the tribes inhabiting adjacent districts are almost always at war with each other; for the social instincts never extend to all the individuals of the same species. Judging from the analogy of the majority of the Quadrumana, it is probable that the early ape-like progenitors of man were likewise social; but this is not of much importance for us. Although man, as he now exists, has few special instincts, having lost any which his early progenitors may have possessed, this is no reason why he should not have retained from an extremely remote period some degree of instinctive love and sympathy for his fellows. We are indeed all conscious that we do possess such sympathetic feelings;[23] but our consciousness does not tell us whether they are instinctive, having originated long ago in the same manner as with the lower animals, or whether they have been acquired by each of us during our early years. As man is a social animal, it is almost certain that he would inherit a tendency to be faithful to his comrades, and obedient to the leader of his tribe; for these qualities are common to most social animals. He would consequently possess some capacity for self-command. He would from an inherited tendency be willing to defend, in concert with others, his fellow-men; and would be ready to aid them in any way, which did not too greatly interfere with his own welfare or his own strong desires.

The social animals which stand at the bottom of the scale are guided almost exclusively, and those which stand higher in the scale are largely guided, by special instincts in the aid which they give to the members of the same community; but they are

23. Hume remarks ('An Enquiry Concerning the Principles of Morals', edit. of 1751, p. 132), 'There seems a necessity for confessing that the happiness and misery of others are not spectacles altogether indifferent to us, but that the view of the former . . . communicates a secret joy; the appearance of the latter . . . throws a melancholy damp over the imagination.'

likewise in part impelled by mutual love and sympathy, assisted apparently by some amount of reason. Although man, as just remarked, has no special instincts to tell him how to aid his fellow-men, he still has the impulse, and with his improved intellectual faculties would naturally be much guided in this respect by reason and experience. Instinctive sympathy would also cause him to value highly the approbation of his fellows; for, as Mr Bain has clearly shewn,[24] the love of praise and the strong feeling of glory, and the still stronger horror of scorn and infamy, 'are due to the workings of sympathy'. Consequently man would be influenced in the highest degree by the wishes, approbation, and blame of his fellow-men, as expressed by their gestures and language. Thus the social instincts, which must have been acquired by man in a very rude state, and probably even by his early ape-like progenitors, still give the impulse to some of his best actions; but his actions are in a higher degree determined by the expressed wishes and judgment of his fellow-men, and unfortunately very often by his own strong selfish desires. But as love, sympathy and self-command become strengthened by habit, and as the power of reasoning becomes clearer, so that man can value justly the judgments of his fellows, he will feel himself impelled, apart from any transitory pleasure or pain, to certain lines of conduct. He might then declare—not that any barbarian or uncultivated man could thus think—I am the supreme judge of my own conduct, and in the words of Kant, I will not in my own person violate the dignity of humanity.

The more enduring Social Instincts conquer the less persistent Instincts— We have not, however, as yet considered the main point, on which, from our present point of view, the whole question of the moral sense turns. Why should a man feel that he ought to obey

24. 'Mental and Moral Science', 1868, p. 254.

one instinctive desire rather than another? Why is he bitterly regretful, if he has yielded to a strong sense of self-preservation, and has not risked his life to save that of a fellow-creature? or why does he regret having stolen food from hunger?

It is evident in the first place, that with mankind the instinctive impulses have different degrees of strength; a savage will risk his own life to save that of a member of the same community, but will be wholly indifferent about a stranger: a young and timid mother urged by the maternal instinct will, without a moment's hesitation, run the greatest danger for her own infant, but not for a mere fellow-creature. Nevertheless many a civilized man, or even boy, who never before risked his life for another, but full of courage and sympathy, has disregarded the instinct of self-preservation, and plunged at once into a torrent to save a drowning man, though a stranger. In this case man is impelled by the same instinctive motive, which made the heroic little American monkey, formerly described, save his keeper, by attacking the great and dreaded baboon. Such actions as the above appear to be the simple result of the greater strength of the social or maternal instincts than that of any other instinct or motive; for they are performed too instantaneously for reflection, or for pleasure or pain to be felt at the time; though, if prevented by any cause, distress or even misery might be felt. In a timid man, on the other hand, the instinct of self-preservation might be so strong, that he would be unable to force himself to run any such risk, perhaps not even for his own child.

I am aware that some persons maintain that actions performed impulsively, as in the above cases, do not come under the dominion of the moral sense, and cannot be called moral. They confine this term to actions done deliberately, after a victory over opposing desires, or when prompted by some exalted motive. But it appears scarcely possible to draw any clear line of distinction of this kind.[25] As far as exalted motives are concerned, many instances have been recorded of savages, destitute of any feeling of general benevolence towards mankind, and not guided by any religious motive, who have deliberately sacrificed their lives as prisoners,[26] rather than betray their comrades; and surely their conduct ought to be considered as moral. As far as deliberation, and the victory over opposing motives are concerned, animals may be seen doubting between opposed instincts, in rescuing their offspring or comrades from danger; yet their actions, though done for the good of others, are not called moral. Moreover, anything performed very often by us, will at last be done without deliberation or hesitation, and can then hardly be distinguished from an instinct; yet surely no one will pretend that such an action ceases to be moral. On the contrary, we all feel that an act cannot be considered as perfect, or as performed in the most noble manner, unless it be done impulsively, without deliberation or effort, in the same manner as by a man in whom the requisite qualities are innate. He who is forced to overcome his fear or want of sympathy before he acts, deserves, however, in one way higher credit than the man whose innate disposition leads him to a good act without effort. As we cannot distinguish between motives, we rank all actions of a certain class as moral, if performed by a moral being. A moral being is one who is capable of comparing his past and future actions or motives, and of approving or disapproving of them. We have no reason to suppose that any of the lower animals have this capacity; therefore, when

25. I refer here to the distinction between what has been called *material* and *formal* morality. I am glad to find that Prof. Huxley ('Critiques and Addresses', 1873, p. 287) takes the same view on this subject as I do. Mr Leslie Stephen remarks ('Essays on Freethinking and Plain Speaking', 1873, p. 83), 'the metaphysical distinction between material and formal morality is as irrelevant as other such distinctions'.

26. I have given one such case, namely of three Patagonian Indians who preferred being shot, one after the other, to betraying the plans of their companions in war ('Journal of Researches', 1845, p. 103).

a Newfoundland dog drags a child out of the water, or a monkey faces danger to rescue its comrade, or takes charge of an orphan monkey, we do not call its conduct moral. But in the case of man, who alone can with certainty be ranked as a moral being, actions of a certain class are called moral, whether performed deliberately, after a struggle with opposing motives, or impulsively through instinct, or from the effects of slowly-gained habit.

But to return to our more immediate subject. Although some instincts are more powerful than others, and thus lead to corresponding actions, yet it is untenable, that in man the social instincts (including the love of praise and fear of blame) possess greater strength, or have, through long habit, acquired greater strength than the instincts of self-preservation, hunger, lust, vengeance, &c. Why then does man regret, even though trying to banish such regret, that he has followed the one natural impulse rather than the other; and why does he further feel that he ought to regret his conduct? Man in this respect differs profoundly from the lower animals. Nevertheless we can, I think, see with some degree of clearness the reason of this difference.

Man, from the activity of his mental faculties, cannot avoid reflection: past impressions and images are incessantly and clearly passing through his mind. Now with those animals which live permanently in a body, the social instincts are ever present and persistent. Such animals are always ready to utter the danger-signal, to defend the community, and to give aid to their fellows in accordance with their habits; they feel at all times, without the stimulus of any special passion or desire, some degree of love and sympathy for them; they are unhappy if long separated from them, and always happy to be again in their company. So it is with ourselves. Even when we are quite alone, how often do we think with pleasure or pain of what others think of us—of their imagined approbation or disapprobation; and this all follows from sympathy, a fundamental element of the social instincts. A man who possessed no trace of such instincts would be an unnatural monster. On the other hand, the desire to satisfy hunger, or any passion such as vengeance, is in its nature temporary, and can for a time be fully satisfied. Nor is it easy, perhaps hardly possible, to call up with complete vividness the feeling, for instance, of hunger; nor indeed, as has often been remarked, of any suffering. The instinct of self-preservation is not felt except in the presence of danger; and many a coward has thought himself brave until he has met his enemy face to face. The wish for another man's property is perhaps as persistent a desire as any that can be named; but even in this case the satisfaction of actual possession is generally a weaker feeling than the desire: many a thief, if not a habitual one, after success has wondered why he stole some article.[27]

A man cannot prevent past impressions often repassing through his mind; he will thus be driven to make a comparison between the impressions of past hunger, vengeance satisfied, or danger shunned at other men's cost, with the almost ever-present instinct of sympathy, and with his early knowledge of what others consider as praiseworthy or blameable. This knowledge cannot be banished from his mind, and from instinctive sympathy is

27. Enmity or hatred seems also to be a highly persistent feeling, perhaps more so than any other that can be named. Envy is defined as hatred of another for some excellence or success; and Bacon insists (Essay ix.), 'Of all other affections envy is the most importune and continual.' Dogs are very apt to hate both strange men and strange dogs, especially if they live near at hand, but do not belong to the same family, tribe, or clan; this feeling would thus seem to be innate, and is certainly a most persistent one. It seems to be the complement and converse of the true social instinct. From what we hear of savages, it would appear that something of the same kind holds good with them. If this be so, it would be a small step in any one to transfer such feelings to any member of the same tribe if he had done him an injury and had become his enemy. Nor is it probable that the primitive conscience would reproach a man for injuring his enemy: rather it would reproach him, if he had not revenged himself. To do good in return for evil, to love your enemy, is a height of morality to which it may be doubted whether the social instincts would, by themselves, have ever led us. It is necessary that these instincts, together with sympathy, should have been highly cultivated and extended by the aid of reason, instruction, and the love or fear of God, before any such golden rule would ever be thought of and obeyed.

esteemed of great moment. He will then feel as if he had been baulked in following a present instinct or habit, and this with all animals causes dissatisfaction, or even misery.

The above case of the swallow affords an illustration, though of a reversed nature, of a temporary though for the time strongly persistent instinct conquering another instinct, which is usually dominant over all others. At the proper season these birds seem all day long to be impressed with the desire to migrate; their habits change; they become restless, are noisy, and congregate in flocks. Whilst the mother-bird is feeding, or brooding over her nestlings, the maternal instinct is probably stronger than the migratory; but the instinct which is the more persistent gains the victory, and at last, at a moment when her young ones are not in sight, she takes flight and deserts them. When arrived at the end of her long journey, and the migratory instinct has ceased to act, what an agony of remorse the bird would feel, if, from being endowed with great mental activity, she could not prevent the image constantly passing through her mind, of her young ones perishing in the bleak north from cold and hunger.

At the moment of action, man will no doubt be apt to follow the stronger impulse; and though this may occasionally prompt him to the noblest deeds, it will more commonly lead him to gratify his own desires at the expense of other men. But after their gratification, when past and weaker impressions are judged by the ever-enduring social instinct, and by his deep regard for the good opinion of his fellows, retribution will surely come. He will then feel remorse, repentance, regret, or shame, this latter feeling, however, relates almost exclusively to the judgment of others. He will consequently resolve more or less firmly to act differently for the future; and this is conscience; for conscience looks backwards, and serves as a guide for the future.

The nature and strength of the feelings which we call regret, shame, repentance or remorse, depend apparently not only on the strength of the violated instinct, but partly on the strength of the temptation, and often still more on the judgment of our fellows. How far each man values the appreciation of others, depends on the strength of his innate or acquired feeling of sympathy; and on his own capacity for reasoning out the remote consequences of his acts. Another element is most important, although not necessary, the reverence or fear of the Gods, or Spirits believed in by each man: and this applies especially in cases of remorse. Several critics have objected that though some slight regret or repentance may be explained by the view advocated in this chapter, it is impossible thus to account for the soul-shaking feeling of remorse. But I can see little force in this objection. My critics do not define what they mean by remorse, and I can find no definition implying more than an overwhelming sense of repentance. Remorse seems to bear the same relation to repentance, as rage does to anger, or agony to pain. It is far from strange that an instinct so strong and so generally admired, as maternal love, should, if disobeyed, lead to the deepest misery, as soon as the impression of the past cause of disobedience is weakened. Even when an action is opposed to no special instinct, merely to know that our friends and equals despise us for it is enough to cause great misery. Who can doubt that the refusal to fight a duel through fear has caused many men an agony of shame? Many a Hindoo, it is said, has been stirred to the bottom of his soul by having partaken of unclean food. Here is another case of what must, I think, be called remorse. Dr Landor acted as a magistrate in West Australia, and relates,[28] that a native on his farm, after losing one of his wives from disease, came and said that 'he was going to a distant tribe to spear a woman, to satisfy his sense of duty to his wife. I told him that if he did so, I would send him to prison for life. He remained about the farm for some months,

28. 'Insanity in Relation to Law', Ontario, United States, 1871, p. 14.

but got exceedingly thin, and complained that he could not rest or eat, that his wife's spirit was haunting him, because he had not taken a life for hers. I was inexorable, and assured him that nothing should save him if he did.' Nevertheless the man disappeared for more than a year, and then returned in high condition; and his other wife told Dr Landor that her husband had taken the life of a woman belonging to a distant tribe; but it was impossible to obtain legal evidence of the act. The breach of a rule held sacred by the tribe, will thus, as it seems, give rise to the deepest feelings,—and this quite apart from the social instincts, excepting in so far as the rule is grounded on the judgment of the community. How so many strange superstitions have arisen throughout the world we know not; nor can we tell how some real and great crimes, such as incest, have come to be held in an abhorrence (which is not however quite universal) by the lowest savages. It is even doubtful whether in some tribes incest would be looked on with greater horror, than would the marriage of a man with a woman bearing the same name, though not a relation. 'To violate this law is a crime which the Australians hold in the greatest abhorrence, in this agreeing exactly with certain tribes of North America. When the question is put in either district, is it worse to kill a girl of a foreign tribe, or to marry a girl of one's own, an answer just opposite to ours would be given without hesitation.'[29] We may, therefore, reject the belief, lately insisted on by some writers, that the abhorrence of incest is due to our possessing a special God-implanted conscience. On the whole it is intelligible, that a man urged by so powerful a sentiment as remorse, though arising as above explained, should be led to act in a manner, which he has been taught to believe serves as an expiation, such as delivering himself up to justice.

Man prompted by his conscience, will through long habit

29. E. B. Tylor in 'Contemporary Review', April 1873, p. 707.

acquire such perfect self-command, that his desires and passions will at last yield instantly and without a struggle to his social sympathies and instincts, including his feeling for the judgment of his fellows. The still hungry, or the still revengeful man will not think of stealing food, or of wreaking his vengeance. It is possible, or as we shall hereafter see, even probable, that the habit of self-command may, like other habits, be inherited. Thus at last man comes to feel, through acquired and perhaps inherited habit, that it is best for him to obey his more persistent impulses. The imperious word *ought* seems merely to imply the consciousness of the existence of a rule of conduct, however it may have originated. Formerly it must have been often vehemently urged that an insulted gentleman *ought* to fight a duel. We even say that a pointer *ought* to point, and a retriever to retrieve game. If they fail to do so, they fail in their duty and act wrongly.

If any desire or instinct leading to an action opposed to the good of others still appears, when recalled to mind, as strong as, or stronger than, the social instinct, a man will feel no keen regret at having followed it; but he will be conscious that if his conduct were known to his fellows, it would meet with their disapprobation; and few are so destitute of sympathy as not to feel discomfort when this is realised. If he has no such sympathy, and if his desires leading to bad actions are at the time strong, and when recalled are not over-mastered by the persistent social instincts, and the judgment of others, then he is essentially a bad man;[30] and the sole restraining motive left is the fear of punishment, and the conviction that in the long run it would be best for his own selfish interests to regard the good of others rather than his own.

It is obvious that every one may with an easy conscience

---

30. Dr Prosper Despine, in his 'Psychologie Naturelle', 1868 (tom. i, p. 243; tom. ii, p. 169) gives many curious cases of the worst criminals, who apparently have been entirely destitute of conscience.

gratify his own desires, if they do not interfere with his social instincts, that is with the good of others; but in order to be quite free from self-reproach, or at least of anxiety, it is almost necessary for him to avoid the disapprobation, whether reasonable or not, of his fellow-men. Nor must he break through the fixed habits of his life, especially if these are supported by reason; for if he does, he will assuredly feel dissatisfaction. He must likewise avoid the reprobation of the one God or gods in whom, according to his knowledge or superstition, he may believe; but in this case the additional fear of divine punishment often supervenes.

The strictly Social Virtues at first alone regarded—The above view of the origin and nature of the moral sense, which tells us what we ought to do, and of the conscience which reproves us if we disobey it, accords well with what we see of the early and undeveloped condition of this faculty in mankind. The virtues which must be practised, at least generally, by rude men, so that they may associate in a body, are those which are still recognised as the most important. But they are practised almost exclusively in relation to the men of the same tribe; and their opposites are not regarded as crimes in relation to the men of other tribes. No tribe could hold together if murder, robbery, treachery, &c., were common; consequently such crimes within the limits of the same tribe 'are branded with everlasting infamy';[31] but excite no such sentiment beyond these limits. A North-American Indian is well pleased with himself, and is honoured by others, when he scalps a man of another tribe; and a Dyak cuts off the head of an unoffending person, and dries it as a trophy. The murder of infants has prevailed on the largest scale throughout the

---

31. See an able article in the 'North British Review', 1867, p. 395. See also Mr W. Bagehot's articles on the Importance of Obedience and Coherence to Primitive Man, in the 'Fortnightly Review', 1867, p. 529, and 1868, p. 457, &c..

world,[32] and has met with no reproach; but infanticide, especially of females, has been thought to be good for the tribe, or at least not injurious. Suicide during former times was not generally considered as a crime,[33] but rather, from the courage displayed, as an honourable act; and it is still practised by some semi-civilised and savage nations without reproach, for it does not obviously concern others of the tribe. It has been recorded that an Indian Thug conscientiously regretted that he had not robbed and strangled as many travellers as did his father before him. In a rude state of civilisation the robbery of strangers is, indeed, generally considered as honourable.

Slavery, although in some ways beneficial during ancient times,[34] is a great crime; yet it was not so regarded until quite recently, even by the most civilized nations. And this was especially the case, because the slaves belonged in general to a race different from that of their masters. As barbarians do not regard the opinion of their women, wives are commonly treated like slaves. Most savages are utterly indifferent to the sufferings of strangers, or even delight in witnessing them. It is well known that the women and children of the North-American Indians aided in torturing their enemies. Some savages take a horrid pleasure in cruelty to animals,[35] and humanity is an unknown virtue. Nevertheless, besides the family affections, kindness is common, es-

---

32. The fullest account which I have met with is by Dr Gerland, in his 'Ueber dan Aussterben der Naturvölker', 1868; but I shall have to recur to the subject of infanticide in a future chapter.
33. See the very interesting discussion on Suicide in Lecky's 'History of European Morals', vol. i, 1869, p. 223. With respect to savages, Mr Winwood Reade informs me that the negroes of West Africa often commit suicide. It is well known how common it was amongst the miserable aborigines of South America, after the Spanish conquest. For New Zealand, see the voyage of the 'Novara', and for the Aleutian Islands, Müller, as quoted by Houzeau, 'Les Facultés Mentales', &c., tom. ii, p. 136.
34. See Mr Bagehot, 'Physics and Politics', 1872, p. 72.
35. See, for instance, Mr Hamilton's account of the Kaffirs, 'Anthropological Review', 1870, p. xv.

pecially during sickness, between the members of the same tribe, and is sometimes extended beyond these limits. Mungo Park's touching account of the kindness of the negro women of the interior to him is well known. Many instances could be given of the noble fidelity of savages towards each other, but not to strangers; common experience justifies the maxim of the Spaniard, 'Never, never trust an Indian.' There cannot be fidelity without truth; and this fundamental virtue is not rare between the members of the same tribe: thus Mungo Park heard the negro women teaching their young children to love the truth. This, again, is one of the virtues which becomes so deeply rooted in the mind, that it is sometimes practised by savages, even at a high cost, towards strangers; but to lie to your enemy has rarely been thought a sin, as the history of modern diplomacy too plainly shews. As soon as a tribe has a recognised leader, disobedience becomes a crime, and even abject submission is looked at as a sacred virtue.

As during rude times no man can be useful or faithful to his tribe without courage, this quality has universally been placed in the highest rank; and although in civilised countries a good yet timid man may be far more useful to the community than a brave one, we cannot help instinctively honouring the latter above a coward, however benevolent. Prudence, on the other hand, which does not concern the welfare of others, though a very useful virtue, has never been highly esteemed. As no man can practise the virtues necessary for the welfare of his tribe without self-sacrifice, self-command, and the power of endurance, these qualities have been at all times highly and most justly valued. The American savage voluntarily submits to the most horrid tortures without a groan, to prove and strengthen his fortitude and courage; and we cannot help admiring him, or even an Indian Fakir, who, from a foolish religious motive, swings suspended by a hook buried in his flesh.

The other so called self-regarding virtues, which do not obviously, though they may really, affect the welfare of the tribe, have never been esteemed by savages, though now highly appreciated by civilised nations. The greatest intemperance is no reproach with savages. Utter licentiousness, and unnatural crimes, prevail to an astounding extent.[36] As soon, however, as marriage, whether polygamous, or monogamous, becomes common, jealousy will lead to the inculcation of female virtue; and this, being honoured, will tend to spread to the unmarried females. How slowly it spreads to the male sex, we see at the present day. Chastity eminently requires self-command; therefore it has been honoured from a very early period in the moral history of civilised man. As a consequence of this, the senseless practice of celibacy has been ranked from a remote period as a virtue.[37] The hatred of indecency, which appears to us so natural as to be thought innate, and which is so valuable an aid to chastity, is a modern virtue, appertaining exclusively, as Sir G. Staunton remarks,[38] to civilised life. This is shewn by the ancient religious rites of various nations, by the drawings on the walls of Pompeii, and by the practices of many savages.

We have now seen that actions are regarded by savages, and were probably so regarded by primeval man, as good or bad, solely as they obviously affect the welfare of the tribe—not that of the species, nor that of an individual member of the tribe. This conclusion agrees well with the belief that the so-called moral sense is aboriginally derived from the social instincts, for both relate at first exclusively to the community. The chief causes of the low morality of savages, as judged by our standard, are, firstly, the confinement of sympathy to the same tribe. Secondly, pow-

36. Mr M'Lennan has given ('Primitive Marriage', 1865, p. 176) a good collection of facts on this head.
37. Lecky, 'History of European Morals', vol. i, 1869, p. 109.
38. 'Embassy to China', vol. ii, p. 348.

ers of reasoning insufficient to recognise the bearing of many virtues, especially of the self-regarding virtues, on the general welfare of the tribe. Savages, for instance, fail to trace the multiplied evils consequent on a want of temperance, chastity, &c. And, thirdly, weak power of self-command; for this power has not been strengthened through long-continued, perhaps inherited, habit, instruction and religion.

I have entered into the above details on the immorality of savages,[39] because some authors have recently taken a high view of their moral nature, or have attributed most of their crimes to mistaken benevolence.[40] These authors appear to rest their conclusion on savages possessing those virtues which are serviceable, or even necessary, for the existence of the family and of the tribe—qualities which they undoubtedly do possess, and often in a high degree.

Concluding Remarks—It was assumed formerly by philosophers of the derivative[41] school of morals that the foundation of morality lay in a form of Selfishness; but more recently the 'Greatest happiness principle' has been brought prominently forward. It is, however, more correct to speak of the latter principle as the standard, and not as the motive of conduct. Nevertheless, all the authors whose works I have consulted, with a few exceptions,[42]

---

39. See on this subject copious evidence in Chap. vii of Sir J. Lubbock, 'Origin of Civilisation', 1870.
40. For instance Lecky, 'Hist. European Morals', vol. i. p. 124.
41. This term is used in an able article in the 'Westminster Review', Oct. 1869, p. 498. For the 'Greatest happiness principle', see J. S. Mill, 'Utilitarianism', p. 17.
42. Mill recognises ('System of Logic', vol. ii., p. 422) in the clearest manner, that actions may be performed through habit without the anticipation of pleasure. Mr H. Sidgwick also, in his Essay on Pleasure and Desire ('The Contemporary Review', April 1872, p. 671), remarks: 'To sum up, in contravention of the doctrine that our conscious active impulses are always directed towards the production of agreeable sensations in ourselves, I would maintain that we find everywhere in consciousness extra-regarding impulse, directed towards something that is not pleasure; that in many cases the impulse is so far incompatible with the self-regarding that the two do not easily co-exist in the same moment of conscious-

write as if there must be a distinct motive for every action, and that this must be associated with some pleasure or displeasure. But man seems often to act impulsively, that is from instinct or long habit, without any consciousness of pleasure, in the same manner as does probably a bee or ant, when it blindly follows its instincts. Under circumstances of extreme peril, as during a fire, when a man endeavours to save a fellow-creature without a moment's hesitation, he can hardly feel pleasure; and still less has he time to reflect on the dissatisfaction which he might subsequently experience if he did not make the attempt. Should he afterwards reflect over his own conduct, he would feel that there lies within him an impulsive power widely different from a search after pleasure or happiness; and this seems to be the deeply planted social instinct.

In the case of the lower animals it seems much more appropriate to speak of their social instincts, as having been developed for the general good rather than for the general happiness of the species. The term, general good, may be defined as the rearing of the greatest number of individuals in full vigour and health, with all their faculties perfect, under the conditions to which they are subjected. As the social instincts both of man and the lower animals have no doubt been developed by nearly the same steps, it would be advisable, if found practicable, to use the same definition in both cases, and to take as the standard of morality, the general good or welfare of the community, rather than the general happiness; but this definition would perhaps require some limitation on account of political ethics.

When a man risks his life to save that of a fellow-creature, it

---

ness.' A dim feeling that our impulses do not by any means always arise from any contemporaneous or anticipated pleasure, has, I cannot but think, been one chief cause of the acceptance of the intuitive theory of morality, and of the rejection of the utilitarian or 'Greatest happiness' theory. With respect to the latter theory, the standard and the motive of conduct have no doubt often been confused, but they are really in some degree blended.

seems also more correct to say that he acts for the general good, rather than for the general happiness of mankind. No doubt the welfare and the happiness of the individual usually coincide; and a contented, happy tribe will flourish better than one that is discontented and unhappy. We have seen that even at an early period in the history of man, the expressed wishes of the community will have naturally influenced to a large extent the conduct of each member; and as all wish for happiness, the 'greatest happiness principle' will have become a most important secondary guide and object; the social instinct, however, together with sympathy (which leads to our regarding the approbation and disapprobation of others), having served as the primary impulse and guide. Thus the reproach is removed of laying the foundation of the noblest part of our nature in the base principle of selfishness; unless, indeed, the satisfaction which every animal feels, when it follows its proper instincts, and the dissatisfaction felt when prevented, be called selfish.

The wishes and opinions of the members of the same community, expressed at first orally, but later by writing also, either form the sole guides of our conduct, or greatly reinforce the social instincts; such opinions, however, have sometimes a tendency directly opposed to these instincts. This latter fact is well exemplified by the *Law of Honour*, that is, the law of the opinion of our equals, and not of all our countrymen. The breach of this law, even when the breach is known to be strictly accordant with true morality, has caused many a man more agony than a real crime. We recognise the same influence in the burning sense of shame which most of us have felt, even after the interval of years, when calling to mind some accidental breach of a trifling, though fixed, rule of etiquette. The judgment of the community will generally be guided by some rude experience of what is best in the long run for all the members; but this judgment will not rarely err from ignorance and weak powers of reasoning. Hence

the strangest customs and superstitions, in complete opposition
to the true welfare and happiness of mankind, have become all-
powerful throughout the world. We see this in the horror felt by
a Hindoo who breaks his caste, and in many other such cases.
It would be difficult to distinguish between the remorse felt by
a Hindoo who has yielded to the temptation of eating unclean
food, from that felt after committing a theft; but the former
would probably be the more severe.

How so many absurd rules of conduct, as well as so many
absurd religious beliefs, have originated, we do not know; nor
how it is that they have become, in all quarters of the world, so
deeply impressed on the mind of men; but it is worthy of remark
that a belief constantly inculcated during the early years of life,
whilst the brain is impressible, appears to acquire almost the nature of an instinct; and the very essence of an instinct is that it is
followed independently of reason. Neither can we say why certain admirable virtues, such as the love of truth, are much more
highly appreciated by some savage tribes than by others;[43] nor,
again, why similar differences prevail even amongst highly civilised nations. Knowing how firmly fixed many strange customs
and superstitions have become, we need feel no surprise that the
self-regarding virtues, supported as they are by reason, should
now appear to us so natural as to be thought innate, although
they were not valued by man in his early condition.

Notwithstanding many sources of doubt, man can generally and readily distinguish between the higher and lower moral
rules. The higher are founded on the social instincts, and relate
to the welfare of others. They are supported by the approbation
of our fellow-men and by reason. The lower rules, though some
of them when implying self-sacrifice hardly deserve to be called

43. Good instances are given by Mr Wallace in 'Scientific Opinion', Sept. 15, 1869;
    and more fully in his 'Contributions to the Theory of Natural Selection', 1870,
    p. 353.

lower, relate chiefly to self, and arise from public opinion, matured by experience and cultivation; for they are not practised by rude tribes.

As man advances in civilisation, and small tribes are united into larger communities, the simplest reason would tell each individual that he ought to extend his social instincts and sympathies to all the members of the same nation, though personally unknown to him. This point being once reached, there is only an artificial barrier to prevent his sympathies extending to the men of all nations and races. If, indeed, such men are separated from him by great differences in appearance or habits, experience unfortunately shews us how long it is, before we look at them as our fellow-creatures. Sympathy beyond the confines of man, that is, humanity to the lower animals seems to be one of the latest moral acquistions. It is apparently unfelt by savages, except towards their pets. How little the old Romans knew of it is shewn by their abhorrent gladiatorial exhibitions. The very idea of humanity, as far as I could observe, was new to most of the Gauchos of the Pampas. This virtue, one of the noblest with which man is endowed, seems to arise incidentally from our sympathies becoming more tender and more widely diffused, until they are extended to all sentient beings. As soon as this virtue is honoured and practised by some few men, it spreads through instruction and example to the young, and eventually becomes incorporated in public opinion.

The highest possible stage in moral culture is when we recognise that we ought to control our thoughts, and 'not even in inmost thought to think again the sins that made the past so pleasant to us'.[44] Whatever makes any bad action familiar to the mind, renders its performance by so much the easier. As Marcus Aurelius long ago said, 'Such as are thy habitual thoughts, such

44. Tennyson, 'Idylls of the King', p. 244.

also will be the character of thy mind; for the soul is dyed by the thoughts.'[45]

Our great philosopher, Herbert Spencer, has recently explained his views on the moral sense. He says,[46] 'I believe that the experiences of utility organised and consolidated through all past generations of the human race, have been producing corresponding modifications, which, by continued transmission and accumulation, have become in us certain faculties of moral intuition—certain emotions responding to right and wrong conduct, which have no apparent basis in the individual experiences of utility.' There is not the least inherent improbability, as it seems to me, in virtuous tendencies being more or less strongly inherited; for, not to mention the various dispositions and habits transmitted by many of our domestic animals to their offspring, I have heard of authentic cases in which a desire to steal and a tendency to lie appeared to run in families of the upper ranks; and as stealing is a rare crime in the wealthy classes, we can hardly account by accidental coincidence for the tendency occurring in two or three members of the same family. If bad tendencies are transmitted, it is probable that good ones are likewise transmitted. That the state of the body by affecting the brain, has great influence on the moral tendencies is known to most of those who have suffered from chronic derangements of the digestion or liver. The same fact is likewise shewn by the 'perversion or destruction of the moral sense being often one of the earliest symptoms of mental derangement';[47] and insanity is notoriously often inherited. Except through the principle of the transmission of moral tendencies, we cannot understand the differences believed to exist in this respect between the various races of mankind.

45. 'The Thoughts of the Emperor M. Aurelius Antoninus', Eng. translat., 2nd edit., 1869, p. 112. Marcus Aurelius was born AD 121.
46. Letter to Mr Mill in Bain's 'Mental and Moral Science', 1868, p. 722.
47. Maudsley, 'Body and Mind', 1870, p. 60.

Even the partial transmission of virtuous tendencies would be an immense assistance to the primary impulse derived directly and indirectly from the social instincts. Admitting for a moment that virtuous tendencies are inherited, it appears probable, at least in such cases as chastity, temperance, humanity to animals, &c., that they become first impressed on the mental organization through habit, instruction and example, continued during several generations in the same family, and in a quite subordinate degree, or not at all, by the individuals possessing such virtues having succeeded best in the struggle for life. My chief source of doubt with respect to any such inheritance, is that senseless customs, superstitions, and tastes, such as the horror of a Hindoo for unclean food, ought on the same principle to be transmitted. I have not met with any evidence in support of the transmission of superstitious customs or senseless habits, although in itself it is perhaps not less probable than that animals should acquire inherited tastes for certain kinds of food or fear of certain foes.

Finally the social instincts, which no doubt were acquired by man as by the lower animals for the good of the community, will from the first have given to him some wish to aid his fellows, some feeling of sympathy, and have compelled him to regard their approbation and disapprobation. Such impulses will have served him at a very early period as a rude rule of right and wrong. But as man gradually advanced in intellectual power, and was enabled to trace the more remote consequences of his actions; as he acquired sufficient knowledge to reject baneful customs and superstitions; as he regarded more and more, not only the welfare, but the happiness of his fellow-men; as from habit, following on beneficial experience, instruction and example, his sympathies became more tender and widely diffused, extending to men of all races, to the imbecile, maimed, and other useless members of society, and finally to the lower animals—so would the standard of his morality rise higher and higher. And it is admitted by moralists of the derivative school and by some intuitionists, that the standard of morality has risen since an early period in the history of man.[48]

As a struggle may sometimes be seen going on between the various instincts of the lower animals, it is not surprising that there should be a struggle in man between his social instincts, with their derived virtues, and his lower, though momentarily stronger impulses or desires. This, as Mr Galton[49] has remarked, is all the less surprising, as man has emerged from a state of barbarism within a comparatively recent period. After having yielded to some temptation we feel a sense of dissatisfaction, shame, repentance, or remorse, analogous to the feelings caused by other powerful instincts or desires, when left unsatisfied or baulked. We compare the weakened impression of a past temptation with the ever present social instincts, or with habits, gained in early youth and strengthened during our whole lives, until they have become almost as strong as instincts. If with the temptation still before us we do not yield, it is because either the social instinct or some custom is at the moment predominant, or because we have learnt that it will appear to us hereafter the stronger, when compared with the weakened impression of the temptation, and we realise that its violation would cause us suffering. Looking to future generations, there is no cause to fear that the social instincts will grow weaker, and we may expect that virtuous habits will grow stronger, becoming perhaps fixed by inheritance. In

48. A writer in the 'North British Review' (July 1869, p. 531), well capable of forming a sound judgment, expresses himself strongly in favour of this conclusion. Mr Lecky ('Hist. of Morals', vol. i. p.143) seems to a certain extent to coincide therein.

49. See his remarkable work on 'Hereditary Genius', 1869, p. 349. The Duke of Argyll ('Primeval Man', 1869, p. 188) has some good remarks on the contest in man's nature between right and wrong.

this case the struggle between our higher and lower impulses will be less severe, and virtue will be triumphant.

Summary of the last two Chapters—There can be no doubt that the difference between the mind of the lowest man and that of the highest animal is immense. An anthropomorphous ape, if he could take a dispassionate view of his own case, would admit that though he could form an artful plan to plunder a garden— though he could use stones for fighting or for breaking open nuts, yet that the thought of fashioning a stone into a tool was quite beyond his scope. Still less, as he would admit, could he follow out a train of metaphysical reasoning, or solve a mathematical problem, or reflect on God, or admire a grand natural scene. Some apes, however, would probably declare that they could and did admire the beauty of the coloured skin and fur of their partners in marriage. They would admit, that though they could make other apes understand by cries some of their perceptions and simpler wants, the notion of expressing definite ideas by definite sounds had never crossed their minds. They might insist that they were ready to aid their fellow-apes of the same troop in many ways, to risk their lives for them, and to take charge of their orphans; but they would be forced to acknowledge that disinterested love for all living creatures, the most noble attribute of man, was quite beyond their comprehension.

Nevertheless the difference in mind between man and the higher animals, great as it is, certainly is one of degree and not of kind. We have seen that the senses and intitutions, the various emotions and faculties, such as love, memory, attention, curiosity, imitation, reason, &c., of which man boasts, may be found in an incipient, or even sometimes in a well-developed condition, in the lower animals. They are also capable of some inherited improvement, as we see in the domestic dog compared with the wolf or jackal. If it could be proved that certain high mental powers, such as the formation of general concepts, self-consciousness, &c., were absolutely peculiar to man, which seems extremely doubtful, it is not improbable that these qualities are merely the incidental results of other highly-advanced intellectual faculties; and these again mainly the result of the continued use of a perfect language. At what age does the new-born infant possess the power of abstraction, or become self-conscious, and reflect on its own existence? We cannot answer; nor can we answer in regard to the ascending organic scale. The half-art, half-instinct of language still bears the stamp of its gradual evolution. The ennobling belief in God is not universal with man; and the belief in spiritual agencies naturally follows from other mental powers. The moral sense perhaps affords the best and highest distinction between man and the lower animals; but I need say nothing on this head, as I have so lately endeavoured to shew that the social instincts—the prime principle of man's moral constitution[50]— with the aid of active intellectual powers and the effects of habit, naturally lead to the golden rule, 'As ye would that men should do to you, do ye to them likewise', and this lies at the foundation of morality.

In the next chapter I shall make some few remarks on the probable steps and means by which the several mental and moral faculties of man have been gradually evolved. That such evolution is at least possible, ought not to be denied, for we daily see these faculties developing in every infant; and we may trace a perfect gradation from the mind of an utter idiot, lower than that of an animal low in the scale, to the mind of a Newton.

---

50. 'The Thoughts of Marcus Aurelius', &c., p. 139.

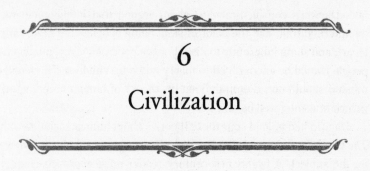

6

Civilization

Two months after the publication of *The Origin of Species*, Charles Darwin wrote a rueful letter to Charles Lyell. "I have received, in a Manchester newspaper, rather a good squib, showing that I have proved 'might is right,' and therefore that Napoleon is right, and every cheating tradesman is also right," he wrote.

It was difficult enough for Darwin to cope with the critics who rejected evolution. But some readers who embraced it seemed to read messages into his books that Darwin had never intended.

Darwin's ideas did not fall into a vacuum. People had been musing about evolution for sixty years, and they had already begun to speculate about the laws that governed the development of human society. They had been inspired by Lamarck and envisioned history as an inexorable march of progress. In *The Origin of Species* those writers saw a new mechanism for history's progress: natural selection. Adversity and competition strengthened societies, while indolence and coddling weakened them.

Archaeologists and anthropologists gave this view of humanity prehistoric roots. Rather than the creation and fall of man as described in the Bible, they envisioned early man much like "savages" of their day, hunting and gathering food and living in wandering bands. Later some peoples developed agriculture, invented writing and the other tools of civilization, and increased their wealth and knowledge.

Some writers called for society to seize control of its evolutionary fate. Darwin's cousin, Francis Galton, argued that intelligence was an inherited trait, like the color of skin. Some people were inherently smart, and some inherently foolish. For the good of society, intelligent people should be encouraged to marry and have children. The feeble-minded should not. Galton dubbed his theory of human breeding eugenics (meaning well-born).

Darwin had helped spur these theories about human societies with *The Origin of Species*. In *The Descent of Man*, he offered his own views on the subject. A twenty-first-century reader must approach chapter five with great care. In the decades after Darwin published *The Descent of Man*, eugenics would be invoked to support some profoundly evil social policies. In the United States, many leading doctors and scientists argued that "feeblemindedness" and "criminality" were carried by genes. For the sake of society, the eugenicists argued, people with these genes must be sterilized. This policy would claim sixty thousand victims, and eugenic laws were still being enforced as late as the 1960s. German scientists embraced American eugenics in the 1920s as they were developing a monstrous Nazi ideology of race, one that would eventually lead to the murder of millions.

It would be a mistake to look back at Darwin's own ideas through this horrifying haze. Nowhere in *The Descent of Man* does Darwin call for murder or sterilization. We should not whitewash *The Descent of Man,* either. Darwin shared many of the notions about human nature held by his Victorian contemporaries, notions we reject today. Darwin maintained that natural selection not only gave rise to the first humans, but also continued to shape humanity. "It is, therefore, highly probable that with mankind the intellectual faculties have been mainly and gradually perfected through natural selection," he wrote in *The Descent of Man*. The fact that brilliant scholars don't have enormous families did not overturn his claim, Darwin argues, because evolution is not based on extreme cases. Instead, evolution is a matter of averages. Overall, intelligence provides a slight evolutionary advantage, which gradually nudges the average intelligence of a society upward.

While mankind could progress from barbarism to civilization, Darwin did not see this progress as inevitable. Any improvement through natural selection could be undone if conditions changed. In fact, Darwin believed that the conditions he lived in—the growing prosperity of nineteenth-century England—might undo previous improvements. The sick, who would have died in earlier years, could now survive and pass on their defective traits to future generations. Darwin therefore argues that society will thrive if intelligent people are encouraged to have families. He also agrees that the less fit should be discouraged.

But Darwin goes no further in his prescriptions. After all, Darwin points out, a profound sympathy for our fellow humans evolved in us long ago, and full-blown eugenics would turn us into monsters. For Darwin, every cheating tradesman was most certainly *not* right.

Scientists can now point to many examples in which natural selection has shaped our species in just the past few thousand years. About ten thousand years ago, for example, some populations of humans began to domesticate cows. This allowed them to drink milk into adulthood. But early cattle herders did not have a physiology that was well-suited to milk-drinking. As infants, most mammals digest their mothers' milk with the help of a protein they produce, called lactase. Acting as a set of molecular scissors, lactase snips apart a sugar found in milk, called lactose. Once it is cut into pieces, lactose can be absorbed into the bloodstream. But as mammals grow older, they stop producing lactase, and as a result, they can no longer digest milk.

Humans started out the same way, and many humans today remain lactose intolerant. They get indigestion when they try to drink milk or eat cheese. But many people who descend from traditional cattle herders can still digest lactose. Scientists have traced the source of their ability to a mutation to a gene called LCT. Scientists are not yet sure how the mutant version of LCT works, but it shows clear signs of having undergone natural selection in the past few thousand years. In fact, it's the strongest selection yet measured in humans.

It appears that thousands of years ago, the LCT mutation randomly emerged in cattle-herding people. It somehow disabled the off-switch for lactase production, and allowed people who carried it to drink milk into adulthood. The extra calories and protein from the milk ultimately translated into extra children for people who had the mutation. Mutations to LCT may have cropped up among people who did not herd cattle, but they offered no benefit because there was no cow's milk for them to drink.

Culture has shaped human evolution not only with the benefits it has brought, but also with new threats. Malaria, for example, has thrived since humans invented agriculture. Malaria is caused by a single-celled mosquito-dwelling parasite called *Plasmodium falciparum*. Originally *Plasmodium*-carrying mosquitoes lived in African forests and drank the blood of forest animals. But over the past few thousand years they moved into the fields that African farmers had started to clear for crops. They laid their eggs in the stagnant water that puddled in the fields and feasted on the blood of farmers as they slept. The parasite's numbers exploded, and eventually it hitchhiked with mosquitoes into southern Europe and southern Asia. Finally humans brought it to the New World. Malaria still causes more than a million deaths a year, mostly in Africa where it began.

Nothing speeds up evolution faster than disease, because any mutation that can provide some resistance may save an animal from death. We humans are no exception. By unwittingly unleashing malaria, our ancestors triggered an explosion of evolution not only in *Plasmodium* but in our own species as well. Throughout malaria's range, human populations have acquired new defenses against the parasite, blocking its entry into blood cells and breaking its life cycle.

Darwin—and many others—were much more interested in natural selection's effects on intelligence than on digestion or disease resistance. But intelligence has proven far more difficult for scientists to analyze. Most experts agree that intelligence is a partially inherited trait. Family members tend to have similar scores on intelligence tests, and their scores are correlated on different kinds of tests. Adopted children show no correlation with their adoptive parents or even their adopted siblings.

It's one thing to identify an inherited trait—it's quite another to identify the genes that produce it. Scientists have been able to pinpoint genes that are associated with lactose digestion and malaria resistance because changing a single gene or a few of them produces dramatic changes in these features. Intelligence is proving to be a very different matter. The candidate genes that have been tied to intelligence can only account for a tiny fraction of the variation in test scores. It looks as if there may be thousands of intelligence-related genes. A mutation to any of those genes may produce a tiny change in a person's performance on intelligence tests.

Making matters even more complex is the fact that intelligence is only partly inherited. About half of the variation in test scores can be linked to genes; the other half is due to other sources of variation, which could be everything from the chemistry of the womb in which children developed to the level of income of their parents. Children who grow up poor in modern cities also experience high levels of stress, which has been shown to interfere with the growth of neurons. The actual role that a gene may have in shaping a person's intelligence may depend on that person's experiences. When scientists study the heritability of intelligence in affluent people, they find a very strong correlation. But when they look at the poor, the correlation collapses. Eugenics is not only immoral, in other words, but impossible to put into practice.

CHAPTER 5:
On the Development of the Intellectual and Moral Faculties during Primeval and Civilised Times

THE SUBJECTS TO BE discussed in this chapter are of the highest interest, but are treated by me in an imperfect and fragmentary manner. Mr Wallace, in an admirable paper before referred to,[1] argues that man, after he had partially acquired those intellectual and moral faculties which distinguish him from the lower animals, would have been but little liable to bodily modifications through natural selection or any other means. For man is enabled through his mental faculties 'to keep with an unchanged body in harmony with the changing universe'. He has great power of adapting his habits to new conditions of life. He invents weapons, tools, and various stratagems to procure food and to defend himself. When he migrates into a colder climate he uses clothes, builds sheds, and makes fires; and by the aid of fire cooks food otherwise indigestible. He aids his fellow-men in many ways, and anticipates future events. Even at a remote period he practised some division of labour.

The lower animals, on the other hand, must have their bodily structure modified in order to survive under greatly changed conditions. They must be rendered stronger, or acquire more effective teeth or claws, for defence against new enemies; or they must be reduced in size, so as to escape detection and danger.

1. 'Anthropological Review', May 1864, p. clviii.

When they migrate into a colder climate, they must become clothed with thicker fur, or have their constitutions altered. If they fail to be thus modified, they will cease to exist.

The case, however, is widely different, as Mr Wallace has with justice insisted, in relation to the intellectual and moral faculties of man. These faculties are variable; and we have every reason to believe that the variations tend to be inherited. Therefore, if they were formerly of high importance to primeval man and to his ape-like progenitors, they would have been perfected or advanced through natural selection. Of the high importance of the intellectual faculties there can be no doubt, for man mainly owes to them his predominant position in the world. We can see, that in the rudest state of society, the individuals who were the most sagacious, who invented and used the best weapons or traps, and who were best able to defend themselves, would rear the greatest number of offspring. The tribes, which included the largest number of men thus endowed, would increase in number and supplant other tribes. Numbers depend primarily on the means of subsistence, and this depends partly on the physical nature of the country, but in a much higher degree on the arts which are there practised. As a tribe increases and is victorious, it is often still further increased by the absorption of other tribes.[2] The stature and strength of the men of a tribe are likewise of some importance for its success, and these depend in part on the nature and amount of the food which can be obtained. In Europe the men of the Bronze period were supplanted by a race more powerful, and, judging from their sword-handles, with larger hands;[3] but their success was probably still more due to their superiority in the arts.

2. After a time the members or tribes which are absorbed into another tribe assume, as Sir Henry Maine remarks ('Ancient Law', 1861, p. 131), that they are the co-descendants of the same ancestors.
3. Morlot, 'Soc. Vaud. Sc. Nat.', 1860, p. 294.

All that we know about savages, or may infer from their traditions and from old monuments, the history of which is quite forgotten by the present inhabitants, shew that from the remotest times successful tribes have supplanted other tribes. Relics of extinct or forgotten tribes have been discovered throughout the civilised regions of the earth, on the wild plains of America, and on the isolated islands in the Pacific Ocean. At the present day civilised nations are everywhere supplanting barbarous nations, excepting where the climate opposes a deadly barrier; and they succeed mainly, though not exclusively, through their arts, which are the products of the intellect. It is, therefore, highly probable that with mankind the intellectual faculties have been mainly and gradually perfected through natural selection; and this conclusion is sufficient for our purpose. Undoubtedly it would be interesting to trace the development of each separate faculty from the state in which it exists in the lower animals to that in which it exists in man; but neither my ability nor knowledge permits the attempt.

It deserves notice that, as soon as the progenitors of man became social (and this probably occurred at a very early period), the principle of imitation, and reason, and experience would have increased, and much modified the intellectual powers in a way, of which we see only traces in the lower animals. Apes are much given to imitation, as are the lowest savages; and the simple fact previously referred to, that after a time no animal can be caught in the same place by the same sort of trap, shews that animals learn by experience, and imitate the caution of others. Now, if some one man in a tribe, more sagacious than the others, invented a new snare or weapon, or other means of attack or defence, the plainest self-interest, without the assistance of much reasoning power, would prompt the other members to imitate him; and all would thus profit. The habitual practice of each new art must likewise in some slight degree strengthen the intellect. If the new invention were an important one, the tribe would increase in number, spread, and supplant other tribes. In a tribe thus rendered more numerous there would always be a rather greater chance of the birth of other superior and inventive members. If such men left children to inherit their mental superiority, the chance of the birth of still more ingenious members would be somewhat better, and in a very small tribe decidedly better. Even if they left no children, the tribe would still include their blood-relations; and it has been ascertained by agriculturists[4] that by preserving and breeding from the family of an animal, which when slaughtered was found to be valuable, the desired character has been obtained.

Turning now to the social and moral faculties. In order that primeval men, or the ape-like pregenitors of man, should become social, they must have acquired the same instinctive feelings, which impel other animals to live in a body; and they no doubt exhibited the same general disposition. They would have felt uneasy when separated from their comrades, for whom they would have felt some degree of love; they would have warned each other of danger, and have given mutual aid in attack or defence. All this implies some degree of sympathy, fidelity, and courage. Such social qualities, the paramount importance of which to the lower animals is disputed by no one, were no doubt acquired by the progenitors of man in a similar manner, namely, through natural selection, aided by inherited habit. When two tribes of primeval man, living in the same country, came into competition, if (other circumstances being equal) the one tribe included a great number of courageous, sympathetic and faithful members, who were always ready to warn each other of danger, to

---

4. I have given instances in my 'Variation of Animals under Domestication', vol. ii. p. 196.

aid and defend each other, this tribe would succeed better and conquer the other. Let it be borne in mind how all-important in the never-ceasing wars of savages, fidelity and courage must be. The advantage which disciplined soldiers have over undisciplined hordes follows chiefly from the confidence which each man feels in his comrades. Obedience, as Mr Bagehot has well shewn,[5] is of the highest value, for any form of government is better than none. Selfish and contentious people will not cohere, and without coherence nothing can be effected. A tribe rich in the above qualities would spread and be victorious over other tribes: but in the course of time it would, judging from all past history, be in its turn overcome by some other tribe still more highly endowed. Thus the social and moral qualities would tend slowly to advance and be diffused throughout the world.

But it may be asked, how within the limits of the same tribe did a large number of members first become endowed with these social and moral qualities, and how was the standard of excellence raised? It is extremely doubtful whether the offspring of the more sympathetic and benevolent parents, or of those who were the most faithful to their comrades, would be reared in greater numbers than the children of selfish and treacherous parents belonging to the same tribe. He who was ready to sacrifice his life, as many a savage has been, rather than betray his comrades, would often leave no offspring to inherit his noble nature. The bravest men, who were always willing to come to the front in war, and who freely risked their lives for others, would on an average perish in larger numbers than other men. Therefore it hardly seems probable, that the number of men gifted with such virtues, or that the standard of their excellence, could be increased through natural selection, that is, by the survival of

---

5. See a remarkable series of articles on 'Physics and Politics' in the 'Fortnightly Review', Nov. 1867; April 1, 1868; July 1, 1869, since separately published.

the fittest; for we are not here speaking of one tribe being victorious over another.

Although the circumstances, leading to an increase in the number of those thus endowed within the same tribe, are too complex to be clearly followed out, we can trace some of the probable steps. In the first place, as the reasoning powers and foresight of the members became improved, each man would soon learn that if he aided his fellow-men, he would commonly receive aid in return. From this low motive he might acquire the habit of aiding his fellows; and the habit of performing benevolent actions certainly strengthens the feeling of sympathy which gives the first impulse to benevolent actions. Habits, moreover, followed during many generations probably tend to be inherited.

But another and much more powerful stimulus to the development of the social virtues, is afforded by the praise and the blame of our fellow-men. To the instinct of sympathy, as we have already seen, it is primarily due, that we habitually bestow both praise and blame on others, whilst we love the former and dread the latter when applied to ourselves; and this instinct no doubt was originally acquired, like all the other social instincts, through natural selection. At how early a period the progenitors of man in the course of their development, became capable of feeling and being impelled by, the praise or blame of their fellow-creatures, we cannot of course say. But it appears that even dogs appreciate encouragement, praise, and blame. The rudest savages feel the sentiment of glory, as they clearly show by preserving the trophies of their prowess, by their habit of excessive boasting, and even by the extreme care which they take of their personal appearance and decorations; for unless they regarded the opinion of their comrades, such habits would be senseless.

They certainly feel shame at the breach of some of their lesser rules, and apparently remorse, as shewn by the case of the Australian who grew thin and could not rest from having delayed to murder some other woman, so as to propitiate his dead wife's spirit. Though I have not met with any other recorded case, it is scarcely credible that a savage, who will sacrifice his life rather than betray his tribe, or one who will deliver himself up as a prisoner rather than break his parole,[6] would not feel remorse in his inmost soul, if he had failed in a duty, which he held sacred.

We may therefore conclude that primeval man, at a very remote period, was influenced by the praise and blame of his fellows. It is obvious, that the members of the same tribe would approve of conduct which appeared to them to be for the general good, and would reprobate that which appeared evil. To do good unto others—to do unto others as ye would they should do unto you—is the foundation-stone of morality. It is, therefore, hardly possible to exaggerate the importance during rude times of the love of praise and the dread of blame. A man who was not impelled by any deep, instinctive feeling, to sacrifice his life for the good of others, yet was roused to such actions by a sense of glory, would by his example excite the same wish for glory in other men, and would strengthen by exercise the noble feeling of admiration. He might thus do far more good to his tribe than by begetting offspring with a tendency to inherit his own high character.

With increased experience and reason, man perceives the more remote consequences of his actions, and the self-regarding virtues, such as temperance, chastity, &c., which during early times are, as we have before seen, utterly disregarded, come to be highly esteemed or even held sacred. I need not, however, repeat what I have said on this head in the fourth chapter. Ultimately our moral sense or conscience becomes a highly complex

---

6. Mr Wallace gives cases in his 'Contributions to the Theory of Natural Selection', 1870, p. 354.

sentiment—originating in the social instincts, largely guided by the approbation of our fellow-men, ruled by reason, self-interest, and in later times by deep religious feelings, and confirmed by instruction and habit.

It must not be forgotten that although a high standard of morality gives but a slight or no advantage to each individual man and his children over the other men of the same tribe, yet that an increase in the number of well-endowed men and an advancement in the standard of morality will certainly give an immense advantage to one tribe over another. A tribe including many members who, from possessing in a high degree the spirit of patriotism, fidelity, obedience, courage, and sympathy, were always ready to aid one another, and to sacrifice themselves for the common good, would be victorious over most other tribes; and this would be natural selection. At all times throughout the world tribes have supplanted other tribes; and as morality is one important element in their success, the standard of morality and the number of well-endowed men will thus everywhere tend to rise and increase.

It is, however, very difficult to form any judgment why one particular tribe and not another has been successful and has risen in the scale of civilisation. Many savages are in the same condition as when first discovered several centuries ago. As Mr Bagehot has remarked, we are apt to look at progress as normal in human society; but history refutes this. The ancients did not even entertain the idea, nor do the Oriental nations at the present day. According to another high authority, Sir Henry Maine,[7] 'the greatest part of mankind has never shewn a particle of desire that its civil institutions should be improved'. Progress seems to depend on many concurrent favourable conditions, far too com-

7. 'Ancient Law', 1861, p. 22. For Mr Bagehot's remarks, 'Fortnightly Review,' April 1, 1868, p. 452.

plex to be followed out. But it has often been remarked, that a cool climate, from leading to industry and to the various arts, has been highly favourable thereto. The Esquimaux, pressed by hard necessity, have succeeded in many ingenious inventions, but their climate has been too severe for continued progress. Nomadic habits, whether over wide plains, or through the dense forests of the tropics, or along the shores of the sea, have in every case been highly detrimental. Whilst observing the barbarous inhabitants of Tierra del Fuego, it struck me that the possession of some property, a fixed abode, and the union of many families under a chief, were the indispensable requisites for civilisation. Such habits almost necessitate the cultivation of the ground; and the first steps in cultivation would probably result, as I have elsewhere shewn,[8] from some such accident as the seeds of a fruit-tree falling on a heap of refuse, and producing an unusually fine variety. The problem, however, of the first advance of savages towards civilisation is at present much too difficult to be solved.

Natural Selection as affecting Civilised Nations—I have hitherto only considered the advancement of man from a semi-human condition to that of the modern savage. But some remarks on the action of natural selection on civilised nations may be worth adding. This subject has been ably discussed by Mr W. R. Greg,[9] and previously by Mr Wallace and Mr Galton.[10] Most of my remarks are taken from these three authors. With savages, the

8. 'The Variation of Animals and Plants under Domestication', vol. i. p. 309.
9. 'Fraser's Magazine', Sept. 1868, p. 353. This article seems to have struck many persons, and has given rise to two remarkable essays and a rejoinder in the 'Spectator', Oct. 3rd and 17th, 1868. It has also been discussed in the 'Q. Journal of Science', 1869, p. 152, and by Mr Lawson Tait in the 'Dublin Q. Journal of Medical Science', Feb. 1869, and by Mr E. Ray Lankester in his 'Comparative Longevity', 1870, p. 128. Similar views appeared previously in the 'Australasian', July 13, 1867. I have borrowed ideas from several of these writers.
10. For Mr Wallace, see 'Anthropolog. Review', as before cited. Mr Galton in 'Macmillan's Magazine', Aug. 1865, p. 318; also his great work, 'Hereditary Genius', 1870.

weak in body or mind are soon eliminated; and those that survive commonly exhibit a vigorous state of health. We civilised men, on the other hand, do our utmost to check the process of elimination; we build asylums for the imbecile, the maimed, and the sick; we institute poor-laws; and our medical men exert their utmost skill to save the life of every one to the last moment. There is reason to believe that vaccination has preserved thousands, who from a weak constitution would formerly have succumbed to small-pox. Thus the weak members of civilised societies propagate their kind. No one who has attended to the breeding of domestic animals will doubt that this must be highly injurious to the race of man. It is surprising how soon a want of care, or care wrongly directed, leads to the degeneration of a domestic race; but excepting in the case of man himself, hardly any one is so ignorant as to allow his worst animals to breed.

The aid which we feel impelled to give to the helpless is mainly an incidental result of the instinct of sympathy, which was originally acquired as part of the social instincts, but subsequently rendered, in the manner previously indicated, more tender and more widely diffused. Nor could we check our sympathy, even at the urging of hard reason, without deterioration in the noblest part of our nature. The surgeon may harden himself whilst performing an operation, for he knows that he is acting for the good of his patient; but if we were intentionally to neglect the weak and helpless, it could only be for a contingent benefit, with an overwhelming present evil. We must therefore bear the undoubtedly bad effects of the weak surviving and propagating their kind; but there appears to be at least one check in steady action, namely that the weaker and inferior members of society do not marry so freely as the sound; and this check might be indefinitely increased by the weak in body or mind refraining from marriage, though this is more to be hoped for than expected.

In every country in which a large standing army is kept up,

the finest young men are taken by the conscription or are enlisted. They are thus exposed to early death during war, are often tempted into vice, and are prevented from marrying during the prime of life. On the other hand the shorter and feebler men, with poor constitutions, are left at home, and consequently have a much better chance of marrying and propagating their kind.[11]

Man accumulates property and bequeaths it to his children, so that the children of the rich have an advantage over the poor in the race for success, independently of bodily or mental superiority. On the other hand, the children of parents who are short-lived, and are therefore on an average deficient in health and vigour, come into their property sooner than other children, and will be likely to marry earlier, and leave a larger number of offspring to inherit their inferior constitutions. But the inheritance of property by itself is very far from an evil; for without the accumulation of capital the arts could not progress; and it is chiefly through their power that the civilised races have extended, and are now everywhere extending their range, so as to take the place of the lower races. Nor does the moderate accumulation of wealth interfere with the process of selection. When a poor man becomes moderately rich, his children enter trades or professions in which there is struggle enough, so that the able in body and mind succeed best. The presence of a body of well-instructed men, who have not to labour for their daily bread, is important to a degree which cannot be over-estimated; as all high intellectual work is carried on by them, and on such work, material progress of all kinds mainly depends, not to mention other and higher advantages. No doubt wealth when very great tends to convert men into useless drones, but their number is never large; and some degree of elimination here occurs, for

---

11. Prof. H. Fick ('Einfluss der Naturwissenschaft auf das Recht', June, 1872) has some good remarks on this head, and on other such points.

we daily see rich men, who happen to be fools or profligate, squandering away their wealth.

Primogeniture with entailed estates is a more direct evil, though it may formerly have been a great advantage by the creation of a dominant class, and any government is better than none. Most eldest sons, though they may be weak in body or mind, marry, whilst the younger sons, however superior in these respects, do not so generally marry. Nor can worthless eldest sons with entailed estates squander their wealth. But here, as elsewhere, the relations of civilised life are so complex that some compensatory checks intervene. The men who are rich through primogeniture are able to select generation after generation the more beautiful and charming women; and these must generally be healthy in body and active in mind. The evil consequences, such as they may be, of the continued preservation of the same line of descent, without any selection, are checked by men of rank always wishing to increase their wealth and power; and this they effect by marrying heiresses. But the daughters of parents who have produced single children, are themselves, as Mr Galton[12] has shewn, apt to be sterile; and thus noble families are continually cut off in the direct line, and their wealth flows into some side channel; but unfortunately this channel is not determined by superiority of any kind.

Although civilisation thus checks in many ways the action of natural selection, it apparently favours the better development of the body, by means of good food and the freedom from occasional hardships. This may be inferred from civilised men having been found, wherever compared, to be physically stronger than savages.[13] They appear also to have equal powers of endurance, as has been proved in many adventurous expeditions. Even the great

12. 'Hereditary Genius', 1870, pp. 132–140.
13. Quatrefages, 'Revue des Cours Scientifiques', 1867–68, p. 659.

luxury of the rich can be but little detrimental; for the expectation of life of our aristocracy, at all ages and of both sexes, is very little inferior to that of healthy English lives in the lower classes.[14]

We will now look to the intellectual faculties. If in each grade of society the members were divided into two equal bodies, the one including the intellectually superior and the other the inferior, there can be little doubt that the former would succeed best in all occupations, and rear a greater number of children. Even in the lowest walks of life, skill and ability must be of some advantage; though in many occupations, owing to the great division of labour, a very small one. Hence in civilised nations there will be some tendency to an increase both in the number and in the standard of the intellectually able. But I do not wish to assert that this tendency may not be more than counterbalanced in other ways, as by the multiplication of the reckless and improvident; but even to such as these, ability must be some advantage.

It has often been objected to views like the foregoing, that the most eminent men who have ever lived have left no offspring to inherit their great intellect. Mr Galton says,[15] 'I regret I am unable to solve the simple question whether, and how far, men and women who are prodigies of genius are infertile. I have, however, shewn that men of eminence are by no means so.' Great lawgivers, the founders of beneficent religions, great philosophers and discoverers in science, aid the progress of mankind in a far higher degree by their works than by leaving a numerous progeny. In the case of corporeal structures, it is the selection of the slightly better-endowed and the elimination of the slightly less well-endowed individuals, and not the preservation of strongly-marked and rare anomalies, that leads to the advancement of a

14. See the fifth and sixth columns, compiled from good authorities, in the table given in Mr E. R. Lankester's 'Comparative Longevity', 1870, p. 115.
15. 'Hereditary Genius', 1870, p. 330.

species.[16] So it will be with the intellectual faculties, since the somewhat abler men in each grade of society succeed rather better than the less able, and consequently increase in number, if not otherwise prevented. When in any nation the standard of intellect and the number of intellectual men have increased, we may expect from the law of the deviation from an average, that prodigies of genius will, as shewn by Mr Galton, appear somewhat more frequently than before.

In regard to the moral qualities, some elimination of the worst dispositions is always in progress even in the most civilised nations. Malefactors are executed, or imprisoned for long periods, so that they cannot freely transmit their bad qualities. Melancholic and insane persons are confined, or commit suicide. Violent and quarrelsome men often come to a bloody end. The restless who will not follow any steady occupation—and this relic of barbarism is a great check to civilisation[17]—emigrate to newly-settled countries, where they prove useful pioneers. Intemperance is so highly destructive, that the expectation of life of the intemperate, at the age of thirty for instance, is only 13.8 years; whilst for the rural labourers of England at the same age it is 40.59 years.[18] Profligate women bear few children, and profligate men rarely marry; both suffer from disease. In the breeding of domestic animals, the elimination of those individuals, though few in number, which are in any marked manner inferior, is by no means an unimportant element towards success. This especially holds good with injurious characters which tend to reappear through reversion, such as blackness in sheep; and with mankind some of the worst dispositions, which occasionally without any assignable cause make their appearance

16. 'Origin of Species' (fifth edition, 1869), p. 104.
17. 'Hereditary Genius', 1870, p. 347.
18. E. Ray Lankester, 'Comparative Longevity', 1870, p. 115. The table of the intemperate is from Neison's 'Vital Statistics'. In regard to profligacy, see Dr Farr, 'Influence of Marriage on Mortality', 'Nat. Assoc. for the Promotion of Social Science', 1858.

in families, may perhaps be reversions to a savage state, from which we are not removed by very many generations. This view seems indeed recognised in the common expression that such men are the black sheep of the family.

With civilised nations, as far as an advanced standard of morality, and an increased number of fairly good men are concerned, natural selection apparently effects but little; though the fundamental social instincts were originally thus gained. But I have already said enough, whilst treating of the lower races, on the causes which lead to the advance of morality, namely, the approbation of our fellow-men—the strengthening of our sympathies by habit—example and imitation—reason—experience, and even self-interest—instruction during youth, and religious feelings.

A most important obstacle in civilised countries to an increase in the number of men of a superior class has been strongly insisted on by Mr Greg and Mr Galton,[19] namely, the fact that the very poor and reckless, who are often degraded by vice, almost invariably marry early, whilst the careful and frugal, who are generally otherwise virtuous, marry late in life, so that they may be able to support themselves and their children in comfort. Those who marry early produce within a given period not only a greater number of generations, but, as shewn by Dr Duncan,[20] they produce many more children. The children, moreover, that are born by mothers during the prime of life are heavier and larger, and therefore probably more vigorous, than those born at other periods. Thus the reckless, degraded, and often vicious members of society, tend to increase at a quicker rate than the provident and generally virtuous members. Or as Mr Greg puts

19. 'Fraser's Magazine', Sept. 1868, p. 353. 'Macmillan's Magazine', Aug. 1865, p. 318. The Rev. F. W. Farrar ('Fraser's Mag.', Aug. 1870, p. 264) takes a different view.
20. 'On the Laws of the Fertility of Women', in 'Transact. Royal Soc.', Edinburgh, vol. xxiv. p. 287; now published separately under the title of 'Fecundity, Fertility, and Sterility', 1871. See, also, Mr Galton, 'Hereditary Genius', pp. 352–357, for observations to the above effect.

the case: 'The careless, squalid, unaspiring Irishman multiplies like rabbits: the frugal, foreseeing, self-respecting, ambitious Scot, stern in his morality, spiritual in his faith, sagacious and disciplined in his intelligence, passes his best years in struggle and in celibacy, marries late, and leaves few behind him. Given a land originally peopled by a thousand Saxons and a thousand Celts—and in a dozen generations five-sixths of the population would be Celts, but five-sixths of the property, of the power, of the intellect, would belong to the one-sixth of Saxons that remained. In the eternal "struggle for existence", it would be the inferior and *less* favoured race that had prevailed—and prevailed by virtue not of its good qualities but of its faults.'

There are, however, some checks to this downward tendency. We have seen that the intemperate suffer from a high rate of mortality, and the extremely profligate leave few offspring. The poorest classes crowd into towns, and it has been proved by Dr Stark from the statistics of ten years in Scotland,[21] that at all ages the death-rate is higher in towns than in rural districts, 'and during the first five years of life the town death-rate is almost exactly double that of the rural districts'. As these returns include both the rich and the poor, no doubt more than twice the number of births would be requisite to keep up the number of the very poor inhabitants in the towns, relatively to those in the country. With women, marriage at too early an age is highly injurious; for it has been found in France that, 'twice as many wives under twenty die in the year, as died out of the same number of the unmarried'. The mortality, also, of husbands under twenty is 'excessively high',[22] but what the cause of this may be, seems doubtful.

21. 'Tenth Annual Report of Births, Deaths, &c., in Scotland', 1867, p. xxix.
22. These quotations are taken from our highest authority on such questions, namely, Dr Farr, in his paper 'On the Influence of Marriage on the Mortality of the French People', read before the Nat. Assoc. for the Promotion of Social Science, 1858.

Lastly, if the men who prudently delay marrying until they can bring up their families in comfort, were to select, as they often do, women in the prime of life, the rate of increase in the better class would be only slightly lessened.

It was established from an enormous body of statistics, taken during 1853, that the unmarried men throughout France, between the ages of twenty and eighty, die in a much larger proportion than the married: for instance, out of every 1000 unmarried men, between the ages of twenty and thirty, 11.3 annually died, whilst of the married only 6.5 died.[23] A similar law was proved to hold good, during the years 1863 and 1864, with the entire population above the age of twenty in Scotland: for instance, out of every 1000 unmarried men, between the ages of twenty and thirty, 14.97 annually died, whilst of the married only 7.24 died, that is less than half.[24] Dr Stark remarks on this, 'Bachelorhood is more destructive to life than the most unwholesome trades, or than residence in an unwholesome house or district where there has never been the most distant attempt at sanitary improvement.' He considers that the lessened mortality is the direct result of 'marriage, and the more regular domestic habits which attend that state'. He admits, however, that the intemperate, profligate, and criminal classes, whose duration of life is low, do not commonly marry; and it must likewise be admitted that men with a weak constitution, ill health, or any great infirmity in body or mind, will often not wish to marry, or will be rejected. Dr Stark seems to have come to the conclusion that marriage in itself is a main cause of prolonged life, from finding that aged married men still have a considerable advantage in this respect over the

---

23. Dr Farr, ibid. The quotations given below are extracted from the same striking paper.

24. I have taken the mean of the quinquennial means, given in 'The Tenth Annual Report of Births, Deaths, &c., in Scotland', 1867. The quotation from Dr Stark is copied from an article in the 'Daily News', Oct. 17th, 1868, which Dr Farr considers very carefully written.

unmarried of the same advanced age; but every one must have known instances of men, who with weak health during youth did not marry, and yet have survived to old age, though remaining weak, and therefore always with a lessened chance of life or of marrying. There is another remarkable circumstance which seems to support Dr Stark's conclusion, namely, that widows and widowers in France suffer in comparison with the married a very heavy rate of mortality; but Dr Farr attributes this to the poverty and evil habits consequent on the disruption of the family, and to grief. On the whole we may conclude with Dr Farr that the lesser mortality of married than of unmarried men, which seems to be a general law, 'is mainly due to the constant elimination of imperfect types, and to the skilful selection of the finest individuals out of each successive generation'; the selection relating only to the marriage state, and acting on all corporeal, intellectual, and moral qualities.[25] We may, therefore, infer that sound and good men who out of prudence remain for a time unmarried, do not suffer a high rate of mortality.

If the various checks specified in the two last paragraphs, and perhaps others as yet unknown, do not prevent the reckless, the vicious and otherwise inferior members of society from increasing at a quicker rate than the better class of men, the nation will retrograde, as has too often occurred in the history of the world. We must remember that progress is no invariable rule. It is very difficult to say why one civilised nation rises, becomes more powerful, and spreads more widely, than another; or why the same nation progresses more quickly at one time than at another. We can only say that it depends on an increase in the actual number of the population, on the number of the men endowed with high intellectual and moral faculties, as well as on

25. Dr Duncan remarks ('Fecundity, Fertility', &c., 1871, p. 334) on this subject; 'At every age the healthy and beautiful go over from the unmarried side to the married, leaving the unmarried columns crowded with the sickly and unfortunate.'

their standard of excellence. Corporeal structure appears to have little influence, except so far as vigour of body leads to vigour of mind.

It has been urged by several writers that as high intellectual powers are advantageous to a nation, the old Greeks, who stood some grades higher in intellect than any race that has ever existed,[26] ought, if the power of natural selection were real, to have risen still higher in the scale, increased in number, and stocked the whole of Europe. Here we have the tacit assumption, so often made with respect to corporeal structures, that there is some innate tendency towards continued development in mind and body. But development of all kinds depends on many concurrent favourable circumstances. Natural selection acts only tentatively. Individuals and races may have acquired certain indisputable advantages, and yet have perished from failing in other characters. The Greeks may have retrograded from a want of coherence between the many small states, from the small size of their whole country, from the practice of slavery, or from extreme sensuality; for they did not succumb until 'they were enervated and corrupt to the very core'.[27] The western nations of Europe, who now so immeasurably surpass their former savage progenitors, and stand at the summit of civilisation, owe little or none of their superiority to direct inheritance from the old Greeks, though they owe much to the written works of that wonderful people.

Who can positively say why the Spanish nation, so dominant at one time, has been distanced in the race. The awakening of the nations of Europe from the dark ages is a still more perplexing problem. At that early period, as Mr Galton has remarked, almost all the men of a gentle nature, those given to meditation or culture of the mind, had no refuge except in the bosom of a

26. See the ingenious and original argument on this subject by Mr Galton, 'Hereditary Genius', pp. 340–342.
27. Mr Greg, 'Fraser's Magazine', Sept. 1868, p. 357.

Church which demanded celibacy;[28] and this could hardly fail to have had a deteriorating influence on each successive generation. During this same period the Holy Inquisition selected with extreme care the freest and boldest men in order to burn or imprison them. In Spain alone some of the best men—those who doubted and questioned, and without doubting there can be no progress—were eliminated during three centuries at the rate of a thousand a year. The evil which the Catholic Church has thus effected is incalculable, though no doubt counterbalanced to a certain, perhaps to a large, extent in other ways; nevertheless, Europe has progressed at an unparalleled rate.

The remarkable success of the English as colonists, compared to other European nations, has been ascribed to their 'daring and persistent energy'; a result which is well illustrated by comparing the progress of the Canadians of English and French extraction; but who can say how the English gained their energy? There is apparently much truth in the belief that the wonderful progress of the United States, as well as the character of the people, are the results of natural selection; for the more energetic, restless, and courageous men from all parts of Europe have emigrated during the last ten or twelve generations to that great country, and have there succeeded best.[29] Looking to the distant future, I do not think that the Rev. Mr Zincke takes an exaggrated view when he says:[30] 'All other series of events—as that which resulted in the culture of mind in Greece, and that which resulted in the empire of Rome—only appear to have purpose and value when viewed in connection with, or rather as subsidiary to ... the

28. 'Hereditary Genius', 1870, pp. 357–359. The Rev. F. W. Farrar ('Fraser's Mag', Aug. 1870, p. 257) advances arguments on the other side. Sir C. Lyell had already ('Principles of Geology', vol. ii. 1868, p. 489) in a striking passage called attention to the evil influence of the Holy Inquisition in having, through selection, lowered the general standard of intelligence in Europe.

29. Mr Galton, 'Macmillan's Magazine', August, 1865, p. 325. See also, 'Nature', 'On Darwinism and National Life', Dec. 1869, p. 184.

30. 'Last Winter in the United States', 1868, p. 29.

great stream of Anglo-Saxon emigration to the west.' Obscure as is the problem of the advance of civilisation, we can at least see that a nation which produced during a lengthened period the greatest number of highly intellectual, energetic, brave, patriotic, and benevolent men, would generally prevail over less favoured nations.

Natural selection follows from the struggle for existence; and this from a rapid rate of increase. It is impossible not to regret bitterly, but whether wisely is another question, the rate at which man tends to increase; for this leads in barbarous tribes to infanticide and many other evils, and in civilised nations to abject poverty, celibacy, and to the late marriages of the prudent. But as man suffers from the same physical evils as the lower animals, he has no right to expect an immunity from the evils consequent on the struggle for existence. Had he not been subjected during primeval times to natural selection, assuredly he would never have attained to his present rank. Since we see in many parts of the world enormous areas of the most fertile land capable of supporting numerous happy homes, but peopled only by a few wandering savages, it might be argued that the struggle for existence had not been sufficiently severe to force man upwards to his highest standard. Judging from all that we know of man and the lower animals, there has always been sufficient variability in their intellectual and moral faculties, for a steady advance through natural selection. No doubt such advance demands many favourable concurrent circumstances; but it may well be doubted whether the most favourable would have sufficed, had not the rate of increase been rapid, and the consequent struggle for existence extremely severe. It even appears from what we see, for instance, in parts of S. America, that a people which may be called civilised, such as the Spanish settlers, is liable to become indolent and to retrograde, when the conditions of life are very easy. With highly civilised nations continued progress depends in a subordinate degree on natural selection; for such nations do not supplant and exterminate one another as do savage tribes. Nevertheless the more intelligent members within the same community will succeed better in the long run than the inferior, and leave a more numerous progeny, and this is a form of natural selection. The more efficient causes of progress seem to consist of a good education during youth whilst the brain is impressible, and of a high standard of excellence, inculcated by the ablest and best men, embodied in the laws, customs and traditions of the nation, and enforced by public opinion. It should, however, be borne in mind, that the enforcement of public opinion depends on our appreciation of the approbation and disapprobation of others; and this appreciation is founded on our sympathy, which it can hardly be doubted was originally developed through natural selection as one of the most important elements of the social instincts.[31]

On the evidence that all civilised nations were once barbarous—The present subject has been treated in so full and admirable a manner by Sir J. Lubbock,[32] Mr Tylor, Mr M'Lennan, and others, that I need here give only the briefest summary of their results. The arguments recently advanced by the Duke of Argyll[33] and formerly by Archbishop Whately, in favour of the belief that man came into the world as a civilised being, and that all savages have since undergone degradation, seem to me weak in comparison with those advanced on the other side. Many nations, no doubt, have fallen away in civilisation, and some may have lapsed into utter barbarism, though on this latter head I have met with no evidence. The Fuegians were probably compelled by other con-

31. I am much indebted to Mr John Morley for some good criticisms on this subject: see, also, Broca, 'Les Sélections', 'Revue d'Anthropologie', 1872.
32. 'On the Origin of Civilisation', 'Proc. Ethnological Soc.', Nov. 26, 1867.
33. 'Primeval Man', 1869.

quering hordes to settle in their inhospitable country, and they may have become in consequence somewhat more degraded; but it would be difficult to prove that they have fallen much below the Botocudos, who inhabit the finest parts of Brazil.

The evidence that all civilised nations are the descendants of barbarians, consists, on the one side, of clear traces of their former low condition in still-existing customs, beliefs, language, &c.; and on the other side, of proofs that savages are independently able to raise themselves a few steps in the scale of civilisation, and have actually thus risen. The evidence on the first head is extremely curious, but cannot be here given: I refer to such cases as that of the art of enumeration, which, as Mr Tylor clearly shews by reference to the words still used in some places, originated in counting the fingers, first of one hand and then of the other, and lastly of the toes. We have traces of this in our own decimal system, and in the Roman numerals, where, after the V, which is supposed to be an abbreviated picture of a human hand, we pass on to VI, &c., when the other hand no doubt was used. So again, 'when we speak of three-score and ten, we are counting by the vigesimal system, each score thus ideally made, standing for 20—for "one man" as a Mexican or Carib would put it'.[34] According to a large and increasing school of philologists, every language bears the marks of its slow and gradual evolution. So it is with the art of writing, for letters are rudiments of pictorial representations. It is hardly possible to read Mr M'Lennan's work[35] and not admit that almost all civilised nations still retain traces of such rude habits as the forcible capture of wives.

34. 'Royal Institution of Great Britain', March 15, 1867. Also, 'Researches into the Early History of Mankind', 1865.
35. 'Primitive Marriage', 1865. See, likewise, an excellent article, evidently by the same author, in the 'North British Review', July, 1869. Also, Mr L. H. Morgan, 'A Conjectural Solution of the Origin of the Class, System of Relationship', in 'Proc. American Acad. of Sciences', vol. vii. Feb. 1868. Prof. Schaaffhausen ('Anthropolog. Review', Oct. 1869, p. 373) remarks on 'the vestiges of human sacrifices found both in Homer and the Old Testament'.

What ancient nation, as the same author asks, can be named that was originally monogamous? The primitive idea of justice, as shewn by the law of battle and other customs of which vestiges still remain, was likewise most rude. Many existing superstitions are the remnants of former false religious beliefs. The highest form of religion—the grand idea of God hating sin and loving righteousness—was unknown during primeval times.

Turning to the other kind of evidence: Sir J. Lubbock has shewn that some savages have recently improved a little in some of their simpler arts. From the extremely curious account which he gives of the weapons, tools, and arts, in use amongst savages in various parts of the world, it cannot be doubted that these have nearly all been independent discoveries, excepting perhaps the art of making fire.[36] The Australian boomerang is a good instance of one such independent discovery. The Tahitians when first visited had advanced in many respects beyond the inhabitants of most of the other Polynesian islands. There are no just grounds for the belief that the high culture of the native Peruvians and Mexicans was derived from abroad;[37] many native plants were there culti-vated, and a few native animals domesticated. We should bear in mind that, judging from the small influence of most missionar-ies, a wandering crew from some semi-civilised land, if washed to the shores of America, would not have produced any marked effect on the natives, unless they had already become somewhat advanced. Looking to a very remote period in the history of the world, we find, to use Sir J. Lubbock's well-known terms, a paleolithic and neolithic period; and no one will pretend that the art of grinding rough flint tools was a borrowed one. In all

36. Sir J. Lubbock, 'Prehistoric Times', 2nd edit. 1869, chap. xv and xvi *et passim*. See also the excellent 9th chapter in Tylor's 'Early History of Mankind', 2nd edit., 1870.
37. Dr F. Müller has made some good remarks to this effect in the 'Reise der Novara: Anthropolog. Theil', Abtheil. iii. 1868, s. 127.

parts of Europe, as far east as Greece, in Palestine, India, Japan, New Zealand, and Africa, including Egypt, flint tools have been discovered in abundance; and of their use the existing inhabitants retain no tradition. There is also indirect evidence of their former use by the Chinese and ancient Jews. Hence there can hardly be a doubt that the inhabitants of these countries, which include nearly the whole civilised world, were once in a barbarous condition. To believe that man was aboriginally civilised and then suffered utter degradation in so many regions, is to take a pitiably low view of human nature. It is apparently a truer and more cheerful view that progress has been much more general than retrogression; that man has risen, though by slow and interrupted steps, from a lowly condition to the highest standard as yet attained by him in knowledge, morals and religion.

7

Race

D arwin dedicated much of his career to demolition. He needed to clear away old explanations of biology in order to introduce his own. He directed his intellectual wrecking ball particularly often to traditional concepts of species. Many naturalists saw species as relatively distinct from one another. They believed that species could be classified in turn into relatively distinct groups—families, classes, kingdoms, and other taxonomic ranks. Taxonomists diagnosed groups by the unique traits found in certain organisms and not others. Mammals had a combination of fur, milk, warm-blooded metabolism, and other traits found in no other group of animals. Although mammals all shared those traits, each species of mammal had its own set of traits to set it apart from other mammal species.

In *The Origin of Species*, Darwin argued that species were not so distinct. Many species, in fact, could not be so carefully distinguished from one another because they were regularly interbreeding to produce viable hybrids. Likewise, taxonomists had established species that turned out to contain enormous variability. By some measures, two populations within a single species could be considered a pair of separate species. Trying to draw sharp lines between species or other taxonomic groups was a futile exercise, Darwin argued. Species had not suddenly come into existence as distinct groups of organisms. They had gradually evolved from older species. Naturalists could not see the

full continuity of life because many species and populations within species had become extinct, leaving gaps.

In *The Descent of Man*, Darwin applied these same arguments to our
own species. He argued that humans descended from a primitive inver-
tebrates, whose descendants evolved a skeleton and fins, and eventually
moved onto land. Later, one lineage of these land vertebrates evolved
into mammals, and one lineage of mammals evolved into primates. Humans belonged to the branch of primates that included chimpanzees
and other apes. While humans were distinct from other species, those
differences did not eliminate the signs of our common ancestry with
other species.

Evolution could not only explain the similarities of humans to other
species, but also the differences between various groups of humans.
Anthropologists and naturalists had offered many different accounts for
human races. Some argued that humans represented a single species,
while others argued that races represented separate species. Exactly
how many species they represented was another matter of debate, with
writers creating anywhere from two to sixty-three of them. The number of races (or species) depended entirely on the standards a writer
used to divide humans into their separate groups. As with other species, this confusion was further proof to Darwin that life evolved. All of
the debates over the divisions of humans dwindled away if one viewed
Homo sapiens as a single species descended from a common ancestor.
Over time, variations emerged in different populations of humans, but
those differences did not prevent humans from interbreeding. Competition between different groups of humans had led to local extinctions.
In some cases, the competition took the form of warfare; in other cases,
the form of disease. Those extinctions created gaps in the evolutionary
continuity of our species.

Darwin's account of race in chapter seven of *The Descent of Man*
stands in sharp contrast to many other scientific accounts of race from
the mid-1800s. Some writers claimed that human races were so distant
from one another that interbreeding was impossible. Darwin, on the

other hand, considered many of the differences between peoples to be minor. Humans were simply a variable species like dogs or pigeons. Many differences were not indelible stamps of heredity. The American descendants of Europeans had grown taller than their ancestors (a phenomenon Barry Bogin would later show could happen over just a few years).

Many of the differences between human races were so minor, in fact, that Darwin did not think natural selection had shaped them. In chapter seven he refers to the force he suspects is instead responsible: sexual selection. Darwin will return later to sexual selection in humans, but only after an enormous digression into the animal kingdom, in which he offers evidence for sexual selection in other species. For the time being, he stresses the unity of *Homo sapiens*, even mentioning some of his own encounters with different races. The Fuegans, he writes, impressed him "with the many little traits of character, shewing how similar their minds were to ours; and so it was with a full-blooded negro with whom I happened once to be intimate." That last phrase refers fleetingly to John Edmonston, a freed slave who worked as a taxidermist in Edinburgh. Edmonston taught the teenage Darwin how to stuff birds. It is a sign of how differently races were viewed in the nineteenth century that Darwin would even contemplate mentioning such a fact.

The scientific study of race has had a stormy history since Darwin's day. Some have tried to find an extreme biological basis for the concept of race—perhaps none more extreme than Nazi scientists who created a pseudoscientific justification for the Holocaust. Nazi racism may be a thing of the past, but race still plays a powerful role in the way people categorize themselves and others. People from Ireland, the Ukraine, and Portugal may all identify themselves as Caucasian. People from Mongolia, Japan, and Laos are all Asian. But these categories and the meanings we assign to them are largely social constructions. At one time, Jews were considered a separate race, but no longer. A person considered black in the United States might be considered white in Brazil.

By the mid-1900s, it was becoming clear that the world could not be sliced up neatly into races. A map of skin color did not line up with a map of blood types, for example. In the early 1970s Richard Lewontin of Harvard made one of the earliest surveys of the genetic variation in humans. He found 85 percent of the variation among members of the same group, while only 15 percent of the variation was between groups. Later research showed that our species is less than two hundred thousand years old, and at some points early in its history, it may have included only a few thousand individuals. Thus humans simply haven't had much time to build up a lot of genetic diversity compared to other animals. There's more genetic diversity in the wildebeest of Kenya alone than in all six billion people on Earth. In 1994 The American Anthropological Association declared that race was biologically meaningless, and that "racial myths bear no relationship to the reality of human capabilities or behavior."

Of course, people do vary, and people from one part of the world tend to look more like one another than people who live far away from them. Today scientists are translating old notions of race into more realistic pictures of human diversity. Humans originated in Africa, and as their populations expanded into new territories, they acquired genetic markers that allow scientists to distinguish their branch of the human family tree. As populations of humans spread to new lands, they became isolated enough for natural selection to create distinctive changes in their ranks.

Skin color is the most obvious of those differences. The earliest hominids were probably covered mostly in hair, as living apes are. While ape hair is dark, the skin underneath is actually pale. When hominids evolved into tall, upright species that walked through open landscapes, their body hair became a liability because it absorbed too much heat. Natural selection favored a relatively hairless body, along with lots of sweat glands, to keep the body cool.

But a hairless body brought its own hazards, such as exposure to the sun's ultraviolet rays. Today doctors advise us to lather up with sunblock to keep away ultraviolet rays because they can cause skin cancer. As deadly as skin cancer may be, it may not have had a big role in the evolution of human skin. Melanomas don't develop until late in life, long after a person has finished having children. Instead, other hazards posed by UV may have been important.

For example, UV rays break down an essential nutrient in the skin called folic acid. Without enough folic acid, men may become infertile and women may give birth to babies with fatal neurological defects. Nina Jablonski and George Chaplin at the California Academy of Sciences have argued that people who were born with UV-absorbing pigments in their skin would have been protected from this danger, and their genes may have been favored by natural selection. As a result, humans evolved dark skin in Africa.

When humans began migrating out of Africa and away from the equator, they were exposed to lower levels of ultraviolet radiation. The advantage of black skin may have disappeared in Asia and Europe. Too much ultraviolet radiation can destroy folic acid in the skin, but our bodies need at least some exposure, because UV rays help our bodies produce vitamin D. Without vitamin D, people develop devastating bone deformities. In regions of the world with low levels of ultraviolet rays, black skin might do its job too well. Jablonksi and Chaplin argue that the balance between protecting folic acid and synthesizing vitamin D favored darker skins where UV exposure is high, and paler skins closer to the poles.

Scientists are just starting to identify the mutations that produced different skin colors in humans. But by measuring the color of skin from around the world, Jablonski and Chaplin have already shown that its geographical pattern matches their predictions pretty well. There are exceptions, but most seem to prove the rule rather than undermine it. People with the "wrong" skin color tend to belong to groups who have migrated long distances within the past few thousand years. The light-brown Khoisan people of Southern Africa have lived there for tens of thousands of years; darker Zulu people came from tropical Africa one thousand years ago. Eskimos ought to be as pale as Scandinavians, but they arrived in the Arctic only about five thousand years ago. They may not be evolving toward paler skin because they can get an extra supply of vitamin D from fish.

Natural selection has produced other transformations, such as lactose tolerance and resistance to malaria. But the genes for these traits do not match the world map of skin color. Lactose tolerance is not a racial trait, but a trait of cattle herders in Africa and Europe. Malaria resistance can be found in people from Africa, the Mediterranean, and New Guinea.

Another fact of human evolution that defies race-based thinking is the spread of genes from one population to another. It only takes two people having children to mingle the genes from two populations. If those children inherit an advantageous gene from one parent, they can spread it throughout the population of the other parent. Even if the two populations look different—thanks to a few genes that produce superficial differences in appearance—they can end up sharing many of the same genes. In fact, genes can even break the species barrier. Scientists have long known that genes can flow between species of fruit flies, sunflowers, and other animals and plants. But now scientists have found this species-crossing gene flow in humans.

Humans and Neanderthals branched off from a common ancestor about five hundred thousand years ago. Neanderthals evolved their own distinct genome as they colonized Europe. It would take another three hundred thousand years for our own ancestors to acquire the capacity for abstract thought and full-blown language that makes us uniquely human. About forty thousand years ago their descendants moved from Africa into Europe, where they encountered Neanderthals. By then they had become so different that many scientists would classify them as two separate species: *Homo sapiens* and *Homo neanderthalensis*.

In 2006, scientists offered compelling evidence of a gene that had entered our species by interbreeding with Neanderthals. It is a version of a gene known as microcephalin. This version is the most common one found in humans today, with more than two thirds of all people carrying it. But its history, as revealed by comparisons with other versions of the gene, is peculiar. They all originated from a duplicated microcephalin gene in hominids over a million years ago. But the common version only began to spread rapidly through human populations about thirty-seven thousand years ago. The best explanation that scientists have at the moment is that humans mated with Neanderthals thirty-seven thousand years ago and acquired their version of the microcephalin gene. Once in our species, it was strongly favored by natural selection.

What makes this sweep particularly striking is the fact that micro-cephalin does not encode a protein involved only in some mundane function like growing toenails or ear hairs. It helps build brains. If microcephalin is disabled, people usually grow tiny, chimp-sized brains. No one knows what advantage the new version of microcephalin may provide in the brain, or whether perhaps its advantage comes from another function it serves elsewhere in the body. But it is clear that if we cling too tightly to old-fashioned notions of race, we will end up having to say that most humans are not actually human.

CHAPTER 7:
On the Races of Man

IT IS NOT MY intention here to describe the several so-called races of men; but I am about to enquire what is the value of the differences between them under a classificatory point of view, and how they have originated. In determining whether two or more allied forms ought to be ranked as species or varieties, naturalists are practically guided by the following considerations; namely, the amount of difference between them, and whether such differences relate to few or many points of structure, and whether they are of physiological importance; but more especially whether they are constant. Constancy of character is what is chiefly valued and sought for by naturalists. Whenever it can be shewn, or rendered probable, that the forms in question have remained distinct for a long period, this becomes an argument of much weight in favour of treating them as species. Even a slight degree of sterility between any two forms when first crossed, or in their offspring, is generally considered as a decisive test of their specific distinctness; and their continued persistence without blending within the same area, is usually accepted as sufficient evidence, either of some degree of mutual sterility, or in the case of animals of some mutual repugnance to pairing.

Independently of fusion from intercrossing, the complete absence, in a well-investigated region, of varieties linking together any two closely-allied forms, is probably the most important of all the criterions of their specific distinctness; and this is a somewhat different consideration from mere constancy of character, for two forms may be highly variable and yet not yield intermediate varieties. Geographical distribution is often brought into play unconsciously and sometimes consciously; so that forms living in two widely separated areas, in which most of the other inhabitants are specifically distinct, are themselves usually looked at as distinct; but in truth this affords no aid in distinguishing geographical races from so-called good or true species.

Now let us apply these generally-admitted principles to the races of man, viewing him in the same spirit as a naturalist would any other animal. In regard to the amount of difference between the race, we must make some allowance for our nice powers of discrimination gained by the long habit of observing ourselves. In India, as Elphinstone remarks, although a newly-arrived European cannot at first distinguish the various native races, yet they soon appear to him extremely dissimilar;[1] and the Hindoo cannot at first perceive any difference between the several European nations. Even the most distinct races of man are much more like each other in form than would at first be supposed; certain negro tribes must be excepted, whilst others, as Dr Rohlfs writes to me, and as I have myself seen, have Caucasian features. This general similarity is well shewn by the French photographs in the Collection Anthropologique du Muséum de Paris of the men belonging to various races, the greater number of which might pass for Europeans, as many persons to whom I have shewn them have remarked. Nevertheless, these men, if seen alive, would undoubtedly appear very distinct, so that we are clearly much influenced in our judgment by the mere colour

1. 'History of India', 1841, vol. i. p. 323. Father Ripa makes exactly the same remark with respect to the Chinese.

of the skin and hair, by slight differences in the features, and by expression.

There is, however, no doubt that the various races, when carefully compared and measured, differ much from each other,—as in the texture of the hair, the relative proportions of all parts of the body,[2] the capacity of the lungs, the form and capacity of the skull, and even in the convolutions of the brain.[3] But it would be an endless task to specify the numerous points of difference. The races differ also in constitution, in acclimatisation and in liability to certain diseases. Their mental characteristics are likewise very distinct; chiefly as it would appear in their emotional, but partly in their intellectual faculties. Every one who has had the opportunity of comparison, must have been struck with the contrast between the taciturn, even morose, aborigines of S. America and the light-hearted, talkative negroes. There is a nearly similar contrast between the Malays and the Papuans,[4] who live under the same physical conditions, and are separated from each other only by a narrow space of sea.

We will first consider the arguments which may be advanced in favour of classing the races of man as distinct species, and then the arguments on the other side. If a naturalist, who had never before seen a Negro, Hottentot, Australian, or Mongolian, were to compare them, he would at once perceive that they differed in a multitude of characters, some of slight and some of considerable importance. On enquiry he would find that they were adapted to live under widely different climates, and that they differed somewhat in bodily constitution and mental disposition.

2. A vast number of measurements of Whites, Blacks, and Indians, are given in the 'Investigations in the Military and Anthropolog. Statistics of American Soldiers', by B. A. Gould, 1869, pp. 298–358; 'On the capacity of the lungs,' p. 471. See also the numerous and valuable tables, by Dr Weisbach, from the observations of Dr Scherzer and Dr Schwarz, in the 'Reise der Novara: Anthropolog. Theil', 1867.

3. See, for instance, Mr Marshall's account of the brain of a Bush-woman, in 'Phil. Transact.' 1864, p. 519.

4. Wallace, 'The Malay Archipelago', vol. ii. 1869, p. 178.

If he were then told that hundreds of similar specimens could be brought from the same countries, he would assuredly declare that they were as good species as many to which he had been in the habit of affixing specific names. This conclusion would be greatly strengthened as soon as he had ascertained that these forms had all retained the same character for many centuries; and that negroes, apparently identical with existing negroes, had lived at least 4000 years ago.[5] He would also hear, on the authority of an excellent observer, Dr Lund,[6] that the human skulls found in the caves of Brazil, entombed with many extinct mammals, belonged to the same type as that now prevailing throughout the American Continent.

Our naturalist would then perhaps turn to geographical distribution, and he would probably declare that those forms must be distinct species, which differ not only in appearance, but are fitted for hot, as well as damp or dry countries, and for the Arctic regions. He might appeal to the fact that no species in the group next to man, namely the Quadrumana, can resist a low temperature, or any considerable change of climate; and that the species which come nearest to man have never been

5. With respect to the figures in the famous Egyptian caves of Abou-Simbel, M. Pouchet says ('The Plurality of the Human Races', Eng. translat. 1864, p. 50), that he was far from finding recognisable representations of the dozen or more nations which some authors believe that they can recognise. Even some of the most strongly-marked races cannot be identified with that degree of unanimity which might have been expected from what has been written on the subject. Thus Messrs. Nott and Gliddon ('Types of Mankind', p. 148) state that Rameses II, or the Great, has features superbly European; whereas Knox, another firm believer in the specific distinctness of the races of man ('Races of Man', 1850, p. 201), speaking of young Memnon (the same as Rameses II, as I am informed by Mr Birch), insists in the strongest manner that he is identical in character with the Jews of Antwerp. Again, when I looked at the statue of Amunoph III, I agreed with two officers of the establishment, both competent judges, that he had a strongly marked negro type of features; but Messrs. Nott and Gliddon (ibid. p. 146, fig. 53) describe him as a hybrid, but not of 'negro intermixture'.

6. As quoted by Nott and Gliddon, 'Types of Mankind', 1854, p. 439. They give also corroborative evidence; but C. Vogt thinks that the subject requires further investigation.

reared to maturity, even under the temperate climate of Europe. He would be deeply impressed with the fact, first noticed by Agassiz,[7] that the different races of man are distributed over the world in the same zoological provinces, as those inhabited by undoubtedly distinct species and genera of mammals. This is manifestly the case with the Australian, Mongolian, and Negro races of man; in a less well-marked manner with the Hottentots; but plainly with the Papuans and Malays, who are separated, as Mr Wallace has shewn, by nearly the same line which divides the great Malayan and Australian zoological provinces. The Aborigines of America range throughout the Continent; and this at first appears opposed to the above rule, for most of the productions of the Southern and Northern halves differ widely: yet some few living forms, as the opossum, range from the one into the other, as did formerly some of the gigantic Edentata. The Esquimaux, like other Arctic animals, extend round the whole polar regions. It should be observed that the amount of difference between the mammals of the several zoological provinces does not correspond with the degree of separation between the latter; so that it can hardly be considered as an anomaly that the Negro differs more, and the American much less from the other races of man, than do the mammals of the African and American continents from the mammals of the other provinces. Man, it may be added, does not appear to have aboriginally inhabited any oceanic island; and in this respect he resembles the other members of his class.

In determining whether the supposed varieties of the same kind of domestic animal should be ranked as such, or as specifically distinct, that is, whether any of them are descended from distinct wild species, every naturalist would lay much stress on

7. 'Diversity of Origin of the Human Races', in the 'Christian Examiner', July 1850.

the fact of their external parasites being specifically distinct. All the more stress would be laid on this fact, as it would be an exceptional one; for I am informed by Mr Denny that the most different kinds of dogs, fowls, and pigeons, in England, are infested by the same species of Pediculi or lice. Now Mr A. Murray has carefully examined the Pediculi collected in different countries from the different races of man;[8] and he finds that they differ, not only in colour, but in the structure of their claws and limbs. In every case in which many specimens were obtained the differences were constant. The surgeon of a whaling ship in the Pacific assured me that when the Pediculi, with which some Sandwich Islanders on board swarmed, strayed on to the bodies of the English sailors, they died in the course of three or four days. These Pediculi were darker coloured, and appeared different from those proper to the natives of Chiloe in South America, of which he gave me specimens. These, again, appeared larger and much softer than European lice. Mr Murray procured four kinds from Africa, namely from the Negroes of the Eastern and Western coasts, from the Hottentots and Kaffirs; two kinds from the natives of Australia; two from North and two from South America. In these latter cases it may be presumed that the Pediculi came from natives inhabiting different districts. With insects slight structural differences, if constant, are generally esteemed of specific value: and the fact of the races of man being infested by parasites, which appear to be specifically distinct, might fairly be urged as an argument that the races themselves ought to be classed as distinct species.

Our supposed naturalist having proceeded thus far in his investigation, would next enquire whether the races of men, when crossed, were in any degree sterile. He might consult the

8. 'Transact. R. Soc. of Edinburgh', vol. xxii. 1861, p. 567.

work[9] of Professor Broca, a cautious and philosophical observer, and in this he would find good evidence that some races were quite fertile together, but evidence of an opposite nature in regard to other races. Thus it has been asserted that the native women of Australia and Tasmania rarely produce children to European men; the evidence, however, on this head has now been shewn to be almost valueless. The half-castes are killed by the pure blacks: and an account has lately been published of eleven half-caste youths murdered and burnt at the same time, whose remains were found by the police.[10] Again, it has often been said that when mulattoes intermarry they produce few children; on the other hand, Dr Bachman of Charleston[11] positively asserts that he has known mulatto families which have intermarried for several generations, and have continued on an average as fertile as either pure whites or pure blacks. Enquiries formerly made by Sir C. Lyell on this subject led him, as he informs me, to the same conclusion.[12] In the United States the census for the year 1854 included, according to Dr Bachman, 405,751 mulattoes; and this number, considering all the circumstances of the case, seems small; but it may partly be accounted for by the degraded and anomalous position

9. 'On the Phenomena of Hybridity in the Genus Homo', Eng. translat. 1864.
10. See the interesting letter by Mr T. A. Murray, in the 'Anthropolog. Review', April 1868, p. liii. In this letter Count Strzelecki's statement, that Australian women who have borne children to a white man are afterwards sterile with their own race, is disproved. M. A. de Quatrefages has also collected ('Revue des Cours Scientifiques', March 1869, p. 239) much evidence that Australians and Europeans are not sterile when crossed.
11. 'An Examination of Prof. Agassiz's Sketch of the Nat. Provinces of the Animal World', Charleston, 1855, p. 44.
12. Dr Rohlfs writes to me that he found the mixed races in the Great Sahara, derived from Arabs, Berbers, and Negroes of three tribes, extraordinarily fertile. On the other hand, Mr Winwood Reade informs me that the Negroes on the Gold Coast, though admiring white men and mulattoes, have a maxim that mulattoes should not intermarry, as the children are few and sickly. This belief, as Mr Reade remarks, deserves attention, as white men have visited and resided on the Gold Coast for four hundred years, so that the natives have had ample time to gain knowledge through experience.

of the class, and by the profligacy of the women. A certain amount of absorption of mulattoes into negroes must always be in progress; and this would lead to an apparent diminution of the former. The inferior vitality of mulattoes is spoken of in a trustworthy work[13] as a well-known phenomenon; and this, although a different consideration from their lessened fertility, may perhaps be advanced as a proof of the specific distinctness of the parent races. No doubt both animal and vegetable hybrids, when produced from extremely distinct species, are liable to premature death; but the parents of mulattoes cannot be put under the category of extremely distinct species. The common Mule, so notorious for long life and vigour, and yet so sterile, shews how little necessary connection there is in hybrids between lessened fertility and vitality; other analogous cases could be cited.

Even if it should hereafter be proved that all the races of men were perfectly fertile together, he who was inclined from other reasons to rank them as distinct species, might with justice argue that fertility and sterility are not safe criterions of specific distinctness. We know that these qualities are easily affected by changed conditions of life, or by close inter-breeding, and that they are governed by highly complex laws, for instance, that of the unequal fertility of converse crosses between the same two species. With forms which must be ranked as undoubted species, a perfect series exists from those which are absolutely sterile when crossed, to those which are almost or completely fertile. The degrees of sterility do not coincide strictly with the degrees of difference between the parents in external structure or habits of life. Man in many respects may be compared with those animals which have long been domesticated, and a large body

13. 'Military and Anthropolog. Statistics of American Soldiers', by B. A. Gould, 1869, p. 319.

of evidence can be advanced in favour of the Pallasian doctrine,[14] that domestication tends to eliminate the sterility which is so general a result of the crossing of species in a state of nature. From these several considerations, it may be justly urged that the perfect fertility of the intercrossed races of man, if established, would not absolutely preclude us from ranking them as distinct species.

Independently of fertility, the characters presented by the offspring from a cross have been thought to indicate whether or not the parent-forms ought to be ranked as species or varieties; but after carefully studying the evidence, I have come to the conclusion that no general rules of this kind can be trusted. The ordinary result of a cross is the production of a blended or intermediate form; but in certain cases some of the offspring take closely after one parent-form, and some after the other. This is es-

---

14. 'The Variation of Animals and Plants under Domestication', vol. ii. p. 109. I may here remind the reader that the sterility of species when crossed is not a specially acquired quality, but, like the incapacity of certain trees to be grafted together, is incidental on other acquired differences. The nature of these differences is unknown, but they relate more especially to the reproductive system, and much less so to external structure or to ordinary differences in constitution. One important element in the sterility of crossed species apparently lies in one or both having been long habituated to fixed conditions; for we know that changed conditions have a special influence on the reproductive system, and we have good reason to believe (as before remarked) that the fluctuating conditions of domestication tend to eliminate that sterility which is so general with species, in a natural state, when crossed. It has elsewhere been shewn by me (ibid. vol. ii. p. 185, and 'Origin of Species' 5th edit. p. 317), that the sterility of crossed species has not been acquired through natural selection: we can see that when two forms have already been rendered very sterile, it is scarcely possible that their sterility should be augmented by the preservation or survival of the more and more sterile individuals; for as the sterility increases, fewer and fewer offspring will be produced from which to breed, and at last only single individuals will be produced, at the rarest intervals. But there is even a higher grade of sterility than this. Both Gärtner and Kölreuter have proved that in genera of plants including many species, a series can be formed from species which when crossed yield fewer and fewer seeds, to species which never produce a single seed, but yet are affected by the pollen of the other species, as shewn by the swelling of the germen. It is here manifestly impossible to select the more sterile individuals, which have already ceased to yield seeds; so that the acme of sterility, when the germen alone is affected, cannot have been gained through selection. This acme, and no doubt the other grades of sterility, are the incidental results of certain unknown differences in the constitution of the reproductive system of the species which are crossed.

pecially apt to occur when the parents differ in characters which first appeared as sudden variations or monstrosities.[15] I refer to this point, because Dr Rohlfs informs me that he has frequently seen in Africa the offspring of negroes crossed with members of other races, either completely black or completely white, or rarely piebald. On the other hand, it is notorious that in America mulattoes commonly present an intermediate appearance.

We have now seen that a naturalist might feel himself fully justified in ranking the races of man as distinct species; for he has found that they are distinguished by many differences in structure and constitution, some being of importance. These differences have, also remained nearly constant for very long periods of time. Our naturalist will have been in some degree influenced by the enormous range of man, which is a great anomaly in the class of mammals, if mankind be viewed as a single species. He will have been struck with the distribution of the several so-called races, which accords with that of other undoubtedly distinct species of mammals. Finally, he might urge that the mutual fertility of all the races has not as yet been fully proved, and even if proved would not be an absolute proof of their specific identity.

On the other side of the question, if our supposed naturalist were to enquire whether the forms of man keep distinct like ordinary species, when mingled together in large numbers in the same country, he would immediately discover that this was by no means the case. In Brazil he would behold an immense mongrel population of Negroes and Portuguese; in Chiloe, and other parts of South America, he would behold the whole population consisting of Indians and Spaniards blended in various degrees.[16]

---

15. 'The Variation of Animals', &c., vol. ii. p. 92.
16. M. de Quatrefages has given ('Anthropolog. Review', Jan. 1869, p. 22) an interesting account of the success and energy of the Paulistas in Brazil, who are a much crossed race of Portuguese and Indians, with a mixture of the blood of other races.

In many parts of the same continent he would meet with the most complex crosses between Negroes, Indians, and Europeans; and judging from the vegetable kingdom, such triple crosses afford the severest test of the mutual fertility of the parent-forms. In one island of the Pacific he would find a small population of mingled Polynesian and English blood; and in the Fiji Archipelago a population of Polynesian and Negritos crossed in all degrees. Many analogous cases could be added; for instance, in Africa. Hence the races of man are not sufficiently distinct to inhabit the same country without fusion; and the absence of fusion affords the usual and best test of specific distinctness.

Our naturalist would likewise be much disturbed as soon as he perceived that the distinctive characters of all the races were highly variable. This fact strikes every one on first beholding the negro slaves in Brazil, who have been imported from all parts of Africa. The same remark holds good with the Polynesians, and with many other races. It may be doubted whether any character can be named which is distinctive of a race and is constant. Savages, even within the limits of the same tribe, are not nearly so uniform in character, as has been often asserted. Hottentot women offer certain peculiarities, more strongly marked than those occurring in any other race, but these are known not to be of constant occurrence. In the several American tribes, colour and hairiness differ considerably; as does colour to a certain degree, and the shape of the features greatly, in the Negroes of Africa. The shape of the skull varies much in some races;[17] and so it is with every other character. Now all naturalists have learnt by dearly-bought experience, how rash it is to attempt to define species by the aid of inconstant characters.

17. For instance with the aborigines of America and Australia. Prof. Huxley says ('Transact. Internat. Congress of Prehist. Arch.', 1868, p. 105) that the skulls of many South Germans and Swiss are 'as short and as broad as those of the Tartars', &c.

But the most weighty of all the arguments against treating the races of man as distinct species, is that they graduate into each other, independently in many cases, as far as we can judge, of their having intercrossed. Man has been studied more carefully than any other animal, and yet there is the greatest possible diversity amongst capable judges whether he should be classed as a single species or race, or as two (Virey), as three (Jacquinot), as four (Kant), five (Blumenbach), six (Buffon), seven (Hunter), eight (Agassiz), eleven (Pickering), fifteen (Bory St Vincent), sixteen (Desmoulins), twenty-two (Morton), sixty (Crawfurd), or as sixty-three, according to Burke.[18] This diversity of judgment does not prove that the races ought not to be ranked as species, but it shews that they graduate into each other, and that it is hardly possible to discover clear distinctive characters between them.

Every naturalist who has had the misfortune to undertake the description of a group of highly varying organisms, has encountered cases (I speak after experience) precisely like that of man; and if of a cautious disposition, he will end by uniting all the forms which graduate into each other, under a single species; for he will say to himself that he has no right to give names to objects which he cannot define. Cases of this kind occur in the Order which includes man, namely in certain genera of monkeys; whilst in other genera, as in Cercopithecus, most of the species can be determined with certainty. In the American genus Cebus, the various forms are ranked by some naturalists as species, by others as mere geographical races. Now if numerous specimens of Cebus were collected from all parts of South America, and those forms which at present appear to be specifically distinct, were found to graduate into each other by close steps, they would usually be ranked as mere varieties or races; and

18. See a good discussion on this subject in Waitz, 'Introduct. to Anthropology', Eng. translat. 1863, pp. 198–208, 227. I have taken some of the above statements from H. Tuttle's 'Origin and Antiquity of Physical Man', Boston, 1866, p. 35.

this course has been followed by most naturalists with respect to the races of man. Nevertheless, it must be confessed that there are forms, at least in the vegetable kingdom,[19] which we cannot avoid naming as species, but which are connected together by numberless gradations, independently of intercrossing.

Some naturalists have lately employed the term 'sub-species' to designate forms which possess many of the charactersitics of true species, but which hardly deserve so high a rank. Now if we reflect on the weighty arguments above given, for raising the races of man to the dignity of species, and the insuperable difficulties on the other side in defining them, it seems that the term 'sub-species' might here be used with propriety. But from long habit the term 'race' will perhaps always be employed. The choice of terms is only so far important in that it is desirable to use, as far as possible, the same terms for the same degrees of difference. Unfortunately this can rarely be done: for the larger genera generally include closely-allied forms, which can be distinguished only with much difficulty, whilst the smaller genera within the same family include forms that are perfectly distinct; yet all must be ranked equally as species. So again, species within the same large genus by no means resemble each other to the same degree: on the contrary, some of them can generally be arranged in little groups round other species, like satellites round planets.[20]

The question whether mankind consists of one or several species has of late years been much discussed by anthropologists, who are divided into the two schools of monogenists and polygenists. Those who do not admit the principle of evolution, must

19. Prof. Nägeli has carefully described several striking cases in his 'Botanische Mittheilungen', B ii. 1866; ss. 294–369. Prof. Asa Gray has made analogous remarks on some intermediate forms in the Compositae of N. America.
20. 'Origin of Species', 5th edit. p. 68.

look at species as separate creations, or as in some manner as distinct entities; and they must decide what forms of man they will consider as species by the analogy of the method commonly pursued in ranking other organic beings as species. But it is a hopeless endeavour to decide this point, until some definition of the term 'species' is generally accepted; and the definition must not include an indeterminate element such as an act of creation. We might as well attempt without any definition to decide whether a certain number of houses should be called a village, town, or city. We have a practical illustration of the difficulty in the never-ending doubts whether many closely-allied mammals, birds, insects, and plants, which represent each other respectively in North America and Europe, should be ranked as species or geographical races; and the like holds true of the productions of many islands situated at some little distance from the nearest continent.

Those naturalists, on the other hand, who admit the principle of evolution, and this is now admitted by the majority of rising men, will feel no doubt that all the races of man are descended from a single primitive stock; whether or not they may think fit to designate the races as distinct species, for the sake of expressing their amount of difference.[21] With our domestic animals the question whether the various races have arisen from one or more species is somewhat different. Although it may be admitted that all the races, as well as all the natural species within the same genus, have sprung from the same primitive stock, yet it is a fit subject for discussion, whether all the domestic races of the dog, for instance, have acquired their present amount of difference since some one species was first domesticated by man; or whether they owe some of their characters to inheritance from distinct species, which had already been differentiated in a state

21. See Prof. Huxley to this effect in the 'Fortnightly Review', 1865, p. 275.

of nature. With man no such question can arise, for he cannot be said to have been domesticated at any particular period.

During an early stage in the divergence of the races of man from a common stock, the differences between the races and their number must have been small; consequently as far as their distinguishing characters are concerned, they then had less claim to rank as distinct species than the existing so-called races. Nevertheless, so arbitrary is the term of species, that such early races would perhaps have been ranked by some naturalists as distinct species, if their differences, although extremely slight, had been more constant than they are at present, and had not graduated into each other.

It is however possible, though far from probable, that the early progenitors of man might formerly have diverged much in character, until they became more unlike each other than any now existing races; but that subsequently, as suggested by Vogt,[22] they converged in character. When man selects the offspring of two distinct species for the same object, he sometimes induces a considerable amount of convergence, as far as general appearance is concerned. This is the case, as shewn by Von Nathusius,[23] with the improved breeds of the pig, which are descended from two distinct species; and in a less marked manner with the improved breeds of cattle. A great anatomist, Gratiolet, maintains that the anthropomorphous apes do not form a natural sub-group; but that the orang is a highly developed gibbon or semnopithecus, the chimpanzee a highly developed macacus, and the gorilla a highly developed mandrill. If this conclusion, which rests almost exclusively on brain-characters, be admitted, we should have a case of convergence at least in external characters, for the

22. 'Lectures on Man', Eng. Translat. 1864, p. 468.
23. 'Die Racen des Schweines', 1860, s. 46. 'Vorstudien für Geschichte, &c., Schweineschädel', 1864, s. 104. With respect to cattle, see M. de Quatrefages, 'Unité de l'Espèce Humaine', 1861, p. 119.

anthropomorphous apes are certainly more like each other in many points, than they are to other apes. All analogical resemblances, as of a whale to a fish, may indeed be said to be cases of convergence; but this term has never been applied to superficial and adaptive resemblances. It would, however, be extremely rash to attribute to convergence close similarity of character in many points of structure amongst the modified descendants of widely distinct beings. The form of a crystal is determined solely by the molecular forces, and it is not surprising that dissimilar substances should sometimes assume the same form; but with organic beings we should bear in mind that the form of each depends on an infinity of complex relations, namely on variations, due to causes far too intricate to be followed—on the nature of the variations preserved, these depending on the physical conditions, and still more on the surrounding organisms which compete with each—and lastly, on inheritance (in itself a fluctuating element) from innumerable progenitors, all of which have had their forms determined through equally complex relations. It appears incredible that the modified descendants of two organisms, if these differed from each other in a marked manner, should ever afterwards converge so closely as to lead to a near approach to identity throughout their whole organisation. In the case of the convergent races of pigs above referred to, evidence of their descent from two primitive stocks is, according to Von Nathusius, still plainly retained, in certain bones of their skulls. If the races of man had descended, as is supposed by some naturalists, from two or more species, which differed from each other as much, or nearly as much, as does the orang from the gorilla, it can hardly be doubted that marked differences in the structure of certain bones would still be discoverable in man as he now exists.

Although the existing races of man differ in many respects, as in colour, hair, shape of skull, proportions of the body, &c., yet if their whole structure be taken into consideration they are found to resemble each other closely in a multitude of points. Many of these are of so unimportant or of so singular a nature, that it is extremely improbable that they should have been independently acquired by aboriginally distinct species or races. The same remark holds good with equal or greater force with respect to the numerous points of mental similarity between the most distinct races of man. The American aborigines, Negroes and Europeans are as different from each other in mind as any three races that can be named; yet I was incessantly struck, whilst living with the Fuegians on board the 'Beagle', with the many little traits of character, shewing how similar their minds were to ours; and so it was with a full-blooded negro with whom I happened once to be intimate.

He who will read Mr Tylor's and Sir J. Lubbock's interesting works[24] can hardly fail to be deeply impressed with the close similarity between the men of all races in tastes, dispositions and habits. This is shewn by the pleasure which they all take in dancing, rude music, acting, painting, tattooing, and otherwise decorating themselves; in their mutual comprehension of gesture-language, by the same expression in their features, and by the same inarticulate cries, when excited by the same emotions. This similarity, or rather identity, is striking, when constrasted with the different expressions and cries made by distinct species of monkeys. There is good evidence that the art of shooting with bows and arrows has not been handed down from any common progenitor of mankind, yet as Westropp and Nilsson have remarked,[25] the stone arrow-heads, brought from the most distant

24. Tylor's 'Early History of Mankind', 1865: with respect to gesture-language, see p. 54. Lubbock's 'Prehistoric Times', 2nd edit., 1869.

25. 'On Analogous Forms of Implements', in 'Memoirs of Anthropolog. Soc.', by H. M. Westropp. 'The Primitive Inhabitants of Scandinavia', Eng. translat. edited by Sir J. Lubbock, 1868, p. 104.

parts of the world, and manufactured at the most remote periods, are almost identical; and this fact can only be accounted for by the various races having similar inventive or mental powers. The same observation has been made by archaeologists[26] with respect to certain widely-prevalent ornaments, such as zigzags, &c.; and with respect to various simple beliefs and customs, such as the burying of the dead under megalithic structures. I remember observing in South America,[27] that there, as in so many other parts of the world, men have generally chosen the summits of lofty hills, to throw up piles of stones, either as a record of some remarkable event, or for burying their dead.

Now when naturalists observe a close agreement in numerous small details of habits, tastes, and dispositions between two or more domestic races, or between nearly-allied natural forms, they use this fact as an argument that they are descended from a common progenitor who was thus endowed; and consequently that all should be classed under the same species. The same argument may be applied with much force to the races of man.

As it is improbable that the numerous and unimportant points of resemblance between the several races of man in bodily structure and mental faculties (I do not here refer to similar customs) should all have been independently acquired, they must have been inherited from progenitors who had these same characters. We thus gain some insight into the early state of man, before he had spread step by step over the face of the earth. The spreading of man to regions widely separated by the sea, no doubt, preceded any great amount of divergence of character in the several races; for otherwise we should sometimes meet with the same race in distinct continents; and this is never the case. Sir J. Lubbock, after comparing the arts now practised by savages

26. Westropp, 'On Cromlechs', &c., 'Journal of Ethnological Soc.' as given in 'Scientific Opinion', June 2nd, 1869, p. 3.
27. 'Journal of Researches: Voyage of the "Beagle"', p. 46.

in all parts of the world, specifies those which man could not have known, when he first wandered from his original birth-place; for if once learnt they would never have been forgotten.[28] He thus shews that 'the spear, which is but a development of the knife-point, and the club, which is but a long hammer, are the only things left'. He admits, however, that the art of making fire probably had been already discovered, for it is common to all the races now existing, and was known to the ancient cave-inhabitants of Europe. Perhaps the art of making rude canoes or rafts was likewise known; but as man existed at a remote epoch, when the land in many places stood at a very different level to what it does now, he would have been able, without the aid of canoes, to have spread widely. Sir J. Lubbock further remarks how improbable it is that our earliest ancestors could have 'counted as high as ten, considering that so many races now in existence cannot get beyond four'. Nevertheless, at this early period, the intellectual and social faculties of man could hardly have been inferior in any extreme degree to those possessed at present by the lowest savages; otherwise primeval man could not have been so eminently successful in the struggle for life, as proved by his early and wide diffusion.

From the fundamental differences between certain languages, some philologists have inferred that when man first became widely diffused, he was not a speaking animal; but it may be sus-pected that languages, far less perfect than any now spoken, aided by gestures, might have been used, and yet have left no traces on subsequent and more highly-developed tongues. Without the use of some language, however imperfect, it appears doubtful whether man's intellect could have risen to the standard implied by his dominant position at an early period.

Whether primeval man, when he possessed but few arts, and

28. 'Prehistoric Times', 1869, p.574.

those of the rudest kind, and when his power of language was extremely imperfect, would have deserved to be called man, must depend on the definition which we employ. In a series of forms graduating insensibly from some ape-like creature to man as he now exists, it would be impossible to fix on any definite point when the terms 'man' ought to be used. But this is a matter of very little importance. So again, it is almost a matter of indifference whether the so-called races of man are thus designated, or are ranked as species or sub-species; but the latter term appears the more appropriate. Finally, we may conclude that when the principle of evolution is generally accepted, as it surely will be before long, the dispute between the monogenists and the polygenists will die a silent and unobserved death.

One other question ought not to be passed over without notice, namely, whether, as is sometimes assumed, each sub-species or race of man has sprung from a single pair of progenitors. With our domestic animals a new race can readily be formed by carefully matching the varying offspring from a single pair, or even from a single individual possessing some new character; but most of our races have been formed, not intentionally from a selected pair, but unconsciously by the preservation of many individuals which have varied, however slightly, in some useful or desired manner. If in one country stronger and heavier horses, and in another country lighter and fleeter ones, were habitually preferred, we may feel sure that two distinct sub-breeds would be produced in the course of time, without any one pair having been separated and bred from, in either country. Many races have been thus formed, and their manner of formation is closely analogous to that of natural species. We know, also, that the horses taken to the Falkland Islands have, during successive generations, become smaller and weaker, whilst those which have run wild on the Pampas have acquired larger and coarser heads; and such changes are manifestly due, not to any one pair, but to all the individuals having been subjected to the same conditions, aided, perhaps, by the principle of reversion. The new sub-breeds in such cases are not descended from any single pair, but from many individuals which have varied in different degrees, but in the same general manner; and we may conclude that the races of man have been similarly produced, the modifications being either the direct result of exposure to different conditions, or the indirect result of some form of selection. But to this latter subject we shall presently return.

On the Extinction of the Races of Man—The partial or complete extinction of many races and sub-races of man is historically known. Humboldt saw in South America a parrot which was the sole living creature that could speak a word of the language of a lost tribe. Ancient monuments and stone implements found in all parts of the world, about which no tradition has been preserved by the present inhabitants, indicate much extinction. Some small and broken tribes, remnants of former races, still survive in isolated and generally mountainous districts. In Europe the ancient races were all, according to Schaaffhausen,[29] 'lower in the scale than the rudest living savages'; they must therefore have differed, to a certain extent, from any existing race. The remains described by Professor Broca from Les Eyzies, though they unfortunately appear to have belonged to a single family, indicate a race with a most singular combination of low or simious, and of high characteristics. This race is 'entirely different from any other, ancient or modern, that we have ever heard of'.[30] It differed, therefore, from the quarternary race of the caverns of Belgium.

Man can long resist conditions which appear extremely unfa-

29. Translation in 'Anthropological Review', Oct. 1868, p. 431.
30. 'Transact. Internat. Congress of Prehistoric Arch', 1868, pp. 172–175. See also Broca (translation) in 'Anthropological Review', Oct. 1868, p. 410.

vourable for his existence.[31] He has long lived in the extreme regions of the North, with no wood for his canoes or implements, and with only blubber as fuel, and melted snow as drink. In the southern extremity of America the Fuegians survive without the protection of clothes, or of any building worthy to be called a hovel. In South Africa the aborigines wander over arid plains, where dangerous beasts abound. Man can withstand the deadly influence of the Terai at the foot of the Himalaya, and the pestilential shores of tropical Africa.

Extinction follows chiefly from the competition of tribe with tribe, and race with race. Various checks are always in action, serving to keep down the numbers of each savage tribe—such as periodical famines, nomadic habits and the consequent deaths of infants, prolonged suckling, wars, accidents, sickness, licentiousness, the stealing of women, infanticide, and especially lessened fertility. If any one of these checks increases in power, even slightly, the tribe thus affected tends to decrease; and when of two adjoining tribes one becomes less numerous and less powerful than the other, the contest is soon settled by war, slaughter, cannibalism, slavery, and absorption. Even when a weaker tribe is not thus abruptly swept away, if it once begins to decrease, it generally goes on decreasing until it becomes extinct.[32]

When civilised nations come into contact with barbarians the struggle is short, except where a deadly climate gives its aid to the native race. Of the causes which lead to the victory of civilised nations, some are plain and simple, others complex and obscure. We can see that the cultivation of the land will be fatal in many ways to savages, for they cannot, or will not, change their habits. New diseases and vices have in some cases proved highly destructive; and it appears that a new disease often causes

31. Dr Gerland 'Ueber das Aussterben der Naturvölker', 1868, s. 82.
32. Gerland (ibid. s. 12) gives facts in support of this statement.

much death, until those who are most susceptible to its destructive influence are gradually weeded out;[33] and so it may be with the evil effects from spirituous liquors, as well as with the unconquerably strong taste for them shewn by so many savages. It further appears, mysterious as is the fact that the first meeting of distinct and separated people generates disease.[34] Mr Sproat, who in Vancouver Island closely attended to the subject of extinction, believed that changed habits of life, consequent on the advent of Europeans, induces much ill health. He lays, also, great stress on the apparently trifling cause that the natives become 'bewildered and dull by the new life around them; they lose the motives for exertion, and get no new ones in their place'.[35]

The grade of their civilisation seems to be a most important element in the success of competing nations. A few centuries ago Europe feared the inroads of Eastern barbarians; now any such fear would be ridiculous. It is a more curious fact, as Mr Bagehot has remarked, that savages did not formerly waste away before the classical nations, as they now do before modern civilised nations; had they done so, the old moralists would have mused over the event; but there is no lament in any writer of that period over the perishing barbarians.[36] The most potent of all the causes of extinction, appears in many cases to be lessened fertility and ill-health, especially amongst the children, arising from changed conditions of life, notwithstanding that the new conditions may not be injurious in themselves. I am much indebted to Mr H. H. Howorth for having called my attention to this subject, and for

33. See remarks to this effect in Sir H. Holland's 'Medical Notes and Reflections', 1839, p. 390.
34. I have collected ('Journal of Researches, Voyage of the "Beagle" ', p. 435) a good many cases bearing on this subject: see also Gerland, ibid. s. 8. Poeppig speaks of the 'breath of civilisation as poisonous to savages'.
35. Sproat, 'Scenes and Studies of Savage Life', 1868, p. 284.
36. Bagehot, 'Physics and Politics', 'Fortnightly Review', April 1, 1868, p. 455.

having given me information respecting it. I have collected the
following cases.

When Tasmania was first colonised the natives were roughly
estimated by some at 7000 and by others at 20,000. Their
number was soon greatly reduced, chiefly by fighting with the
English and with each other. After the famous hunt by all the
colonists, when the remaining natives delivered themselves up to
the government, they consisted only of 120 individuals,[37] who
were in 1832 transported to Flinders Island. This island, situated
between Tasmania and Australia, is forty miles long, and from
twelve to eighteen miles broad: it seems healthy, and the natives
were well treated. Nevertheless, they suffered greatly in health.
In 1834 they consisted (Bonwick, p. 250) of forty-seven adult
males, forty-eight adult females, and sixteen children, or in all
of 111 souls. In 1835 only one hundred were left. As they continued rapidly to decrease, and as they themselves thought that
they should not perish so quickly elsewhere, they were removed
in 1847 to Oyster Cove in the southern part of Tasmania. They
then consisted (Dec. 20th, 1847) of fourteen men, twenty-two
women and ten children.[38] But the change of site did no good.
Disease and death still pursued them, and in 1864 one man (who
died in 1869), and three elderly women alone survived. The infertility of the women is even a more remarkable fact than the
liability of all to ill-health and death. At the time when only
nine women were left at Oyster Cove, they told Mr Bonwick (p.
386), that only two had ever borne children: and these two had
together produced only three children!

With respect to the cause of this extraordinary state of things,
Dr Story remarks that death followed the attempts to civilise

---

37. All the statements here given are taken from 'The last of the Tasmanians', by J.
Bonwick, 1870.
38. This is the statement of the Governor of Tasmania, Sir W. Denison, 'Varieties of
Vice-Regal Life', 1870, vol. i. p. 67.

the natives. 'If left to themselves to roam as they were wont and undisturbed, they would have reared more children, and there would have been less mortality.' Another careful observer of the natives, Mr Davis, remarks, 'The births have been few and the deaths numerous. This may have been in a great measure owing to their change of living and food; but more so to their banishment from the mainland of Van Diemen's Land, and consequent depression of spirits' (Bonwick, pp. 388, 390).

Similar facts have been observed in two widely different parts of Australia. The celebrated explorer, Mr Gregory, told Mr Bonwick, that in Queensland 'the want of reproduction was being already felt with the blacks, even in the most recently settled parts, and that decay would set in'. Of thirteen aborigines from Shark's Bay who visited Murchison River, twelve died of consumption within three months.[39]

The decrease of the Maories of New Zealand has been carefully investigated by Mr Fenton, in an admirable Report, from which all the following statements, with one exception, are taken.[40] The decrease in number since 1830 is admitted by every one, including the natives themselves, and is still steadily progressing. Although it has hitherto been found impossible to take an actual census of the natives, their numbers were carefully estimated by residents in many districts. The result seems trustworthy, and shows that during the fourteen years, previous to 1858, the decrease was 19.42 per cent. Some of the tribes, thus carefully examined, lived above a hundred miles apart, some on the coast, some inland; and their means of subsistence and habits differed to a certain extent (p. 28). The total number in 1858 was believed to be 53,700, and in 1872, after a second interval of fourteen

39. For these cases, see Bonwick's 'Daily Life of the Tasmanians', 1870, p. 90; and the 'Last of the Tasmanians', 1870, p. 386.
40. 'Observations on the Aboriginal Inhabitants of New Zealand', published by the Government, 1859.

years, another census was taken, and the number is given as only 36,359, shewing a decrease of 32.29 per cent![41] Mr Fenton, after shewing in detail the insufficiency of the various causes, usually assigned in explanation of this extraordinary decrease, such as new diseases, the profligacy of the women, drunkenness, wars, &c., concludes on weighty grounds that it depends chiefly on the unproductiveness of the women, and on the extraordinary mortality of the young children (pp. 31, 34). In proof of this he shews (p. 33) that in 1844 there was one non-adult for every 2.57 adults; whereas in 1858 there was only one non-adult for every 3.27 adults. The mortality of the adults is also great. He adduces as a further cause of the decrease the inequality of the sexes; for fewer females are born than males. To this latter point, depending perhaps on a widely distinct cause, I shall return in a future chapter. Mr Fenton contrasts with astonishment the decrease in New Zealand with the increase in Ireland; countries not very dissimilar in climate, and where the inhabitants now follow nearly similar habits. The Maories themselves (p. 35) 'attribute their decadence, in some measure, to the introduction of new food and clothing, and the attendant change of habits'; and it will be seen, when we consider the influence of changed conditions on fertility, that they are probably right. The diminution began between the years 1830 and 1840; and Mr Fenton shews (p. 40) that about 1830, the art of manufacturing putrid corn (maize), by long steeping in water, was discovered and largely practised; and this proves that a change of habits was beginning amongst the natives, even when New Zealand was only thinly inhabited by Europeans. When I visited the Bay of Islands in 1835, the dress and food of the inhabitants had already been much modified: they raised potatoes, maize, and other agricultural produce, and exchanged them for English manufactured goods and tobacco.

41. 'New Zealand', by Alex. Kennedy, 1873, p. 47.

It is evident from many statements in the life of Bishop Patteson,[42] that the Melanesians of the New Hebrides and neighbouring archipelagoes, suffered to an extraordinary degree in health, and perished in large numbers, when they were removed to New Zealand, Norfolk Island, and other salubrious places, in order to be educated as missionaries.

The decrease of the native population of the Sandwich Islands is as notorious as that of New Zealand. It has been roughly estimated by those best capable of judging, that when Cook discovered the Islands in 1779, the population amounted to about 300,000. According to a loose census in 1823, the numbers then were 142,050. In 1832, and at several subsequent periods, an accurate census was officially taken, but I have been able to obtain only the following returns:

Year	Native Population (Except during 1832 and 1836, when the few foreigners in the islands were included)	Annual rate of decrease per cent., assuming it to have been uniform between the successive censuses; these censuses being taken at irregular intervals
1832	130,313	4.46
1836	108,579	2.47
1853	71,019	0.81
1860	67,084	2.18
1866	58,765	2.17
1872	51,531	

We here see that in the interval of forty years, between 1832 and 1872, the population has decreased no less than sixty-eight per

42. 'Life of J. C. Patteson', by C. M. Younge, 1874; see more especially vol. i. p. 530.

cent! This has been attributed by most writers to the profligacy of the women, to former bloody wars, and to the severe labour imposed on conquered tribes and to newly introduced diseases, which have been on several occasions extremely destructive. No doubt these and other such causes have been highly efficient, and may account for the extraordinary rate of decrease between the years 1832 and 1836; but the most potent of all the causes seems to be lessened fertility. According to Dr Ruschenberger of the US Navy, who visited these islands between 1835 and 1837, in one district of Hawaii, only twenty-five men out of 1134, and in another district only ten out of 637, had a family with as many as three children. Of eighty married women, only thirty-nine had ever borne children; and 'the official report gives an average of half a child to each married couple in the whole island'. This is almost exactly the same average as with the Tasmanians at Oyster Cove. Jarves, who published his History in 1843, says that 'families who have three children are freed from all taxes; those having more, are rewarded by gifts of land and other encouragements'. This un-paralleled enactment by the government well shews how infertile the race had become. The Rev. A. Bishop stated in the Hawaiian 'Spectator' in 1839, that a large proportion of the children die at early ages, and Bishop Staley informs me that this is still the case, just as in New Zealand. This has been attributed to the neglect of the children by the women, but it is probably in large part due to innate weakness of constitution in the children, in relation to the lessened fertility of their parents. There is, moreover, a further resemblance to the case of New Zealand, in the fact that there is a large excess of male over female births: the census of 1872 gives 31,650 males to 25,247 females of all ages, that is 125.36 males for every 100 females; whereas in all civilised countries the females exceed the males. No doubt the profligacy of the women may in part account for their small fertility; but their changed habits of life is a much more probable cause, and which will at the same time account for the increased mortality especially of the children. The islands were visited by Cook in 1779, by Vancouver in 1794, and often subsequently by whalers. In 1819 missionaries arrived, and found that idolatry had been already abolished, and other changes effected by the king. After this period there was a rapid change in almost all the habits of life of the natives, and they soon became 'the most civilised of the Pacific Islanders'. One of my informants, Mr Coan, who was born on the islands, remarks that the natives have undergone a greater change in their habits of life in the course of fifty years than Englishman during a thousand years. From information received from Bishop Staley, it does not appear that the poorer classes have ever much changed their diet, although many new kinds of fruit have been introduced, and the sugar-cane is in universal use. Owing, however, to their passion for imitating Europeans, they altered their manner of dressing at an early period, and the use of alcoholic drinks became very general. Although these changes appear inconsiderable, I can well believe, from what is known with respect to animals, that they might suffice to lessen the fertility of the natives.[43]

Lastly, Mr Macnamara states[44] that the low and degraded inhabitants of the Andaman Islands, on the eastern side of the Gulf of Bengal, are 'eminently susceptible to any change of climate: in fact, take them away from their island homes, and they are almost certain to die, and that independently of diet or extraneous influences'. He further states that the inhabitants of the Valley of

---

43. The foregoing statements are taken chiefly from the following works: 'Jarves' History of the Hawaiian Islands', 1843, pp. 400–407. Cheever, 'Life in the Sandwich Islands', 1851, p. 277. Ruschenberger is quoted by Bonwick, 'Last of the Tasmanians', 1870, p. 378. Bishop is quoted by Sir E. Belcher, 'Voyage Round the World', 1843, vol. i. p. 272. I owe the census of the several years to the kindness of Mr Coan, at the request of Dr Youmans of New York; and in most cases I have compared the Youmans figures with those given in several of the above-named works. I have omitted the census for 1850, as I have seen two widely different numbers given.
44. 'The Indian Medical Gazette', Nov. 1, 1871, p. 240.

Nepâl, which is extremely hot in summer, and also the various hill-tribes of India, suffer from dysentery and fever when on the plains; and they die if they attempt to pass the whole year there.

We thus see that many of the wilder races of man are apt to suffer much in health when subjected to changed conditions or habits of life, and not exclusively from being transported to a new climate. Mere alterations in habits, which do not appear injurious in themselves, seem to have this same effect; and in several cases the children are particularly liable to suffer. It has often been said, as Mr Macnamara remarks, that man can resist with impunity the greatest diversities of climate and other changes; but this is true only of the civilised races. Man in his wild condition seems to be in this respect almost as susceptible as his nearest allies, the anthropoid apes, which have never yet survived long, when removed from their native country.

Lessened fertility from changed conditions, as in the case of the Tasmanians, Maories, Sandwich Islanders, and apparently the Australians, is still more interesting than their liability to ill-health and death; for even a slight degree of infertility, combined with those other causes which tend to check the increase of every population, would sooner or later lead to extinction. The diminution of fertility may be explained in some cases by the profligacy of the women (as until lately with the Tahitians), but Mr Fenton has shewn that this explanation by no means suffices with the New Zealanders, nor does it with the Tasmanians.

In the paper above quoted, Mr Macnamara gives reasons for believing that the inhabitants of districts subject to malaria are apt to be sterile; but this cannot apply in several of the above cases. Some writers have suggested that the aborigines of islands have suffered in fertility and health from long continued inter-breeding; but in the above cases infertility has coincided too closely with the arrival of Europeans for us to admit this explanation. Nor have we at present any reason to believe that man is highly sensitive to the evil effects of inter-breeding, especially in areas so large as New Zealand, and the Sandwich archipelago with its diversified stations. On the contrary, it is known that the present inhabitants of Norfolk Island are nearly all cousins or near relations, as are the Todas in India, and the inhabitants of some of the Western Islands of Scotland; and yet they seem not to have suffered in fertility.[45]

A much more probable view is suggested by the analogy of the lower animals. The reproductive system can be shewn to be susceptible to an extraordinary degree (though why we know not) to changed conditions of life; and this susceptibility leads both to beneficial and to evil results. A large collection of facts on this subject is given in chap. xviii. of vol. ii. of my 'Variation of Animals and Plants under Domestication', I can here give only the briefest abstract; and every one interested in the subject may consult the above work. Very slight changes increase the health, vigour and fertility of most or all organic beings, whilst other changes are known to render a large number of animals sterile. One of the most familiar cases, is that of tamed elephants not breeding in India; though they often breed in Ava, where the females are allowed to roam about the forests to some extent, and are thus placed under more natural conditions. The case of various American monkeys, both sexes of which have been kept for many years together in their own countries, and yet have very rarely or never bred, is a more apposite instance, because of their relationship to man. It is remarkable how slight a change in the conditions often induces sterility in a wild animal when captured; and this is the more strange as all our domesticated animals have become more fertile than they were in a state of nature; and

45. On the close relationship of the Norfolk Islanders; see Sir W. Denison, 'Varieties of Vice-Regal Life', vol. i. 1870, p. 410. For the Todas, see Col. Marshall's work, 1873, p. 110. For the Western Islands of Scotland, Dr Mitchell, 'Edinburgh Medical Journal', March to June, 1865.

some of them can resist the most unnatural conditions with undiminished fertility.[46] Certain groups of animals are much more liable than others to be affected by captivity; and generally all the species of the same group are affected in the same manner. But sometimes a single species in a group is rendered sterile, whilst the others are not so; on the other hand, a single species may retain its fertility whilst most of the others fail to breed. The males and females of some species when confined, or when allowed to live almost, but not quite free, in their native country, never unite; others thus circumstanced frequently unite but never produce offspring; others again produce some offspring, but fewer than in a state of nature; and as bearing on the above cases of man, it is important to remark that the young are apt to be weak and sickly, or malformed, and to perish at an early age.

Seeing how general is this law of the susceptibility of the reproductive system to changed conditions of life, and that it holds good with our nearest allies, the Quadrumana, I can hardly doubt that it applies to man in his primeval state. Hence if savages of any race are induced suddenly to change their habits of life, they become more or less sterile, and their young offspring suffer in health, in the same manner and from the same cause, as do the elephant and hunting-leopard in India, many monkeys in America, and a host of animals of all kinds, on removal from their natural conditions.

We can see why it is that aborigines, who have long inhabited islands, and who must have been long exposed to nearly uniform conditions, should be specially affected by any change in their habits, as seems to be the case. Civilised races can certainly resist changes of all kinds far better than savages; and in this respect they resemble domesticated animals, for though the latter sometimes suffer in health (for instance European dogs in

46. For the evidence on this head, see 'Variation of Animals' &c., vol. ii. p. 111.

India), yet they are rarely rendered sterile, though a few such instances have been recorded.[47] The immunity of civilised races and domesticated animals is probably due to their having been subjected to a greater extent, and therefore having grown somewhat more accustomed, to diversified or varying conditions, than the majority of wild animals; and to their having formerly immigrated or been carried from country to country, and to different families or sub-races having inter-crossed. It appears that a cross with civilised races at once gives to an aboriginal race an immunity from the evil consequences of changed conditions. Thus the crossed offspring from the Tahitians and English, when settled in Pitcairn Island, increased so rapidly that the island was soon overstocked; and in June 1856 they were removed to Norfolk Island. They then consisted of 60 married persons and 134 children, making a total of 194. Here they likewise increased so rapidly, that although sixteen of them returned to Pitcairn Island in 1859, they numbered in January 1868, 300 souls; the males and females being in exactly equal numbers. What a contrast does this case present with that of the Tasmanians; the Norfolk Islanders *increased* in only twelve and a half years from 194 to 300; whereas the Tasmanians *decreased* during fifteen years from 120 to 46, of which latter number only ten were children.[48]

So again in the interval between the census of 1866 and 1872 the natives of full blood in the Sandwich Islands decreased by 8081, whilst the half-castes, who are believed to be healthier, increased by 847; but I do not know whether the latter number includes the offspring from the half-castes, or only the half-castes of the first generation.

---

47. 'Variation of Animals', &c., vol. ii. p. 16.
48. These details are taken from 'The Mutineers of the "Bounty"', by Lady Belcher, 1870; and from 'Pitcairn Island', ordered to be printed by the House of Commons, May 29th, 1863. The following statements about the Sandwich Islanders are from the 'Honolulu Gazette', and from Mr Coan.

The cases which I have here given all relate to aborigines, who have been subjected to new conditions as the result of the immigration of civilised men. But sterility and ill-health would probably follow, if savages were compelled by any cause, such as the inroad of a conquering tribe, to desert their homes and to change their habits. It is an interesting circumstance that the chief check to wild animals becoming domesticated, which implies the power of their breeding freely when first captured, and one chief check to wild men, when brought into contact with civilisation, surviving to form a civilised race, is the same, namely, sterility from changed conditions of life.

Finally, although the gradual decrease and ultimate extinction of the races of man is a highly complex problem, depending on many causes which differ in different places and at different times; it is the same problem as that presented by the extinction of one of the higher animals—of the fossil horse, for instance, which disappeared from South America, soon afterwards to be replaced, within the same districts, by countless troops of the Spanish horse. The New Zealander seems conscious of this parallelism, for he compares his future fate with that of the native rat now almost exterminated by the European rat. Though the difficulty is great to our imagination, and really great, if we wish to ascertain the precise causes and their manner of action, it ought not to be so to our reason, as long as we keep steadily in mind that the increase of each species and each race is constantly checked in various ways; so that if any new check, even a slight one, be superadded, the race will surely decrease in number; and decreasing numbers will sooner or later lead to extinction; the end, in most cases, being promptly determined by the inroads of conquering tribes.

On the Formation of the Races of Man—In some cases the crossing of distinct races has led to the formation of a new race. The singular fact that Europeans and Hindoos, who belong to the same Aryan stock, and speak a language fundamentally the same, differ widely in appearance, whilst Europeans differ but little from Jews, who belong to the Semitic stock, and speak quite another language, has been accounted for by Broca,[49] through certain Aryan branches having been largely crossed by indigenous tribes during their wide diffusion. When two races in close contact cross, the first result is a heterogeneous mixture: thus Mr Hunter, in describing the Santali or hill-tribes of India, says that hundreds of imperceptible gradations may be traced 'from the black, squat tribes of the mountains to the tall olive-coloured Brahman, with his intellectual brow, calm eyes, and high but narrow head', so that it is necessary in courts of justice to ask the witnesses whether they are Santalis or Hindoos.[50] Whether a heterogeneous people, such as the inhabitants of some of the Polynesian islands, formed by the crossing of two distinct races, with few or no pure members left, would ever become homogeneous, is not known from direct evidence. But as with our domesticated animals, a cross-breed can certainly be fixed and made uniform by careful selection[51] in the course of a few generations, we may infer that the free intercrossing of a heterogeneous mixture during a long descent would supply the place of selection, and overcome any tendency to reversion; so that the crossed race would ultimately become homogeneous, though it might not partake in an equal degree of the characters of the two parent-races.

Of all the differences between the races of man, the colour of the skin is the most conspicuous and one of the best marked. It was formerly thought that differences of this kind could be accounted for by long exposure to different climates; but Pal-

---

49. 'On Anthropology', translation 'Anthropolog. Review', Jan. 1868, p. 38.
50. 'The Annals of Rural Bengal', 1868, p. 134.
51. 'The Variation of Animals and Plants under Domestication', vol. ii. p. 95.

las first shewed that this is not tenable, and he has since been followed by almost all anthropologists.[52] This view has been rejected chiefly because the distribution of the variously coloured races, most of whom must have long inhabited their present homes, does not coincide with corresponding differences of climate. Some little weight may be given to such cases as that of the Dutch families, who, as we hear on excellent authority,[53] have not undergone the least change of colour after residing for three centuries in South Africa. An argument on the same side may likewise be drawn from the uniform appearance in various parts of the world of gipsies and Jews, though the uniformity of the latter has been somewhat exaggerated.[54] A very damp or a very dry atmosphere has been supposed to be more influential in modifying the colour of the skin than mere heat; but as D'Orbigny in South America, and Livingstone in Africa, arrived at diametrically opposite conclusions with respect to dampness and dryness, any conclusion on this head must be considered as very doubtful.[55]

Various facts, which I have given elsewhere, prove that the colour of the skin and hair is sometimes correlated in a surprising manner with a complete immunity from the action of certain vegetable poisons, and from the attacks of certain parasites. Hence it occurred to me, that negroes and other dark races might have acquired their dark tints by the darker individuals escaping from the deadly influence of the miasma of their native countries, during a long series of generations.

I afterwards found that this same idea had long ago occurred

52. Pallas, 'Act. Acad. St Petersburg', 1780, part ii. p. 69. He was followed by Rudolphi, in his 'Beyträge zur Anthropologie', 1812. An excellent summary of the evidence is given by Godron, 'De l'Espèce', 1859, vol. ii. p. 246, &c.
53. Sir Andrew Smith, as quoted by Knox, 'Races of Man', 1850, p. 473.
54. See De Quatrefages on this head, 'Revue des Cours Scientifiques', Oct. 17, 1868, p. 731.
55. Livingstone's 'Travels and Researches in S. Africa', 1857, pp. 338, 329. D'Orbigny, as quoted by Godron, 'De l'Espèce', vol. ii. p. 266.

to Dr Wells.[56] It has long been known that negroes, and even mulattoes, are almost completely exempt from the yellow fever, so destructive in tropical America.[57] They likewise escape to a large extent the fatal intermittent fevers, that prevail along at least 2600 miles of the shores of Africa, and which annually cause one-fifth of the white settlers to die, and another fifth to return home invalided.[58] This immunity in the negro seems to be partly inherent, depending on some unknown peculiarity of constitution, and partly the result of acclimatisation. Pouchet[59] states that the negro regiments recruited near the Soudan, and borrowed from the Viceroy of Egypt for the Mexican war, escaped the yellow-fever almost equally with the negroes originally brought from various parts of Africa and accustomed to the climate of the West Indies. That acclimatisation plays a part, is shewn by the many cases in which negroes have become somewhat liable to tropical fevers, after having resided for some time in a colder climate.[60] The nature of the climate under which the white races have long resided, likewise has some influence on them; for during the fearful epidemic of yellow-fever in Demerara during 1837, Dr Blair found that the death-rate of the immigrants was proportional to the latitude of the country whence they had come. With the negro the immunity, as far as it is the result of acclimatisation, implies exposure during a prodigious length of time; for the aborigines of tropical America who have re-

56. See a paper read before the Royal Soc. in 1813, and published in his Essays in 1818. I have given an account of Dr Wells' views in the Historical Sketch (p. xvi.) to my 'Origin of Species'. Various cases of colour correlated with constitutional peculiarities are given in my 'Variation of Animals under Domestication', vol. ii. pp. 227, 335.

57. See, for instance, Nott and Gliddon, 'Types of Mankind', p. 68.

58. Major Tulloch, in a paper read before the Statistical Society, April 20th, 1840, and given in the 'Athenaeum', 1840, p. 353.

59. 'The Plurality of the Human Race' (translat.), 1864, p. 60.

60. Quatrefages, 'Unité de l'Espèce Humaine', 1861, p. 205. Waitz, 'Introduct. to Anthropology', translat. vol. i. 1863, p. 124. Livingstone gives analogous cases in his 'Travels'.

sided there from time immemorial, are not exempt from yellow fever; and the Rev. H. B. Tristram states, that there are districts in Northern Africa which the native inhabitants are compelled annually to leave, though the negroes can remain with safety.

That the immunity of the negro is in any degree correlated with the colour of his skin is a mere conjecture: it may be correlated with some difference in his blood, nervous system, or other tissues. Nevertheless, from the facts above alluded to, and from some connection apparently existing between complexion and a tendency to consumption, the conjecture seemed to me not improbable. Consequently I endeavoured, with but little success,[61] to ascertain how far it holds good. The late Dr Daniell, who had long lived on the West Coast of Africa, told me that he did not believe in any such relation. He was himself unusually fair, and had withstood the climate in a wonderful manner. When he first arrived as a boy on the coast, an old and experienced

61. In the spring of 1862 I obtained permission from the Director-General of the Medical department of the Army, to transmit to the surgeons of the various regiments on foreign service a blank table, with the following appended remarks, but I have received no returns. 'As several well-marked cases have been recorded with our domestic animals of a relation between the colour of the dermal appendages and the constitution; and it being notorious that there is some limited degree of relation between the colour of the races of man and the climate inhabited by them; the following investigation seems worth consideration. Namely, whether there is any relation in Europeans between the colour of their hair, and their liability to the diseases of tropical countries. If the surgeons of the several regiments, when stationed in unhealthy tropical districts, would be so good as first to count, as a standard of comparison, how many men, in the force whence the sick are drawn, have dark and light-coloured hair, and hair of intermediate or doubtful tints; and if a similar account were kept by the same medical gentlemen, of all the men who suffered from malarious and yellow fevers, or from dysentery, it would soon be apparent, after some thousand cases had been tabulated, whether there exists any relation between the colour of the hair and constitutional liability to tropical diseases. Perhaps no such relation would be discovered, but the investigation is well worth making. In case any positive result were obtained, it might be of some practical use in selecting men for any particular service. Theoretically the result would be of high interest, as indicating one means by which a race of men inhabiting from a remote period an unhealthy tropical climate, might have become dark-coloured by the better preservation of dark-haired or dark-complexioned individuals during a long succession of generations.'

negro chief predicted from his appearance that this would prove the case. Dr Nicholson, of Antigua, after having attended to this subject, writes to me that he does not think that dark-coloured Europeans escape the yellow-fever more than those that are light-coloured. Mr J. M. Harris altogether denies that Europeans with dark hair withstand a hot climate better than other men: on the contrary, experience has taught him in making a selection of men for service on the coast of Africa, to choose those with red hair.[62] As far, therefore, as these slight indications go, there seems no foundation for the hypothesis, that blackness has resulted from the darker and darker individuals having survived better during long exposure to fever-generating miasma.

Dr Sharpe remarks,[63] that a tropical sun, which burns and blisters a white skin, does not injure a black one at all; and, as he adds, this is not due to habit in the individual, for children only six or eight months old are often carried about naked, and are not affected. I have been assured by a medical man, that some years ago during each summer, but not during the winter, his hands became marked with light brown patches, like, although larger than freckles, and that these patches were never affected by sun-burning, whilst the white parts of his skin have on several occasions been much inflamed and blistered. With the lower animals there is, also, a constitutional difference in liability to the action of the sun between those parts of the skin clothed with white hair and other parts.[64] Whether the saving of the skin from being thus

62. 'Anthropological Review', Jan. 1866, p. xxi. Dr Sharpe also says, with respect to India ('Man a Special Creation,' 1873, p. 118), that 'it has been noticed by some medical officers that Europeans with light hair and florid complexions suffer less from diseases of tropical countries than persons with dark hair and sallow complexions; and, so far as I know, there appear to be good grounds for this remark'. On the other hand, Mr Heddle, of Sierra Leone 'who has had more clerks killed under him than any other man', by the climate of the West African Coast (W. Reade, 'African Sketch Book', vol. ii. p. 522), holds a directly opposite view, as does Capt. Burton.

63. 'Man a Special Creation', 1873, p. 119.

64. 'Variation of Animals and Plants under Domestication', vol. ii. pp. 336, 337.

burnt is of sufficient importance to account for a dark tint having been gradually acquired by man through natural selection, I am unable to judge. If it be so, we should have to assume that the natives of tropical America have lived there for a much shorter time than the negroes in Africa, or the Papuans in the southern parts of the Malay archipelago, just as the lighter-coloured Hindoos have resided in India for a shorter time than the darker aborigines of the central and southern parts of the peninsula.

Although with our present knowledge we cannot account for the differences of colour in the races of man, through any advantage thus gained, or from the direct action of climate; yet we must not quite ignore the latter agency, for there is good reason to believe that some inherited effect is thus produced.[65]

We have seen in the second chapter that the conditions of life affect the development of the bodily frame in a direct manner, and that the effects are transmitted. Thus, as is generally admitted, the European settlers in the United States undergo a slight but extraordinarily rapid change of appearance. Their bodies and limbs become elongated; and I hear from Col. Bernys that during the late war in the United States, good evidence was afforded of this fact by the ridiculous appearance presented by the German regiments, when dressed in ready-made clothes manufactured for the American market, and which were much too long for the men in every way. There is, also, a considerable body of evidence shewing that in the Southern States the house-slaves of the third generation present a markedly different appearance from the field-slaves.[66]

---

65. See, for instance, Quatrefages ('Revue des Cours Scientifiques', Oct. 10, 1868, p. 724) on the effects of residence in Abyssinia and Arabia, and other analogous cases. Dr Rolle ('Der Mensch, seine Abstammung', &c., 1865, s. 99) states, on the authority of Khanikof, that the greater number of German families settled in Georgia, have acquired in the course of two generations dark hair and eyes. Mr D. Forbes informs me that the Quichuas in the Andes vary greatly in colour, according to the position of the valleys inhabited by them.

66. Harlan, 'Medical Researches', p. 532. Quatrefages ('Unité de l'Espèce Humaine,' 1861, p. 128) has collected much evidence on this head.

If, however, we look to the races of man as distributed over the world, we must infer that their characteristic differences cannot be accounted for by the direct action of different conditions of life, even after exposure to them for an enormous period of time. The Esquimaux live exclusively on animal food; they are clothed in thick fur, and are exposed to intense cold and to prolonged darkness; yet they do not differ in any extreme degree from the inhabitants of Southern China, who live entirely on vegetable food, and are exposed almost naked to a hot, glaring climate. The unclothed Fuegians live on the marine productions of their inhospitable shores; the Botocudos of Brazil wander about the hot forests of the interior and live chiefly on vegetable productions; yet these tribes resemble each other so closely that the Fuegians on board the 'Beagle' were mistaken by some Brazilians for Botocudos. The Botocudos again, as well as the other inhabitants of tropical America, are wholly different from the Negroes who inhabit the opposite shores of the Atlantic, are exposed to a nearly similar climate, and follow nearly the same habits of life.

Nor can the differences between the races of man be accounted for by the inherited effects of the increased or decreased use of parts, except to a quite insignificant degree. Men who habitually live in canoes, may have their legs somewhat stunted; those who inhabit lofty regions may have their chests enlarged; and those who constantly use certain sense-organs may have the cavities in which they are lodged somewhat increased in size, and their features consequently a little modified. With civilised nations, the reduced size of the jaws from lessened use—the habitual play of different muscles serving to express different emotions—and the increased size of the brain from greater intellectual activity, have together produced a considerable effect on their general appearance when compared with savages.[67] Increased bodily stature,

67. See Prof. Schaaffhausen, translat. in 'Anthropological Review', Oct. 1868, p. 429.

without any corresponding increase in the size of the brain, may (judging from the previously adduced case of rabbits), have given to some races an elongated skull of the dolichocephalic type.

Lastly, the little-understood principle of correlated development has sometimes come into action, as in the case of great muscular development and strongly projecting supra-orbital ridges. The colour of the skin and hair are plainly correlated, as is the texture of the hair with its colour in the Mandans of North America.[68] The colour also of the skin, and the odour emitted by it, are likewise in some manner connected. With the breeds of sheep the number of hairs within a given space and the number of the excretory pores are related.[69] If we may judge from the analogy of our domesticated animals, many modifications of structure in man probably come under this principle of correlated development.

We have now seen that the external characteristic differences between the races of man cannot be accounted for in a satisfactory manner by the direct action of the conditions of life, nor by the effects of the continued use of parts, nor through the principle of correlation. We are therefore led to inquire whether slight individual differences, to which man is eminently liable, may not have been preserved and augmented during a long series of generations through natural selection. But here we are at once met by the objection that beneficial variations alone can be thus preserved; and as far as we are enabled to judge, although always liable to err on this head, none of the differences between

68. Mr Catlin states ('N. American Indians', 3rd edit. 1842, vol. I, p. 49) that in the whole tribe of the Mandans, about one in ten or twelve of the members, of all ages and both sexes, have bright silvery grey hair, which is hereditary. Now this hair is as coarse and harsh as that of a horse's mane, whilst the hair of other colours is fine and soft.

69. On the odour of the skin, Godron, 'Sur l'Espèce,' tom. ii. p. 217. On the pores in the skin, Dr Wilckens, 'Die Aufgaben der Landwirth. Zootechnik', 1869, s. 7.

the races of man are of any direct or special service to him. The intellectual and moral or social faculties must of course be excepted from this remark. The great variability of all the external differences between the races of man, likewise indicates that they cannot be of much importance; for if important, they would long ago have been either fixed and preserved, or eliminated. In this respect man resembles those forms, called by naturalists protean or polymorphic, which have remained extremely variable, owing, as it seems, to such variations being of an indifferent nature, and to their having thus escaped the action of natural selection.

We have thus far been baffled in all our attempts to account for the differences between the races of man; but there remains one important agency, namely Sexual Selection, which appears to have acted powerfully on man, as on many other animals. I do not intend to assert that sexual selection will account for all the differences between the races. An unexplained residuum is left, about which we can only say, in our ignorance, that as individuals are continually born with, for instance, heads a little rounder or narrower, and with noses a little longer or shorter, such slight differences might become fixed and uniform, if the unknown agencies which induced them were to act in a more constant manner, aided by long-continued intercrossing. Such variations come under the provisional class, alluded to in our second chapter, which for the want of a better term are often called spontaneous. Nor do I pretend that the effects of sexual selection can be indicated with scientific precision; but it can be shewn that it would be an inexplicable fact if man had not been modified by this agency, which appears to have acted powerfully on innumerable animals. It can further be shewn that the differences between the races of man, as in colour, hairiness, form of features, &c., are of a kind which might have been expected to come under the influence of sexual selection. But in order to treat this subject properly, I have found it necessary to pass the whole animal kingdom in review. I have therefore devoted to it the Second Part of this work. At the close I shall return to man, and, after attempting to shew how far he has been modified through sexual selection, will give a brief summary of the chapters in this First Part.

8

Sexual Selection in Animals

The full title of Darwin's book is *The Descent of Man, and Selection in Relation to Sex*. The two parts are like conjoined twins—almost separate, and yet inseparable. Having spent the first half of the work investigating the links between humans and animals in their behavior and anatomy, Darwin abruptly turns his back on our species. He presents the reader with hundreds of pages of details on natural history that appear, at first, to have nothing to do with what it means to be human. Only at the end of the book does Darwin make his ultimate strategy clear. Along with natural selection, Darwin recognizes a parallel process of evolution at work throughout the animal kingdom: sexual selection. Darwin wants to persuade his readers of the reality of sexual selection in other species before finally turning to our own. To Darwin, human evolution makes no sense at all without sexual selection.

I have selected just a few passages from Darwin's eleven chapters on sexual selection in animals. This choice does not mean that Darwin's work on sexual selection in animals is unimportant. In fact, it is one of Darwin's greatest triumphs. Darwin recognized that reproductive success was the result of more than mere survival. Survival was necessary for reproduction, but hardly sufficient. An animal had to find a mate and successfully produce offspring. If an animal had some hereditary trait that allowed it to find more mates and to produce more offspring, that trait would become more common in the species. The trait would be, in Darwin's terms, sexually selected.

Darwin emphasized two factors in sexual selection. In some species males competed fiercely with other males for the opportunity to mate. Sexual selection should favor weapons to help males win these battles—in some cases with bigger bodies, and in other cases with wider horns, and so on. But Darwin also argued that females played a powerful role in the evolution of males because they could choose which males to mate with. The elaborate courtship displays that some males perform could be driven by the preferences of females.

Today a number of scientists are exploring sexual selection's deep effects on life. In some cases they have vindicated Darwin's core insights. Females do indeed prefer to mate with some males over others, and in many cases their choices can be tied to male ornamentation. But why should a female have such a preference? Darwin wrote vaguely of a sense of beauty in females, but some scientists today prefer a much more prosaic explanation: Females choose certain males because their genes will boost the survival of the females' young. Elaborate tail feathers or powerful chirps may serve as accurate signals of genetic quality. In a sense, sexual selection may be natural selection by other means.

But the full picture of sexual selection turns out to be far richer than Darwin imagined. By his account, females are coy and relatively passive, waiting for males to put on their displays and then making their choice. They behave like the ideal woman in Victorian England, receiving suitors and choosing the man she will marry. But females are more active and dynamic. In many species of birds, males and females form long-term bonds, mating and rearing their chicks together. Yet quite often the females will sneak off to mate with other birds, with the result that male birds will sometimes rear chicks that are not their own. In some species, females can receive the sperm of many different males and store them in receptacles. Later they can choose which sperm to use to fertilize their eggs.

These adaptations raise the reproductive success of females, and they provide a new challenge for males. Any adaptation that allows males to counter the females' adaptations may increase their own reproductive success. And so a tug of war emerges. Male birds will sometimes abandon females and their chicks, possibly because they pick up on cues that the chicks are not their own. Males have also evolved elaborate adaptations to give their own sperm the best chance of fertilizing eggs. In species where females mate with many males, the males tend to have large testes, which can produce more sperm. The increased number of sperm may raise their chances of competing successfully with the sperm of other males. Some have scrapers to remove the sperm of other males. Others plug their mates up after intercourse. Others infuse their semen with druglike substances that make their mates less receptive to other males. Those substances boost the male's reproductive success at the female's expense—they are so toxic that they shorten her lifespan.

Sexual conflict's most astonishing effects emerge long after mating—as eggs develop into embryos and then into full-grown adults. Animals typically receive two copies of each gene, one from the mother and one from the father. These copies may have different effects on the offspring.

One of the best-studied of these genes is insulin growth factor 2 (Igf2). Produced only in fetal cells, it stimulates rapid growth. Normally, only the father's copy produces proteins. To understand the gene's function, scientists disabled the father's copy in the placenta of fetal mice. The mice were born weighing 40 percent below average.

Mice also carry a gene called Igf2r. It interferes with the growth-spurring activity of Igf2. In the case of Igf2r, it is the father's gene that is silent, perhaps as a way for fathers to speed up the growth of their offspring. If the mother's copy of this second gene is disabled, mouse pups are born 125 percent heavier than average.

This pattern is apparently the result of the unconscious struggle between the sexes. A male mouse's reproductive success will be boosted if his offspring grow as big as possible, because bigger mouse pups are more likely to survive till adulthood. On the other hand, bigger mouse pups can be hard on a mother's health because of the demands they make on her for energy. In the long run, a female mouse's reproductive success may benefit most from producing a lot of not-quite-so-big mouse pups. The insulin genes have evolved through the struggle between males and females over how quickly their offspring should grow.

All of these insights flow from Darwin's initial realization that sexual selection might be very important. But they also undermine the last chapters of his book, where he tries to explain human nature in light of sexual selection. Animals have not evolved according to the rules of propriety in Victorian England. As I'll explain in further detail in the next section, one cannot assume that humans have either.

From CHAPTER 8:
Principles of Sexual Selection; CHAPTER 13: Secondary Sexual Characters of Birds; and CHAPTER 16: Birds Concluded

WITH ANIMALS which have their sexes separated, the males necessarily differ from the females in their organs of reproduction; and these are the primary sexual characters. But the sexes often differ in what Hunter has called secondary sexual characters, which are not directly connected with the act of reproduction; for instance, the male possesses certain organs of sense or locomotion, of which the female is quite destitute, or has them more highly-developed, in order that he may readily find or reach her; or again the male has special organs of prehension for holding her securely. These latter organs, of infinitely diversified kinds, graduate into those which are commonly ranked as primary, and in some cases can hardly be distinguished from them; we see instances of this in the complex appendages at the apex of the abdomen in male insects. Unless indeed we confine the term 'primary' to the reproductive glands, it is scarcely possible to decide which ought to be called primary and which secondary.

The female often differs from the male in having organs for the nourishment or protection of her young, such as the mammary glands of mammals, and the abdominal sacks of the marsupials. In some few cases also the male possesses similar organs, which are wanting in the female, such as the receptacles for the ova in certain male fishes, and those temporarily developed in certain male frogs. The females of most bees are provided with a special apparatus for collecting and carrying pollen, and their ovipositor is modified into a sting for the defence of the larvae and the community. Many similar cases could be given, but they do not here concern us. There are, however, other sexual differences quite unconnected with the primary reproductive organs, and it is with these that we are more especially concerned—such as the greater size, strength, and pugnacity of the male, his weapons of offence or means of defence against rivals, his gaudy colouring and various ornaments, his power of song, and other such characters.

Besides the primary and secondary sexual differences, such as the foregoing, the males and females of some animals differ in structures related to different habits of life, and not at all, or only indirectly, to the reproductive functions. Thus the females of certain flies (Culicidae and Tabanidae) are blood-suckers, whilst the males, living on flowers, have mouths destitute of mandibles.[1] The males of certain moths and of some crustaceans (*e.g.* Tanais) have imperfect, closed mouths, and cannot feed. The complemental males of certain Cirripedes live like epiphytic plants either on the female or the hermaphrodite form, and are destitute of a mouth and of prehensile limbs. In these cases it is the male which has been modified, and has lost certain important organs, which the females possess. In other cases it is the female which has lost such parts; for instance, the female glow-worm is destitute of wings, as also are many female moths, some of which never leave their cocoons. Many female parasitic crustaceans have lost their natatory legs. In some weevil-beetles (Curculionidae) there is a great difference between the male and female in the length of the rostrum or snout;[2] but the meaning of this and of many analogous differences, is not at all understood. Differences of structure between the two

---

1. Westwood, 'Modern Class. of Insects', vol. ii. 1840, p. 541. For the statement about Tanais, mentioned below, I am indebted to Fritz Müller.
2. Kirby and Spence, 'Introduction to Entomology', vol. iii. 1826, p. 309.

sexes in relation to different habits of life are generally confined to the lower animals; but with some few birds the beak of the male differs from that of the female. In the Huia of New Zealand the difference is wonderfully great, and we hear from Dr Buller[3] that the male uses his strong beak in chiselling the larvae of insects out of decayed wood, whilst the female probes the softer parts with her far longer, much curved and pliant beak: and thus they mutually aid each other. In most cases, differences of structure between the sexes are more or less directly connected with the propagation of the species: thus a female, which has to nourish a multitude of ova, requires more food than the male, and consequently requires special means for procuring it. A male animal, which lives for a very short time, might lose its organs for procuring food through disuse, without detriment; but he would retain his locomotive organs in a perfect state, so that he might reach the female. The female, on the other hand, might safely lose her organs for flying, swimming, or walking, if she gradually acquired habits which rendered such powers useless.

We are, however, here concerned only with sexual selection. This depends on the advantage which certain individuals have over others of the same sex and species solely in respect of reproduction. When, as in the cases above mentioned, the two sexes differ in structure in relation to different habits of life, they have no doubt been modified through natural selection, and by inheritance limited to one and the same sex. So again the primary sexual organs, and those for nourishing or protecting the young, come under the same influence; for those individuals which generated or nourished their offspring best, would leave, *caeteris paribus*, the greatest number to inherit their superiority; whilst those which generated or nourished their offspring badly, would leave but few to inherit their weaker powers. As the male has

3. 'Birds of New Zealand', 1872, p. 66.

to find the female, he requires organs of sense and locomotion, but if these organs are necessary for the other purposes of life, as is generally the case, they will have been developed through natural selection. When the male has found the female, he some-times absolutely requires prehensile organs to hold her; thus Dr Wallace informs me that the males of certain moths cannot unite with the females if their tarsi or feet are broken. The males of many oceanic crustaceans, when adult, have their legs and antennae modified in an extraordinary manner for the prehension of the female; hence we may suspect that it is because these animals are washed about by the waves of the open sea, that they require these organs in order to propagate their kind, and if so, their development has been the result of ordinary or natural selection. Some animals extremely low in the scale have been modified for this same purpose; thus the males of certain parasitic worms, when fully grown, have the lower surface of the terminal part of their bodies roughened like a rasp, and with this they coil round and permanently hold the females.[4]

When the two sexes follow exactly the same habits of life, and the male has the sensory or locomotive organs more highly developed than those of the female, it may be that the perfec-tion of these is indispensable to the male for finding the female; but in the vast majority of cases, they serve only to give one male an advantage over another, for with sufficient time, the less well-endowed males would succeed in pairing with the females;

4. M. Perrier advances this case ('Revue Scientifique', Feb. 1, 1873, p. 865) as one fatal to the belief in sexual selection, inasmuch as he supposes that I attribute all the differences between the sexes to sexual selection. This distinguished naturalist, therefore, like so many other Frenchmen, has not taken the trouble to under-stand even the first principles of sexual selection. An English naturalist insists that the claspers of certain male animals could not have been developed through the choice of the female! Had I not met with this remark, I should not have thought it possible for any one to have read this chapter and to have imagined that I main-tain that the choice of the female had anything to do with the development of the prehensile organs in the male.

and judging from the structure of the female, they would be in all other respects equally well adapted for their ordinary habits of life. Since in such cases the males have acquired their present structure, not from being better fitted to survive in the struggle for existence, but from having gained an advantage over other males, and from having transmitted this advantage to their male offspring alone, sexual selection must here have come into action. It was the importance of this distinction which led me to designate this form of selection as Sexual Selection. So again, if the chief service rendered to the male by his prehensile organs is to prevent the escape of the female before the arrival of other males, or when assaulted by them, these organs will have been perfected through sexual selection, that is by the advantage acquired by certain individuals over their rivals. But in most cases of this kind it is impossible to distinguish between the effects of natural and sexual selection. Whole chapters could be filled with details on the differences between the sexes in their sensory, locomotive, and prehensile organs. As, however, these structures are not more interesting than others adapted for the ordinary purposes of life I shall pass them over almost entirely, giving only a few instances under each class.

There are many other structures and instincts which must have been developed through sexual selection—such as the weapons of offence and the means of defence of the males for fighting with and driving away their rivals—their courage and pugnacity—their various ornaments—their contrivances for producing vocal or instrumental music—and their glands for emitting odours, most of these latter structures serving only to allure or excite the female. It is clear that these characters are the result of sexual and not of ordinary selection, since unarmed, unornamented, or unattractive males would succeed equally well in the battle for life and in leaving a numerous progeny, but for the presence of better endowed males. We may infer that this would be the case, because the females, which are unarmed and unornamented, are able to survive and procreate their kind. Secondary sexual characters of the kind just referred to, will be fully discussed in the following chapters, as being in many respects interesting, but especially as depending on the will, choice, and rivalry of the individuals of either sex. When we behold two males fighting for the possession of the female, or several male birds displaying their gorgeous plumage, and performing strange antics before an assembled body of females, we cannot doubt that, though led by instinct, they know what they are about, and consciously exert their mental and bodily powers.

Just as man can improve the breed of his game-cocks by the selection of those birds which are victorious in the cockpit, so it appears that the strongest and most vigorous males, or those provided with the best weapons, have prevailed under nature, and have led to the improvement of the natural breed or species. A slight degree of variability leading to some advantage, however slight, in reiterated deadly contests would suffice for the work of sexual selection; and it is certain that secondary sexual characters are eminently variable. Just as man can give beauty, according to his standard of taste, to his male poultry, or more strictly can modify the beauty originally acquired by the parent species, can give to the Sebright bantam a new and elegant plumage, an erect and peculiar carriage—so it appears that female birds in a state of nature, have by a long selection of the more attractive males, added to their beauty or other attractive qualities. No doubt this implies powers of discrimination and taste on the part of the female which will at first appear extremely improbable; but by the facts to be adduced hereafter, I hope to be able to shew that the females actually have these powers. When, however, it is said that the lower animals have a sense of beauty, it must not be supposed that such sense is comparable with that of a cultivated man, with his multiform and complex associated ideas. A more just comparison would be between the taste for the beautiful in animals, and that in the lowest savages, who admire and deck themselves with any brilliant, glittering, or curious object.

From our ignorance on several points, the precise manner in which sexual selection acts is somewhat uncertain. Nevertheless if those naturalists who already believe in the mutability of species, will read the following chapters, they will, I think, agree with me, that sexual selection has played an important part in the history of the organic world. It is certain that amongst almost all animals there is a struggle between the males for the possession of the female. This fact is so notorious that it would be superfluous to give instances. Hence the females have the opportunity of selecting one out of several males, on the supposition that their mental capacity suffices for the exertion of a choice. In many cases special circumstances tend to make the struggle between the males particularly severe. Thus the males of our migratory birds generally arrive at their places of breeding before the females, so that many males are ready to contend for each female. I am informed by Mr Jenner Weir, that the bird-catchers assert that this is invariably the case with the nightingale and blackcap, and with respect to the latter he can himself confirm the statement.

Mr Swaysland of Brighton has been in the habit, during the last forty years, of catching our migratory birds on their first arrival, and he has never known the females of any species to arrive before their males. During one spring he shot thirty-nine males of Ray's wagtail (*Budytes Raii*) before he saw a single female. Mr Gould has ascertained by the dissection of those snipes which arrive the first in this country, that the males come before the females. And the like holds good with most of the migratory birds of the United States.[5] The majority of the male salmon in

5. J. A. Allen, on the 'Mammals and Winter Birds of Florida', Bull. Comp. Zoology, Harvard College, p. 268.

our rivers, on coming up from the sea, are ready to breed before the females. So it appears to be with frogs and toads. Throughout the great class of insects the males almost always are the first to emerge from the pupal state, so that they generally abound for a time before any females can be seen.[6] The cause of this difference between the males and females in their periods of arrival and maturity is sufficiently obvious. Those males which annually first migrated into any country, or which in the spring were first ready to breed, or were the most eager, would leave the largest number of offspring; and these would tend to inherit similar instincts and constitutions. It must be borne in mind that it would have been impossible to change very materially the time of sexual maturity in the females, without at the same time interfering with the period of the production of the young—a period which must be determined by the seasons of the year. On the whole there can be no doubt that with almost all animals, in which the sexes are separate, there is a constantly recurrent struggle between the males for the possession of the females.

Our difficulty in regard to sexual selection lies in understanding how it is that the males which conquer other males, or those which prove the most attractive to the females, leave a greater number of offspring to inherit their superiority than their beaten and less attractive rivals. Unless this result does follow, the characters which give to certain males an advantage over others, could not be perfected and augmented through sexual selection. When the sexes exist in exactly equal numbers, the worst-endowed males will (except where polygamy prevails), ultimately find females, and leave as many offspring, as

6. Even with those plants in which the sexes are separate, the male flowers are generally mature before the female. As first shewn by C. K. Sprengel, many hermaphrodite plants are dichogamous; that is, their male and female organs are not ready at the same time, so that they cannot be self-fertilised. Now in such flowers, the pollen is in general matured before the stigma, though there are exceptional cases in which the female organs are beforehand.

well fitted for their general habits of life, as the best-endowed males. From various facts and considerations, I formerly inferred that with most animals, in which secondary sexual characters are well developed, the males considerably exceeded the females in number; but this is not by any means always true. If the males were to the females as two to one, or as three to two, or even in a somewhat lower ratio, the whole affair would be simple; for the better-armed or more attractive males would leave the largest number of offspring. But after investigating, as far as possible, the numerical proportion of the sexes, I do not believe that any great inequality in number commonly exists. In most cases sexual selection appears to have been effective in the following manner.

Let us take any species, a bird for instance, and divide the females inhabiting a district into two equal bodies, the one consisting of the more vigorous and better-nourished individuals, and the other of the less vigorous and healthy. The former, there can be little doubt, would be ready to breed in the spring before the others; and this is the opinion of Mr Jenner Weir, who has carefully attended to the habits of birds during many years. There can also be no doubt that the most vigorous, best-nourished and earliest breeders would on an average succeed in rearing the largest number of fine offspring.[7] The males, as we have seen, are generally ready to breed before the females; the strongest, and with some species the best armed of the males, drive away the weaker; and the former would then unite with the more vigorous and better-nourished females, because they are the first to

7. Here is excellent evidence on the character of the offspring from an experienced ornithologist. Mr J. A. Allen, in speaking ('Mammals and Winter Birds of E. Florida', p. 229) of the later broods, after the accidental destruction of the first, says, that these 'are found to be smaller and paler-coloured than those hatched earlier in the season. In cases where several broods are reared each year, as a general rule the birds of the earlier broods seem in all respects the most perfect and vigorous'.

breed.[8] Such vigorous pairs would surely rear a larger number of offspring than the retarded females, which would be compelled to unite with the conquered and less powerful males, supposing the sexes to be numerically equal; and this is all that is wanted to add, in the course of successive generations, to the size, strength and courage of the males, or to improve their weapons.

But in very many cases the males which conquer their rivals, do not obtain possession of the females, independently of the choice of the latter. The courtship of animals is by no means so simple and short an affair as might be thought. The females are most excited by, or prefer pairing with, the more ornamented males, or those which are the best songsters, or play the best antics; but it is obviously probable that they would at the same time prefer the more vigorous and lively males, and this has in some cases been confirmed by actual observation.[9] Thus the more vigorous females, which are the first to breed, will have the choice of many males; and though they may not always select the strongest or best armed, they will select those which are vigorous and well armed, and in other respects the most attractive. Both sexes, therefore, of such early pairs would as above explained, have an advantage over others in rearing offspring; and this apparently has sufficed during a long course of generations to add not only to the strength and fighting powers of the males, but likewise to their various ornaments or other attractions.

In the converse and much rarer case of the males selecting particular females, it is plain that those which were the most vigorous and had conquered others, would have the freest choice; and it is

8. Hermann Müller has come to this same conclusion with respect to those female bees which are the first to emerge from the pupa each year. See his remarkable essay, 'Anwendung den Darwin'schen Lehre auf Bienen', 'Verh. d. V. Jahrg.', xxix. p. 45.

9. With respect to poultry, I have received information, hereafter to be given, to this effect. Even with birds, such as pigeons, which pair for life, the female, as I hear from Mr Jenner Weir, will desert her mate if he is injured or grows weak.

almost certain that they would select vigorous as well as attractive females. Such pairs would have an advantage in rearing offspring, more especially if the male had the power to defend the female during the pairing-season as occurs with some of the higher animals, or aided her in providing for the young. The same principles would apply if each sex preferred and selected certain individuals of the opposite sex; supposing that they selected not only the more attractive, but likewise the more vigorous individuals.

Numerical Proportion of the Two Sexes—I have remarked that sexual selection would be a simple affair if the males were considerably more numerous than the females. Hence I was led to investigate, as far as I could, the proportions between the two sexes of as many animals as possible; but the materials are scanty. I will here give only a brief abstract of the results, retaining the details for a supplementary discussion, so as not to interfere with the course of my argument. Domesticated animals alone afford the means of ascertaining the proportional numbers at birth; but no records have been specially kept for this purpose. By indirect means, however, I have collected a considerable body of statistics, from which it appears that with most of our domestic animals the sexes are nearly equal at birth. Thus 25,560 births of race-horses have been recorded during twenty-one years, and the male births were to the female births as 99.7 to 100. In greyhounds the inequality is greater than with any other animal, for out of 6878 births during twelve years, the male births were to the female as 110.1 to 100. It is, however, in some degree doubtful whether it is safe to infer that the proportion would be the same under natural conditions as under domestication; for slight and unknown differences in the conditions affect the proportion of the sexes. Thus with mankind, the male births in England are as 104.5, in Russia as 108.9, and with the Jews of Livonia as 120, to 100 female births. But I shall recur to this curious point of the excess of male births in the supplement to this chapter. At the Cape of Good Hope, however, male children of European extraction have been born during several years in the proportion of between 90 and 99 to 100 female children.

For our present purpose we are concerned with the proportion of the sexes, not only at birth, but also at maturity, and this adds another element of doubt; for it is a well-ascertained fact that with man the number of males dying before or during birth, and during the first few years of infancy, is considerably larger than that of females. So it almost certainly is with male lambs, and probably with some other animals. The males of some species kill one another by fighting; or they drive one another about until they become greatly emaciated. They must also be often exposed to various dangers, whilst wandering about in eager search for the females. In many kinds of fish the males are much smaller than the females, and they are believed often to be devoured by the latter, or by other fishes. The females of some birds appear to die earlier than the males; they are also liable to be destroyed on their nests, or whilst in charge of their young. With insects the female larvae are often larger than those of the males, and would consequently be more likely to be devoured. In some cases the mature females are less active and less rapid in their movements than the males, and could not escape so well from danger. Hence, with animals in a state of nature, we must rely on mere estimation, in order to judge of the proportions of the sexes at maturity; and this is but little trustworthy, except when the inequality is strongly marked. Nevertheless, as far as a judgment can be formed, we may conclude from the facts given in the supplement, that the males of some few mammals, of many birds, of some fish and insects, are considerably more numerous than the females.

The proportion between the sexes fluctuates slightly during successive years: thus with race-horses, for every 100 mares born the stallions varied from 107.1 in one year to 92.6 in another year, and with greyhounds from 116.3 to 95.3. But had larger numbers been tabulated throughout an area more extensive than England, these fluctuations would probably have disappeared; and such as they are, would hardly suffice to lead to effective sexual selection in a state of nature. Nevertheless, in the cases of some few wild animals, as shewn in the supplement, the proportions seem to fluctuate either during different seasons or in different localities in a sufficient degree to lead to such selection. For it should be observed that any advantage gained during certain years or in certain localities by those males which were able to conquer their rivals, or were the most attractive to the females, would probably be transmitted to the offspring, and would not subsequently be eliminated. During the succeeding seasons, when, from the equality of the sexes, every male was able to procure a female, the stronger or more attractive males previously produced would still have at least as good a chance of leaving offspring as the weaker or less attractive.

Polygamy—The practice of polygamy leads to the same results as would follow from an actual inequality in the number of the sexes; for if each male secures two or more females, many males cannot pair; and the latter assuredly will be the weaker or less attractive individuals. Many mammals and some few birds are polygamous, but with animals belonging to the lower classes I have found no evidence of this habit. The intellectual powers of such animals are, perhaps, not sufficient to lead them to collect and guard a harem of females. That some relation exists between polygamy and the development of secondary sexual characters, appears nearly certain; and this supports the view that a numerical preponderance of males would be eminently favourable to the action of sexual selection. Nevertheless many animals, which are strictly monogamous, especially birds, display strongly-marked secondary sexual characters; whilst some few animals, which are polygamous, do not have such characters.

We will first briefly run through the mammals, and then turn to birds. The gorilla seems to be polygamous, and the male differs considerably from the female; so it is with some baboons, which live in herds containing twice as many adult females as males. In South America the *Mycetes caraya* presents well-marked sexual differences, in colour, beard, and vocal organs; and the male generally lives with two or three wives: the male of the *Cebus capucinus* differs somewhat from the female, and appears to be polygamous.[10] Little is known on this head with respect to most other monkeys, but some species are strictly monogamous. The ruminants are eminently polygamous, and they present sexual differences more frequently than almost any other group of mammals; this holds good, especially in their weapons, but also in other characters. Most deer, cattle, and sheep are polygamous; as are most antelopes, though some are monogamous. Sir Andrew Smith, in speaking of the antelopes of South Africa, says that in herds of about a dozen there was rarely more than one mature male. The Asiatic *Antilope saiga* appears to be the most inordinate polygamist in the world; for Pallas[11] states that the male drives away all rivals, and collects a herd of about a hundred females and kids together; the female is hornless and has softer hair, but does not otherwise differ much from the male. The wild horse of the Falkland Islands and of the Western States of N. America is polygamous, but, except in his greater size and in the proportions of his body, differs but little from the mare. The wild boar presents well-marked sexual characters, in his great

10. On the Gorilla, Savage and Wyman. 'Boston Journal of Nat. Hist.', vol. v. 1845–47, p. 423. On Cynocephalus, Brehm, 'Illust. Thierleben', B. i. 1864, s. 77. On Mycetes, Rengger, 'Naturgesch.: Säugethiere von Paraguay', 1830, s. 14, 20. Cebus, Brehm, ibid. s. 108.

11. Pallas, 'Spicilegia Zoolog.', Fasc. xii. 1777, p. 29. Sir Andrew Smith, 'Illustrations of the Zoology of S. Africa', 1849, pl. 29, on the Kobus. Owen, in his 'Anatomy of Vertebrates' (vol. iii. 1868, p. 633) gives a table shewing incidentally which species of antelopes are gregarious.

tusks and some other points. In Europe and in India he leads a solitary life, except during the breeding-season; but as is believed by Sir W. Elliot, who has had many opportunities in India of observing this animal, he consorts at this season with several females. Whether this holds good in Europe is doubtful, but it is supported by some evidence. The adult male Indian elephant, like the boar, passes much of his time in solitude; but as Dr Campbell states, when with others, 'it is rare to find more than one male with a whole herd of females', the larger males expelling or killing the smaller and weaker ones. The male differs from the female in his immense tusks, greater size, strength, and endurance; so great is the difference in these respects, that the males when caught are valued at one-fifth more than the females.[12] The sexes of other pachydermatous animals differ very little or not at all, and, as far as known, they are not polygamists. Nor have I heard of any species in the Orders of Cheiroptera, Edentata, Insectivora and Rodents being polygamous, excepting that amongst the Rodents, the common rat, according to some rat-catchers, lives with several females. Nevertheless the two sexes of some sloths (Edentata) differ in the character and colour of certain patches of hair on their shoulders.[13] And many kinds of bats (Cheiroptera) present well-marked sexual differences, chiefly in the males possessing odoriferous glands and pouches, and by their being of a lighter colour.[14] In the great order of Rodents, as far as I can learn, the sexes rarely differ, and when they do so, it is but slightly in the tint of the fur.

As I hear from Sir Andrew Smith, the lion in South Africa sometimes lives with a single female, but generally with more, and, in one case, was found with as many as five females; so that he is polygamous. As far as I can discover, he is the only

---

12. Dr Campbell, in 'Proc. Zoolog. Soc.', 1869, p. 138. See also an interesting paper, by Lieut. Johnstone, in 'Proc. Asiatic Soc. of Bengal', May, 1868.
13. Dr Gray, in 'Annals and Mag. of Nat. Hist.', 1871, p. 302.
14. See Dr Dobson's excellent paper, in 'Proc. Zoolog. Soc.', 1873, p. 241.

polygamist amongst all the terrestrial Carnivora, and he alone presents well-marked sexual characters. If, however, we turn to the marine Carnivora, as we shall hereafter see, the case is widely different; for many species of seals offer extraordinary sexual differences, and they are eminently polygamous. Thus, according to Péron, the male sea-elephant of the Southern Ocean always possesses several females, and the sea-lion of Forster is said to be surrounded by from twenty to thirty females. In the North, the male sea-bear of Steller is accompanied by even a greater number of females. It is an interesting fact, as Dr Gill remarks,[15] that in the monogamous species, 'or those living in small communities, there is little difference in size between the males and females; in the social species, or rather those of which the males have harems, the males are vastly larger than the females'.

Amongst birds, many species, the sexes of which differ greatly from each other, are certainly monogamous. In Great Britain we see well-marked sexual differences, for instance, in the wild-duck which pairs with a single female, the common blackbird, and the bullfinch which is said to pair for life. I am informed by Mr Wallace that the like is true of the Chatterers or Cotingidae of South America, and of many other birds. In several groups I have not been able to discover whether the species are polygamous or monogamous. Lesson says that birds of paradise, so remarkable for their sexual differences, are polygamous, but Mr Wallace doubts whether he had sufficient evidence. Mr Salvin tells me he has been led to believe that humming-birds are polygamous. The male widow-bird, remarkable for his caudal plumes, certainly seems to be a polygamist.[16] I have been assured by Mr Jen-

---

15. The Eared Seals, 'American Naturalist', vol. iv. Jan. 1871.
16. 'The Ibis', vol. iii. 1861, p. 133, on the Progne Widow-bird. See also on the *Vidua axillaris*, ibid. vol. ii. 1860, p. 211. On the polygamy of the Capercailzie and Great Bustard, see L. Lloyd, 'Game Birds of Sweden', 1867, pp. 19, and 182. Montagu and Selby speak of the Black Grouse as polygamous and of the Red Grouse as monogamous.

ner Weir and by others, that it is somewhat common for three starlings to frequent the same nest; but whether this is a case of polygamy or polyandry has not been ascertained.

The Gallinaceae exhibit almost as strongly marked sexual differences as birds of paradise or humming-birds, and many of the species are, as is well known, polygamous; others being strictly monogamous. What a contrast is presented between the sexes of the polygamous peacock or pheasant, and the monogamous guinea-fowl or partridge! Many similar cases could be given, as in the grouse tribe, in which the males of the polygamous capercailzie and black-cock differ greatly from the females; whilst the sexes of the monogamous red grouse and ptarmigan differ very little. In the Cursores, except amongst the bustards, few species offer strongly-marked sexual differences, and the great bustard (*Otis tarda*) is said to be polygamous. With the Grallatores, extremely few species differ sexually, but the ruff (*Machetes pugnax*) affords a marked exception, and this species is believed by Montagu to be a polygamist. Hence it appears that amongst birds there often exists a close relation between polygamy and the development of strongly-marked sexual differences. I asked Mr Bartlett, of the Zoological Gardens, who has had very large experience with birds, whether the male tragopan (one of the Gallinaceae) was polygamous, and I was struck by his answering, 'I do not know, but should think so from his splendid colours.'

It deserves notice that the instinct of pairing with a single female is easily lost under domestication. The wild-duck is strictly monogamous, the domestic-duck highly polygamous. The Rev. W. D. Fox informs me that out of some half-tamed wild-ducks, on a large pond in his neighbourhood, so many mallards were shot by the gamekeeper that only one was left for every seven or eight females; yet unusually large broods were reared. The guinea-fowl is strictly monogamous; but Mr Fox finds that his birds succeed best when he keeps one cock to two or three hens. Canary-birds pair in a state of nature, but the breeders in England successfully put one male to four or five females. I have noticed these cases, as rendering it probable that wild monogamous species might readily become either temporarily or permanently polygamous.

Too little is known of the habits of reptiles and fishes to enable us to speak of their marriage arrangements. The stickle-back (Gasterosteus), however, is said to be a polygamist;[17] and the male during the breeding season differs conspicuously from the female.

To sum up on the means through which, as far as we can judge, sexual selection has led to the development of secondary sexual characters. It has been shewn that the largest number of vigorous offspring will be reared from the pairing of the strongest and best-armed males, victorious in contests over other males, with the most vigorous and best-nourished females, which are the first to breed in the spring. If such females select the more attractive, and at the same time vigorous males, they will rear a larger number of offspring than the retarded females, which must pair with the less vigorous and less attractive males. So it will be if the more vigorous males select the more attractive and at the same time healthy and vigorous females; and this will especially hold good if the male defends the female, and aids in providing food for the young. The advantage thus gained by the more vigorous pairs in rearing a larger number of offspring has apparently sufficed to render sexual selection efficient. But a large numerical preponderance of males over females will be still more efficient; whether the preponderance is only occasional and local, or permanent; whether it occurs at birth, or afterwards from the greater destruction of the females; or whether it indirectly follows from the practice of polygamy.

The Male generally more modified than the Female—Throughout the animal kingdom, when the sexes differ in external appearance, it is,

17. Noel Humphreys, 'River Gardens', 1857.

with rare exceptions, the male which has been the more modified; for, generally, the female retains a closer resemblance to the young of her own species, and to other adult members of the same group. The cause of this seems to lie in the males of almost all animals having stronger passions than the females. Hence it is the males that fight together and sedulously display their charms before the females; and the victors transmit their superiority to their male offspring. Why both sexes do not thus acquire the characters of their fathers, will be considered hereafter. That the males of all mammals eagerly pursue the females is notorious to every one. So it is with birds; but many cock birds do not so much pursue the hen, as display their plumage, perform strange antics, and pour forth their song in her presence. The male in the few fish observed seems much more eager than the female; and the same is true of alligators, and apparently of Batrachians. Throughout the enormous class of insects, as Kirby remarks,[18] 'the law is, that the male shall seek the female'. Two good authorities, Mr Blackwall and Mr C. Spence Bate, tell me that the males of spiders and crustaceans are more active and more erratic in their habits than the females. When the organs of sense or locomotion are present in the one sex of insects and crustaceans and absent in the other, or when, as is frequently the case, they are more highly developed in the one than in the other, it is, as far as I can discover, almost invariably the male which retains such organs, or has them most developed; and this shews that the male is the more active member in the courtship of the sexes.[19]

18. Kirby and Spence, 'Introduction to Entomology', vol. iii. 1826, p. 342.
19. One parasitic Hymenopterous insect (Westwood, 'Modern Class. of Insects', vol. ii. p. 160) forms an exception to the rule, as the male has rudimentary wings, and never quits the cell in which it is born, whilst the female has well-developed wings. Audouin believes that the females of this species are impregnated by the males which are born in the same cells with them; but it is much more probable that the females visit other cells, so that close interbreeding is thus avoided. We shall hereafter meet in various classes, with a few exceptional cases, in which the female, instead of the male, is the seeker and wooer.

The female, on the other hand, with the rarest exceptions, is less eager than the male. As the illustrious Hunter[20] long ago observed, she generally 'requires to be courted'; she is coy, and may often be seen endeavouring for a long time to escape from the male. Every observer of the habits of animals will be able to call to mind instances of this kind. It is shown by various facts, given hereafter, and by the results fairly attributable to sexual selection, that the female, though comparatively passive, generally exerts some choice and accepts one male in preference to others. Or she may accept, as appearances would sometimes lead us to believe, not the male which is the most attractive to her, but the one which is the least distasteful. The exertion of some choice on the part of the female seems a law almost as general as the eagerness of the male.

We are naturally led to enquire why the male, in so many and such distinct classes, has become more eager than the female, so that he searches for her, and plays the more active part in court-ship. It would be no advantage and some loss of power if each sex searched for the other; but why would the male almost always be the seeker? The ovules of plants after fertilisation have to be nour-ished for a time; hence the pollen is necessarily brought to the female organs—being placed on the stigma, by means of insects or the wind, or by the spontaneous movements of the stamens; and in the Algae, &c., by the locomotive power of the antherozooids. With lowly-organised aquatic animals, permanently affixed to the same spot and having their sexes separate, the male element is invariably brought to the female; and of this we can see the reason, for even if the ova were detached before fertilisation, and did not require sub-sequent nourishment or protection, there would yet be greater dif-ficulty in transporting them than the male element, because, being larger than the latter, they are produced in far smaller numbers. So

20. 'Essays and Observations', edited by Owen, vol. i. 1861, p. 194.

that many of the lower animals are, in this respect, analogous with plants.[21] The males of affixed and aquatic animals having been led to emit their fertilising element in this way, it is natural that any of their descendants, which rose in the scale and became locomotive, should retain the same habit; and they would approach the female as closely as possible, in order not to risk the loss of the fertilising element in a long passage of it through the water. With some few of the lower animals, the females alone are fixed, and the males of these must be the seekers. But it is difficult to understand why the males of species, of which the progenitors were primordially free, should invariably have acquired the habit of approaching the females, instead of being approached by them. But in all cases, in order that the males should seek efficiently, it would be necessary that they should be endowed with strong passions; and the acquirement of such passions would naturally follow from the more eager leaving a larger number of offspring than the less eager.

The great eagerness of the males has thus indirectly led to their much more frequently developing secondary sexual characters than the females. But the development of such characters would be much aided, if the males were more liable to vary than the females—as I concluded they were—after a long study of domesticated animals. Von Nathusius, who has had very wide experience, is strongly of the same opinion.[22] Good evidence also in favour of this conclusion can be produced by a comparison of the two sexes in mankind. During the Novara Expedition[23] a vast number of measurements was made of various parts of the body

21. Prof. Sachs ('Lehrbuch der Botanik', 1870, s. 633) in speaking of the male and female reproductive cells, remarks, 'verhält sich die eine bei der Vereinigung activ, . . . die andere erscheint bei der Vereinigung passiv' ['the one is active in the union while the other appears to play a passive role'].
22. 'Vortrage über Viehzucht', 1872, p. 63.
23. 'Reise der Novara: Anthropolog. Theil', 1867, s. 216–269. The results were calculated by Dr Weisbach from measurements made by Drs K. Scherzer and Schwarz. On the greater variability of the males of domesticated animals, see my 'Variation of Animals and Plants under Domestication', vol, ii. 1868, p. 75.

in different races, and the men were found in almost every case to present a greater range of variation than the women; but I shall have to recur to this subject in a future chapter. Mr J. Wood,[24] who has carefully attended to the variation of the muscles in man, puts in italics the conclusion that 'the greatest number of abnormalities in each subject is found in the males'. He had previously remarked that 'altogether in 102 subjects, the varieties of redundancy were found to be half as many again as in females, contrasting widely with the greater frequency of deficiency in females before described'. Professor Macalister likewise remarks[25] that variations in the muscles 'are probably more common in males than females'. Certain muscles which are not normally present in mankind are also more frequently developed in the male than in the female sex, although exceptions to this rule are said to occur. Dr Burt Wilder[26] has tabulated the cases of 152 individuals with supernumerary digits, of which 86 were males, and 39, or less than half, females, the remaining 27 being of unknown sex. It should not, however, be overlooked that women would more frequently endeavour to conceal a deformity of this kind than men. Again, Dr L. Meyer asserts that the ears of man are more variable in form than those of woman.[27] Lastly the temperature is more variable in man than in woman.[28]

The cause of the greater general variability in the male sex, than in the female is unknown, except in so far as secondary sexual characters are extraordinarily variable, and are usually confined to the males; and, as we shall presently see, this fact is to a certain extent, intelligible. Through the action of sexual and natural selection male animals have been rendered in very many instances widely

24. 'Proceedings Royal Soc.', vol. xvi. July 1868, pp. 519 and 524.
25. 'Proc. Royal Irish Academy', vol. x. 1868, p. 123.
26. 'Massachusetts Medical Soc.', vol. ii. No. 3, 1868, p. 9.
27. 'Archiv für Path. Anat. und Phys.', 1871, p. 488.
28. The conclusions recently arrived at by Dr J. Stockton Hough, on the temperature of man, are given in the 'Pop. Science Review', Jan. 1st, 1874, p. 97.

different from their females; but independently of selection the two sexes, from differing constitutionally, tend to vary in a somewhat different manner. The female has to expend much organic matter in the formation of her ova, whereas the male expends much force in fierce contests with his rivals, in wandering about in search of the female, in exerting his voice, pouring out odoriferous secretions, &c.: and this expenditure is generally concentrated within a short period. The great vigour of the male during the season of love seems often to intensify his colours, independently of any marked difference from the female.[29] In mankind, and even as low down in the organic scale as in the Lepidoptera, the temperature of the body is higher in the male than in the female, accompanied in the case of man by a slower pulse.[30] On the whole the expenditure of matter and force by the two sexes is probably nearly equal, though effected in very different ways and at different rates.

From the causes just specified the two sexes can hardly fail to differ somewhat in constitution, at least during the breeding season; and, although they may be subjected to exactly the same conditions, they will tend to vary in a different manner. If such variations are of no service to either sex, they will not be accumulated and increased by sexual or natural selection. Nevertheless, they may become permanent if the exciting cause acts permanently; and in accordance with a frequent form of inheritance they may be transmitted to that sex alone in which they first appeared. In this case the two sexes will come to present permanent, yet unimportant, differences of character. For instance, Mr Allen shews that with a large number of birds inhabiting the

29. Prof. Mantegazza is inclined to believe ('Lettera a Carlo Darwin', 'Archivio per l'Anthropologia', 1871, p. 306) that the bright colours, common in so many male animals, are due to the presence and retention by them of the spermatic fluid; but this can hardly be the case; for many male birds, for instance young pheasants, become brightly coloured in the autumn of their first year.

30. For mankind, see Dr J. Stockton Hough, whose conclusions are given in the 'Pop. Science Review', 1874, p. 97. See Girard's observations on the Lepidoptera, as given in the 'Zoological Record', 1869, p. 347.

northern and southern United States, the specimens from the south are darker-coloured than those from the north; and this seems to be the direct result of the difference in temperature, light, &c., between the two regions. Now, in some few cases, the two sexes of the same species appear to have been differently affected; in the *Agelaeus phoeniceus* the males have had their colours greatly intensified in the south; whereas with *Cardinalis virginianus* it is the females which have been thus affected; with *Quiscalus major* the females have been rendered extremely variable in tint, whilst the males remain nearly uniform.[31]

A few exceptional cases occur in various classes of animals, in which the females instead of the males have acquired well pronounced secondary sexual characters, such as brighter colours, greater size, strength, or pugnacity. With birds there has sometimes been a complete transposition of the ordinary characters proper to each sex; the females having become the more eager in courtship, the males remaining comparatively passive, but apparently selecting the more attractive females, as we may infer from the results. Certain hen birds have thus been rendered more highly coloured or otherwise ornamented, as well as more powerful and pugnacious than the cocks; these characters being transmitted to the female offspring alone.

It may be suggested that in some cases a double process of selection has been carried on; that the males have selected the more attractive females, and the latter the more attractive males. This process, however, though it might lead to the modification of both sexes, would not make the one sex different from the other, unless indeed their tastes for the beautiful differed; but this is a supposition too improbable to be worth considering in the case of any animal, excepting man. There are, however, many animals in which the sexes resemble each other, both being furnished with

31. 'Mammals and Birds of E. Florida', pp. 234, 280, 295.

the same ornaments, which analogy would lead us to attribute to the agency of sexual selection. In such cases it may be suggested with more plausibility, that there has been a double or mutual process of sexual selection; the more vigorous and precocious females selecting the more attractive and vigorous males, the latter rejecting all except the more attractive females. But from what we know of the habits of animals, this view is hardly probable, for the male is generally eager to pair with any female. It is more probable that the ornaments common to both sexes were acquired by one sex, generally the male, and then transmitted to the offspring of both sexes. If, indeed, during a lengthened period the males of any species were greatly to exceed the females in number, and then during another lengthened period, but under different conditions, the reverse were to occur, a double, but not simultaneous, process of sexual selection might easily be carried on, by which the two sexes might be rendered widely different.

We shall hereafter see that many animals exist, of which neither sex is brilliantly coloured or provided with special ornaments, and yet the members of both sexes or of one alone have probably acquired simple colours, such as white or black, through sexual selection. The absence of bright tints or other ornaments may be the result of variations of the right kind never having occurred, or of the animals themselves having preferred plain black or white. Obscure tints have often been developed through natural selection for the sake of protection, and the acquirement through sexual selection of conspicuous colours, appears to have been sometimes checked from the danger thus incurred. But in other cases the males during long ages may have struggled together for the possession of the females, and yet no effect will have been produced, unless a larger number of offspring were left by the more successful males to inherit their superiority, than by the less successful: and this, as previously shewn, depends on many complex contingencies.

Sexual selection acts in a less rigorous manner than natural selection. The latter produces its effects by the life or death at all ages of the more or less successful individuals. Death, indeed, not rarely ensues from the conflicts of rival males. But generally the less successful male merely fails to obtain a female, or obtains a retarded and less vigorous female later in the season, or, if polygamous, obtains fewer females; so that they leave fewer, less vigorous, or no offspring. In regard to structures acquired through ordinary or natural selection, there is in most cases, as long as the conditions of life remain the same, a limit to the amount of advantageous modification in relation to certain special purposes; but in regard to structures adapted to make one male victorious over another, either in fighting or in charming the female, there is no definite limit to the amount of advantageous modification; so that as long as the proper variations arise the work of sexual selection will go on. This circumstance may partly account for the frequent and extraordinary amount of variability presented by secondary sexual characters. Nevertheless, natural selection will determine that such characters shall not be acquired by the victorious males, if they would be highly injurious, either by expending too much of their vital powers, or by exposing them to any great danger. The development, however, of certain structures—of the horns, for instance, in certain stags—has been carried to a wonderful extreme; and in some cases to an extreme which, as far as the general conditions of life are concerned, must be slightly injurious to the male. From this fact we learn that the advantages which favoured males derive from conquering other males in battle or courtship, and thus leaving a numerous progeny, are in the long run greater than those derived from rather more perfect adaptation to their conditions of life. We shall further see, and it could never have been anticipated, that the power to charm the female has sometimes been more important than the power to conquer other males in battle.

CHAPTER 13:
Secondary Sexual Characters of Birds

SECONDARY SEXUAL CHARACTERS are more diversified and conspicuous in birds, though not perhaps entailing more important
changes of structure, than in any other class of animals. I shall,
therefore, treat the subject at considerable length. Male birds
sometimes, though rarely, possess special weapons for fighting
with each other. They charm the female by vocal or instrumental
music of the most varied kinds. They are ornamented by all sorts
of combs, wattles, protuberances, horns, air-distended sacks, top-
knots, naked shafts, plumes and lengthened feathers gracefully
springing from all parts of the body. The beak and naked skin
about the head, and the feathers are often gorgeously coloured.
The males sometimes pay their court by dancing, or by fantastic antics performed either on the ground or in the air. In one
instance, at least, the male emits a musky odour, which we may
suppose serves to charm or excite the female; for that excellent
observer, Mr Ramsay,[1] says of the Australian musk-duck (*Biziura
lobata*) that 'the smell which the male emits during the summer months is confined to that sex, and in some individuals is
retained throughout the year; I have never, even in the breeding-
season, shot a female which had any smell of musk'. So powerful

1. 'Ibis', vol. iii. (new series) 1867, p. 414.

is this odour during the pairing-season, that it can be detected long before the bird can be seen.[2] On the whole, birds appear to be the most aesthetic of all animals, excepting of course man, and they have nearly the same taste for the beautiful as we have. This is shewn by our enjoyment of the singing of birds, and by our women, both civilised and savage, decking their heads with borrowed plumes, and using gems which are hardly more brilliantly coloured than the naked skin and wattles of certain birds. In man, however, when cultivated, the sense of beauty is manifestly a far more complex feeling, and is associated with various intellectual ideas.

Before treating of the sexual characters with which we are here more particularly concerned, I may just allude to certain differences between the sexes which apparently depend on differences in their habits of life; for such cases, though common in the lower, are rare in the higher classes. Two humming-birds belonging to the genus Eustephanus, which inhabit the island of Juan Fernandez, were long thought to be specifically distinct, but are now known, as Mr Gould informs me, to be the male and female of the same species, and they differ slightly in the form of the beak. In another genus of humming-birds (*Grypus*), the beak of the male is serrated along the margin and hooked at the extremity, thus differing much from that of the female. In the Neomorpha of New Zealand, there is, as we have seen, a still wider difference in the form of the beak in relation to the manner of feeding of the two sexes. Something of the same kind has been observed with the goldfinch (*Carduelis elegans*), for I am assured by Mr J. Jenner Weir that the birdcatchers can distinguish the males by their slightly longer beaks. The flocks of males are often found feeding on the seeds of the teazle (Dipsacus), which they can reach with their elongated beaks, whilst the females

2. Gould, 'Handbook to the Birds of Australia', 1865, vol. ii. p. 383.

more commonly feed on the seeds of the betony or Scrophularia. With a slight difference of this kind as a foundation, we can see how the beaks of the two sexes might be made to differ greatly through natural selection. In some of the above cases, however, it is possible that the beaks of the males may have been first modified in relation to their contests with other males; and that this afterwards led to slightly changed habits of life.

Law of Battle—Almost all male birds are extremely pugnacious, using their beaks, wings, and legs for fighting together. We see this every spring with our robins and sparrows. The smallest of all birds, namely the humming-bird, is one of the most quarrelsome. Mr Gosse[3] describes a battle in which a pair seized hold of each other's beaks, and whirled round and round, till they almost fell to the ground; and M. Montes de Oca, in speaking of another genus of humming-bird, says that two males rarely meet without a fierce aërial encounter: when kept in cages 'their fighting has mostly ended in the splitting of the tongue of one of the two, which then surely dies from being unable to feed'.[4] With Waders, the males of the common water-hen (*Gallinula chloropus*) 'when pairing, fight violently for the females: they stand nearly upright in the water and strike with their feet'. Two were seen to be thus engaged for half an hour, until one got hold of the head of the other, which would have been killed, had not the observer interfered; the female all the time looking on as a quiet spectator.[5] Mr Blyth informs me that the males of an allied bird (*Gallicrex cristatus*) are a third larger than the females, and are so pugnacious during the breeding-season, that they are kept by the natives of Eastern Bengal for the sake of fighting. Various other birds are kept in India for the

3. Quoted by Mr Gould, 'Introduction to the Trochilidae', 1861, p. 29.
4. Gould, ibid. p. 52.
5. W. Thompson, 'Nat. Hist. of Ireland: Birds', vol. ii. 1850, p. 327.

Fig. 37. The Ruff or Machetes pugnax (from Brehm's 'Thierleben').

same purpose, for instance, the bulbuls (*Pycnonotus haemorrhous*) which 'fight with great spirit'.[6]

The polygamous ruff (*Machetes pugnax*, fig. 37) is notorious for his extreme pugnacity; and in the spring, the males, which are considerably larger than the females, congregate day after day at a particular spot, where the females propose to lay their eggs. The fowlers discover these spots by the turf being trampled somewhat bare. Here they fight very much like game-cocks, seizing each other with their beaks and striking with their wings. The great ruff of feathers round the neck is then erected, and according to Col. Montagu 'sweeps the ground as a shield to defend the more tender parts', and this is the only instance known to me in the case of birds, of any structure serving as a shield. The ruff of feathers, however, from its varied and rich colours probably serves in chief part as an ornament. Like most pugnacious birds,

6. Jerdon, 'Birds of India', 1863, vol. ii. p. 96.

they seem always ready to fight, and when closely confined often kill each other; but Montagu observed that their pugnacity becomes greater during the spring, when the long feathers on their necks are fully developed; and at this period the least movement by any one bird provokes a general battle.[7] Of the pugnacity of web-footed birds, two instances will suffice: in Guiana 'bloody fights occur during the breeding-season between the males of the wild musk-duck (*Cairina moschata*); and where these fights have occurred the river is covered for some distance with feathers'.[8] Birds which seem ill-adapted for fighting engage in fierce conflicts; thus the stronger males of the pelican drive away the weaker ones, snapping with their huge beaks and giving heavy blows with their wings. Male snipe fight together, 'tugging and pushing each other with their bills in the most curious manner imaginable'. Some few birds are believed never to fight; this is the case, according to Audubon, with one of the woodpeckers of the United States (*Picus auratus*), although 'the hens are followed by even half a dozen of their gay suitors'.[9]

The males of many birds are larger than the females, and this no doubt is the result of the advantage gained by the larger and stronger males over their rivals during many generations. The difference in size between the two sexes is carried to an extreme point in several Australian species; thus the male musk-duck (Biziura) and the male *Cincloramphus cruralis* (allied to our pipits) are by measurement actually twice as large as their respective females.[10] With many other birds the females are larger than the males; and as formerly remarked, the explanation often given, namely, that the females have most of the work in feeding their

---

7. Macgillivray, 'Hist. Brit. Birds', vol. iv. 1852, pp. 177–181.
8. Sir R. Schomburgk, in 'Journal of R. Geograph. Soc.', vol. xiii. 1843, p. 31.
9. 'Ornithological Biography', vol. i. p. 191. For pelicans and snipes, see vol. iii. pp. 138, 477.
10. Gould, 'Handbook of Birds of Australia', vol. i. p. 395, vol. ii. p. 383.

young, will not suffice. In some few cases, as we shall hereafter see, the females apparently have acquired their greater size and strength for the sake of conquering other females and obtaining possession of the males.

The males of many gallinaceous birds, especially of the polygamous kinds, are furnished with special weapons for fighting with their rivals, namely spurs, which can be used with fearful effect. It has been recorded by a trustworthy writer[11] that in Derbyshire a kite struck at a game-hen accompanied by her chickens, when the cock rushed to the rescue, and drove his spur right through the eye and skull of the aggressor. The spur was with difficulty drawn from the skull, and as the kite though dead retained his grasp, the two birds were firmly locked together; but the cock when disentangled was very little injured. The invincible courage of the game-cock is notorious: a gentleman who long ago witnessed the brutal scene, told me that a bird had both its legs broken by some accident in the cockpit, and the owner laid a wager that if the legs could be spliced so that the bird could stand upright, he would continue fighting. This was effected on the spot, and the bird fought with undaunted courage until he received his death-stroke. In Ceylon a closely allied, wild species, the *Gallus Stanleyi*, is known to fight desperately 'in defence of his seraglio', so that one of the combatants is frequently found dead.[12] An Indian partridge (*Ortygornis gularis*), the male of which is furnished with strong and sharp spurs, is so quarrelsome, 'that the scars of former fights disfigure the breast of almost every bird you kill'.[13]

The males of almost all gallinaceous birds, even those which are not furnished with spurs, engage during the breeding-season in fierce conflicts. The Capercailzie and Black-cock (*Tetrao urogallus*

11. Mr Hewitt in the 'Poultry Book by Tegetmeier', 1866, p. 137.
12. Layard, 'Annals and Mag. of Nat. Hist.' vol. xiv. 1854, p. 63.
13. Jerdon, 'Birds of India', vol. iii. p. 574.

and *T. tetrix*), which are both polygamists, have regular appointed places, where during many weeks they congregate in numbers to fight together and to display their charms before the females. Dr W. Kovalevsky informs me that in Russia he has seen the snow all bloody on the arenas where the capercailzie have fought; and the black-cocks 'make the feathers fly in every direction', when several 'engage in a battle royal'. The elder Brehm gives a curious account of the Balz, as the love-dances and love-songs of the Black-cock are called in Germany. The bird utters almost continuously the strangest noises: 'he holds his tail up and spreads it out like a fan, he lifts up his head and neck with all the feathers erect, and stretches his wings from the body. Then he takes a few jumps in different directions, sometimes in a circle, and presses the under part of his beak so hard against the ground that the chin feathers are rubbed off. During these movements he beats his wings and turns round and round. The more ardent he grows the more lively he becomes, until at last the bird appears like a frantic creature'. At such times the black-cocks are so absorbed that they become almost blind and deaf, but less so than the capercailzie: hence bird after bird may be shot on the same spot, or even caught by the hand. After performing these antics the males begin to fight: and the same black-cock, in order to prove his strength over several antagonists, will visit in the course of one morning several Balz-places, which remain the same during successive years.[14]

The peacock with his long train appears more like a dandy than a warrior, but he sometimes engages in fierce contests: the Rev. W. Darwin Fox informs me that at some little distance from Chester two peacocks became so excited whilst fighting, that they flew over the whole city, still engaged, until they alighted on the top of St John's tower.

14. Brehm, 'Illust. Thierleben', 1867, B. iv. s. 351. Some of the foregoing statements are taken from L. Lloyd, 'The Game Birds of Sweden', &c., 1867, p. 79.

The spur, in those gallinaceous birds which are thus provided, is generally single; but Polyplectron (see fig. 51, p. 351 has two or more on each leg; and one of the Blood-pheasants (*Ithaginis cruentus*) has been seen with five spurs. The spurs are generally confined to the male, being represented by mere knobs or rudiments in the female; but the females of the Java peacock (*Pavo muticus*) and, as I am informed by Mr Blyth, of the small fire-backed pheasant (*Euplocamus erythropthalmus*) possess spurs. In Galloperdix it is usual for the males to have two spurs, and for the females to have only one on each leg.[15] Hence spurs may be considered as a masculine structure, which has been occasionally more or less transferred to the females. Like most other second-ary sexual characters, the spurs are highly variable, both in number and development, in the same species.

Various birds have spurs on their wings. But the Egyptian goose (*Chenalopex aegyptiacus*) has only 'bare obtuse knobs', and these probably shew us the first steps by which true spurs have been developed in other species. In the spur-winged goose, *Plectropterus gambensis*, the males have much larger spurs than the females; and they use them, as I am informed by Mr Bartlett, in fighting together, so that, in this case, the wing-spurs serve as sexual weapons; but according to Livingstone, they are chiefly used in the defence of the young. The Palamedea (fig. 38) is armed with a pair of spurs on each wing; and these are such formidable weapons, that a single blow has been known to drive a dog howling away. But it does not appear that the spurs in this case, or in that of some of the spur-winged rails, are larger in the male than in the female.[16] In certain plovers, however, the

15. Jerdon, 'Birds of India': on Ithaginis, vol. iii. p. 523; on Galloperdix, p. 541.
16. For the Egyptian goose, see Macgillivray, 'British Birds', vol. iv. p. 639. For Plectropterus, 'Livingstone's Travels', p. 254. For Palamedea, Brehm's 'Thierleben', B. iv. s. 740. See also on this bird Azara, 'Voyages dans l'Amérique mérid.', tom. iv. 1809, pp. 179, 253.

Fig. 38. Palamedea cornuta (from Brehm), shewing the double wing-spurs, and the filament on the head.

wing-spurs must be considered as a sexual character. Thus in the male of our common peewit (*Vanellus cristatus*) the tubercle on the shoulder of the wing becomes more prominent during the breeding-season, and the males fight together. In some species of Lobivanellus a similar tubercle becomes developed during the breeding-season 'into a short horny spur'. In the Australian *L. lobatus* both sexes have spurs, but these are much larger in the males than in the females. In an allied bird, the *Hoplopterus arma-*

tus, the spurs do not increase in size during the breeding-season; but these birds have been seen in Egypt to fight together, in the same manner as our peewits, by turning suddenly in the air and striking sideways at each other, sometimes with fatal results. Thus also they drive away other enemies.[17]

The season of love is that of battle; but the males of some birds, as of the game-fowl and ruff, and even the young males of the wild turkey and grouse,[18] are ready to fight whenever they meet. The presence of the female is the *teterrima belli causa*. The Bengali baboos make the pretty little males of the amadavat (*Estrelda amandava*) fight together by placing three small cages in a row, with a female in the middle; after a little time the two males are turned loose, and immediately a desperate battle ensues.[19] When many males congregate at the same appointed spot and fight together, as in the case of grouse and various other birds, they are generally attended by the females,[20] which afterwards pair with the victorious combatants. But in some cases the pairing precedes instead of succeeding the combat: thus according to Audubon,[21] several males of the Virginian goat-sucker (*Caprimulgus virginianus*) 'court, in a highly entertaining manner the female, and no sooner has she made her choice, than her approved gives chase to all intruders, and drives them beyond his dominions'. Generally the males try to drive away or kill their rivals before

17. See, on our peewit, Mr R. Carr in 'Land and Water', Aug. 8th, 1868, p. 46. In regard to Lobivanellus, see Jerdon's 'Birds of India', vol. iii. p. 647, and Gould's 'Handbook of Birds of Australia', vol. ii. p. 220. For the Holopterus, see Mr Allen in the 'Ibis', vol. v 1863, p. 156.

18. Audubon, 'Ornith. Biography', vol. ii. p. 492; vol. i. pp. 4–13.

19. Mr Blyth, 'Land and Water', 1867, p. 212.

20. Richardson on *Tetrao umbellus* 'Fauna Bor. Amer.: Birds', 1831, p. 343. L. Lloyd, 'Game Birds of Sweden', 1867, pp. 22, 79, on the capercailzie and black-cock. Brehm, however, asserts ('Thierleben', &c., B. iv. s. 352) that in Germany the grey-hens do not generally attend the Balzen of the black-cocks, but this is an exception to the common rule; possibly the hens may lie hidden in the surrounding bushes, as is known to be the case with the grey-hens in Scandinavia, and with other species in N. America.

21. 'Ornithological Biography', vol. ii. p. 275.

they pair. It does not, however, appear that the females invariably prefer the victorious males. I have indeed been assured by Dr W. Kovalevsky that the female capercailzie sometimes steals away with a young male who has not dared to enter the arena with the older cocks, in the same manner as occasionally happens with the does of the red-deer in Scotland. When two males contend in presence of a single female, the victor, no doubt, commonly gains his desire; but some of these battles are caused by wandering males trying to distract the peace of an already mated pair.[22]

Even with the most pugnacious species it is probable that the pairing does not depend exclusively on the mere strength and courage of the male; for such males are generally decorated with various ornaments, which often become more brilliant during the breeding-season, and which are sedulously displayed before the females. The males also endeavour to charm or excite their mates by love-notes, songs, and antics; and the courtship is, in many instances, a prolonged affair. Hence it is not probable that the females are indifferent to the charms of the opposite sex, or that they are invariably compelled to yield to the victorious males. It is more probable that the females are excited, either before or after the conflict, by certain males, and thus unconsciously prefer them. In the case of *Tetrao umbellus*, a good observer[23] goes so far as to believe that the battles of the males 'are all a sham, performed to show themselves to the greatest advantage before the admiring females who assemble around; for I have never been able to find a maimed hero, and seldom more than a broken feather'. I shall have to recur to this subject, but I may here add that with the *Tetrao cupido* of the United States, about a score of males assemble at a particular spot, and strutting about, make the whole air resound with their extraordinary noises. At the

22. Brehm, 'Thierleben', &c., B. iv. 1867, p. 990. Audubon, 'Ornith. Biography', vol. ii. p. 492.
23. 'Land and Water', July 25th, 1868, p. 14.

first answer from a female the males begin to fight furiously, and the weaker give way; but then, according to Audubon, both the victors and vanquished search for the female, so that the females must either then exert a choice, or the battle must be renewed. So, again, with one of the field-starlings of the United States (*Sturnella ludoviciana*) the males engage in fierce conflicts, 'but at the sight of a female they all fly after her, as if mad'.[24]

Vocal and instrumental music—With birds the voice serves to express various emotions, such as distress, fear, anger, triumph, or mere happiness. It is apparently sometimes used to excite terror, as in the case of the hissing noise made by some nestling-birds. Audubon[25] relates that a night-heron (*Ardea nycticorax*, Linn.) which he kept tame, used to hide itself when a cat approached, and then 'suddenly start up uttering one of the most frightful cries, apparently enjoying the cat's alarm and flight'. The common domestic cock clucks to the hen, and the hen to her chickens, when a dainty morsel is found. The hen, when she has laid an egg, 'repeats the same note very often, and concludes with the sixth above, which she holds for a longer time';[26] and thus she expresses her joy. Some social birds apparently call to each other for aid; and as they flit from tree to tree, the flock is kept together by chirp answering chirp. During the nocturnal migrations of geese and other water-fowl, sonorous clangs from the van may be heard in the darkness overhead, answered by clangs in the rear. Certain cries serve as danger signals, which, as the sportsman knows to his cost, are understood by the same species and by others. The domestic cock crows, and the humming-bird chirps, in triumph over a defeated rival. The true song, however,

---

24. Audubon's 'Ornitholog. Biography'; on *Tetrao cupido*, vol. ii. p. 492; on the Sturnus, vol. ii. p. 219.
25. 'Ornithological Biograph.', vol. v. p. 601.
26. The Hon. Daines Barrington, 'Philosoph. Transact.', 1773, p. 252.

of most birds and various strange cries are chiefly uttered during the breeding-season, and serve as a charm, or merely as a call-note, to the other sex.

Naturalists are much divided with respect to the object of the singing of birds. Few more careful observers ever lived than Montagu, and he maintained that the 'males of song-birds and of many others do not in general search for the female, but, on the contrary, their business in the spring is to perch on some conspicuous spot, breathing out their full and amorous notes, which, by instinct, the female knows, and repairs to the spot to choose her mate'.[27] Mr Jenner Weir informs me that this is certainly the case with the nightingale. Bechstein, who kept birds during his whole life, asserts, 'that the female canary always chooses the best singer, and that in a state of nature the female finch selects that male out of a hundred whose notes please her most'.[28] There can be no doubt that birds closely attend to each other's song. Mr Weir has told me of the case of a bullfinch which had been taught to pipe a German waltz, and who was so good a performer that he cost ten guineas; when this bird was first introduced into a room where other birds were kept and he began to sing, all the others, consisting of about twenty linnets and canaries, ranged themselves on the nearest side of their cages, and listened with the greatest interest to the new performer. Many naturalists believe that the singing of birds is almost exclusively 'the effect of rivalry and emulation', and not for the sake of charming their mates. This was the opinion of Daines Barrington and White of Selborne, who both especially attended to this subject.[29] Barrington, however, admits that 'superiority in

27. 'Ornithological Dictionary', 1833, p. 475.
28. 'Naturgeschichte der Stubenvögel', 1840, s. 4. Mr Harrison Weir likewise writes to me:—'I am informed that the best singing males generally get a mate first, when they are bred in the same room.'
29. 'Philosophical Transactions', 1773, p. 263. White's 'Natural History of Selborne', 1825, vol. i. p. 246.

song gives to birds an amazing ascendancy over others, as is well known to bird-catchers'.

It is certain that there is an intense degree of rivalry between the males in their singing. Bird-fanciers match their birds to see which will sing longest; and I was told by Mr Yarrell that a first-rate bird will sometimes sing till he drops down almost dead, or according to Bechstein,[30] quite dead from rupturing a vessel in the lungs. Whatever the cause may be, male birds, as I hear from Mr Weir, often die suddenly during the season of song. That the habit of singing is sometimes quite independent of love is clear, for a sterile, hybrid canary-bird has been described[31] as singing whilst viewing itself in a mirror, and then dashing at its own image; it likewise attacked with fury a female canary, when put into the same cage. The jealousy excited by the act of singing is constantly taken advantage of by bird-catchers; a male, in good song, is hidden and protected, whilst a stuffed bird, surrounded by limed twigs, is exposed to view. In this manner, as Mr Weir informs me, a man has in the course of a single day caught fifty, and in one instance seventy, male chaffinches. The power and inclination to sing differ so greatly with birds that although the price of an ordinary male chaffinch is only sixpence, Mr Weir saw one bird for which the bird-catcher asked three pounds; the test of a really good singer being that it will continue to sing whilst the cage is swung round the owner's head.

That male birds should sing from emulation as well as for charming the female, is not at all incompatible; and it might have been expected that these two habits would have concurred, like those of display and pugnacity. Some authors, however, argue that the song of the male cannot serve to charm the female, because the females of some few species, such as of the canary, robin,

30. 'Naturgesch. der Stubenvögel', 1840, s. 252.
31. Mr Bold, 'Zoologist', 1843–44, p. 659.

lark, and bullfinch, especially when in a state of widowhood, as Bechstein remarks, pour forth fairly melodious strains. In some of these cases the habit of singing may be in part attributed to the females having been highly fed and confined,[32] for this disturbs all the usual functions connected with the reproduction of the species. Many instances have already been given of the partial transference of secondary masculine characters to the female, so that it is not at all surprising that the females of some species should possess the power of song. It has also been argued, that the song of the male cannot serve as a charm, because the males of certain species, for instance of the robin, sing during the autumn.[33] But nothing is more common than for animals to take pleasure in practising whatever instinct they follow at other times for some real good. How often do we see birds which fly easily, gliding and sailing through the air obviously for pleasure? The cat plays with the captured mouse, and the cormorant with the captured fish. The weaver-bird (Ploceus), when confined in a cage, amuses itself by neatly weaving blades of grass between the wires of its cage. Birds which habitually fight during the breeding-season are generally ready to fight at all times; and the males of the capercailzie sometimes hold their *Balzen* or *leks* at the usual place of assemblage during the autumn.[34] Hence it is not at all surprising that male birds should continue singing for their own amusement after the season for courtship is over.

As shown in a previous chapter, singing is to a certain extent an art, and is much improved by practice. Birds can be taught various tunes, and even the unmelodious sparrow has learnt to sing like a linnet. They acquire the song of their foster parents,[35]

---

32. D. Barrington, 'Phil. Transact.' 1773, p. 262. Bechstein, 'Stubenvögel', 1840, s. 4.
33. This is likewise the case with the water-ouzel, see Mr Hepburn in the 'Zoologist', 1845–1846, p. 1068.
34. L. Lloyd, 'Game Birds of Sweden', 1867, p. 25.
35. Barrington, ibid. p. 264. Bechstein, ibid. s. 5.

and sometimes that of their neighbours.[36] All the common song-sters belong to the Order of Insessores, and their vocal organs are much more complex than those of most other birds; yet it is a singular fact that some of the Insessores, such as ravens, crows, and magpies, possess the proper apparatus,[37] though they never sing, and do not naturally modulate their voices to any great extent. Hunter asserts[38] that with the true songsters the muscles of the larynx are stronger in the males than in the females; but with this slight exception there is no difference in the vocal organs of the two sexes, although the males of most species sing so much better and more continuously than the females.

It is remarkable that only small birds properly sing. The Australian genus Menura, however, must be excepted; for the *Menura Alberti*, which is about the size of a half-grown turkey, not only mocks other birds, but 'its own whistle is exceedingly beautiful and varied'. The males congregate and form '*corroborying* places', where they sing, raising and spreading their tails like peacocks, and drooping their wings.[39] It is also remarkable that birds which sing well are rarely decorated with brilliant colours or other ornaments. Of our British birds, excepting the bullfinch and goldfinch, the best songsters are plain-coloured. The kingfisher, bee-eater, roller, hoopoe, woodpeckers, &c., utter harsh cries; and the brilliant birds of the tropics are hardly ever songsters.[40] Hence bright colours and the power of song seem to replace each other. We can perceive that if the plumage did not vary in brightness, or if bright colours were dangerous to the species, other means

36. Dureau de la Malle gives a curious instance ('Annales des Sc. Nat.', 3rd series, Zoolog. tom. x. p. 118) of some wild blackbirds in his garden in Paris, which naturally learnt a republican air from a caged bird.
37. Bishop, in, 'Todd's Cyclop. of Anat. and Phys.', vol. iv. p. 1496.
38. As stated by Barrington in 'Philosoph. Transact.', 1773, p. 262.
39. Gould, 'Handbook to the Birds of Australia', vol. i. 1865, pp. 308–310. See also Mr T. W. Wood in the 'Student', April 1870, p. 125.
40. See remarks to this effect in Gould's 'Introduction to the Trochilidæ,' 1861, p. 22.

Fig. 39. Tetrao cupido: male. (T. W. Wood)

would be employed to charm the females; and melody of voice offers one such means.

In some birds the vocal organs differ greatly in the two sexes. In the *Tetrao cupido* (fig. 39) the male has two bare, orange-coloured sacks, one on each side of the neck; and these are largely inflated when the male, during the breeding-season, makes his curious hollow sound, audible at a great distance. Audubon proved that the sound was intimately connected with this apparatus (which reminds us of the air-sacks on each side of the mouth of certain male frogs), for he found that the sound was much diminished when one of the sacks of a tame bird was pricked, and when both were pricked it was altogether stopped. The female has 'a somewhat similar, though smaller naked space of skin on the neck; but this is not capable of inflation'.[41] The male of another

41. 'The Sportsman and Naturalist in Canada', by Major W. Ross King, 1866, pp. 144–146. Mr T. W. Wood gives in the 'Student' (April, 1870, p. 116) an excellent account of the attitude and habits of this bird during its courtship. He states that the ear-tufts or neck-plumes are erected, so that they meet over the crown of the head. See his drawing, fig. 39.

kind of grouse (*Tetrao urophasianus*), whilst courting the female, has his 'bare yellow oesophagus inflated to a prodigious size, fully half as large as the body'; and he then utters various grating, deep, hollow tones. With his neck-feathers erect, his wings lowered, and buzzing on the ground, and his long pointed tail spread out like a fan, he displays a variety of grotesque attitudes. The oesophagus of the female is not in any way remarkable.[42]

It seems now well made out that the great throat pouch of the European male bustard (*Otis tarda*), and of at least four other species, does not, as was formerly supposed, serve to hold water, but is connected with the utterance during the breeding-season of a peculiar sound resembling 'oak'.[43] A crow-like bird inhabiting South America (*Cephalopterus ornatus*, fig. 40) is called the umbrella-bird, from its immense top knot, formed of bare white quills surmounted by dark-blue plumes, which it can elevate into a great dome no less than five inches in diameter, covering the whole head. This bird has on its neck a long, thin, cylindrical fleshy appendage, which is thickly clothed with scale-like blue feathers. It probably serves in part as an ornament, but likewise as a resounding apparatus; for Mr Bates found that it is connected 'with an unusual development of the trachea and vocal organs'. It is dilated when the bird utters its singularly deep, loud and long sustained fluty note. The head-crest and neck-appendage are rudimentary in the female.[44]

The vocal organs of various web-footed and wading birds

42. Richardson, 'Fauna Bor. American: Birds', 1831, p. 359. Audubon ibid. vol. iv. p. 507.

43. The following papers have been lately written on this subject: Prof. A. Newton, in the 'Ibis', 1862, p. 107, Dr Cullen, ibid. 1865, p. 145; Mr Flower, in 'Proc. Zool. Soc.', 1865, p. 747; and Dr Murie, in 'Proc. Zool. Soc.', 1868, p. 471. In this latter paper an excellent figure is given of the male Australian Bustard in full display with the sack distended. It is a singular fact that the sack is not developed in all the males of the same species.

44. Bates, 'The Naturalist on the Amazons', 1863, vol. ii. p. 284; Wallace, in 'Proc. Zool. Soc.', 1850, p. 206. A new species, with a still larger neck-appendage (*C. penduliger*), has lately been discovered, see 'Ibis', vol. i. p. 457.

Fig. 40. The Umbrella-bird or Cephalopterus ornatus male (from Brehm).

are extraordinarily complex, and differ to a certain extent in the two sexes. In some cases the trachea is convoluted, like a French horn, and is deeply embedded in the sternum. In the wild swan (*Cygnus ferus*) it is more deeply embedded in the adult male, than in the adult female or young male. In the male Merganser the enlarged portion of the trachea is furnished with an additional pair of muscles.[45] In one of the ducks, however, namely *Anas punctata*, the bony enlargement is only a little more developed in the male than in the female.[46] But the meaning of these differences in the trachea of the two sexes of the Anatidae is not understood; for the male is not always the more vociferous; thus with the common duck, the male hisses, whilst the female ut-

45. Bishop, in Todd's 'Cyclop. of Anat. and Phys.', vol. iv. p. 1499.
46. Prof. Newton, 'Proc. Zoolog. Soc.', 1871, p. 651.

ters a loud quack.[47] In both sexes of one of the cranes (*Grus virgo*) the trachea penetrates the sternum, but presents 'certain sexual modifications'. In the male of the black stork there is also a well-marked sexual difference in the length and curvature of the bronchi.[48] Highly important structures have, therefore, in these cases been modified according to sex.

It is often difficult to conjecture whether the many strange cries and notes uttered by male birds during the breeding-season, serve as a charm or merely as a call to the female. The soft cooing of the turtle-dove and of many pigeons, it may be presumed, pleases the female. When the female of the wild turkey utters her call in the morning, the male answers by a note which differs from the gobbling noise made, when with erected feathers, rustling wings and distended wattles, he puffs and struts before her.[49] The *spel* of the black-cock certainly serves as a call to the female, for it has been known to bring four or five females from a distance to a male under confinement; but as the black-cock continues his *spel* for hours during successive days, and in the case of the capercailzie 'with an agony of passion', we are led to suppose that the females which are present are thus charmed.[50] The voice of the common rook is known to alter during the breeding-season, and is therefore in some way sexual.[51] But what shall we say about the harsh screams of, for instance, some kinds of macaws; have these birds as bad taste for musical sounds as they apparently have for colour, judging by the inharmonious

47. The spoonbill (Platalea) has its trachea convoluted into a figure of eight, and yet this bird (Jerdon, 'Birds of India', vol. iii. p. 763) is mute; but Mr Blyth informs me that the convolutions are not constantly present, so that perhaps they are now tending towards abortion.
48. 'Elements of Comp. Anat.' by R. Wagner, Eng. translat. 1845, p. 111. With respect to the swan, as given above, Yarrell's 'Hist. of British Birds', 2nd edit. 1845, vol. iii. p. 193.
49. C. L. Bonaparte, quoted in the 'Naturalist Library: Birds', vol. xiv. p. 126.
50. L. Lloyd, 'The Game Birds of Sweden', &c., 1867, pp. 22, 81.
51. Jenner, 'Philosoph. Transactions', 1824, p. 20.

contrast of their bright yellow and blue plumage? It is indeed possible that without any advantage being thus gained, the loud voices of many male birds may be the result of the inherited effects of the continued use of their vocal organs, when excited by the strong passions of love, jealousy and rage; but to this point we shall recur when we treat of quadrupeds.

We have as yet spoken only of the voice, but the males of various birds practise, during their courtship, what may be called instrumental music. Peacocks and Birds of Paradise rattle their quills together. Turkey-cocks scrape their wings against the ground, and some kinds of grouse thus produce a buzzing sound. Another North American grouse, the *Tetrao umbellus*, when with his tail erect, his ruffs displayed, 'he shows off his finery to the females, who lie hid in the neighbourhood', drums by rapidly striking his wings together above his back, according to Mr R. Haymond, and not, as Audubon thought, by striking them against his sides. The sound thus produced is compared by some to distant thunder, and by others to the quick roll of a drum. The female never drums, 'but flies directly to the place where the male is thus engaged'. The male of the Kalij-pheasant, in the Himalayas, 'often makes a singular drumming noise with his wings, not unlike the sound produced by shaking a stiff piece of cloth'. On the west coast of Africa the little black-weavers (Ploceus?) congregate in a small party on the bushes round a small open space, and sing and glide through the air with quivering wings, 'which make a rapid whirring sound like a child's rattle'. One bird after another thus performs for hours together, but only during the courting-season. At this season and at no other time, the males of certain night-jars (Caprimulgus) make a strange booming noise with their wings. The various species of wood-peckers strike a sonorous branch with their beaks, with so rapid a vibratory movement that 'the head appears to be in two places at once'. The sound thus produced is audible at a considerable distance, but cannot be described; and I feel sure that its source would never be conjectured by any one hearing it for the first time. As this jarring sound is made chiefly during the breeding-season, it has been considered as a love-song; but it is perhaps more strictly a love-call. The female, when driven from her nest, has been observed thus to call her mate, who answered in the same manner and soon appeared. Lastly, the male Hoopoe (*Upupa epops*) combines vocal and instrumental music; for during the breeding-season this bird, as Mr Swinhoe observed, first draws in air, and then taps the end of its beak perpendicularly down against a stone or the trunk of a tree, 'when the breath being forced down the tubular bill produces the correct sound'. If the beak is not thus struck against some object, the sound is quite different. Air is at the same time swallowed, and the oesophagus thus becomes much swollen; and this probably acts as a resonator, not only with the hoopoe, but with pigeons and other birds.[52]

In the foregoing cases sounds are made by the aid of structures already present and otherwise necessary; but in the following cases certain feathers have been specially modified for the express purpose of producing sounds. The drumming, bleating, neighing, or thundering noise (as expressed by different observers) made by the common snipe (*Scolopax gullinago*) must have surprised every one who has ever heard it. This bird, during the pairing-season, flies to 'perhaps a thousand feet in height',

52. For the foregoing facts see, on Birds of Paradise, Brehm, 'Thierleben', Band iii. s. 325. On Grouse, Richardson, 'Fauna Bor. Americ.: Birds', pp. 343 and 359; Major W. Ross King, 'The Sportsman in Canada', 1866, p. 156; Mr Haymond, in Prof. Cox's 'Geol. Survey of Indiana', p. 227; Audubon, 'American Ornitholog. Biograph.', vol. i. p. 216. On the Kalij-pheasant, Jerdon, 'Birds of India', vol. iii. p. 533. On the Weavers, 'Livingstone's Expedition to the Zambesi', 1865, p. 425. On Woodpeckers, Macgillivray, 'Hist. of British Birds', vol. iii. 1840, pp. 84, 88, 89, and 95. On the Hoopoe, Mr Swinhoe, in 'Proc. Zoolog. Soc.', June 23, 1863 and 1871, p. 348. On the Night-jar, Audubon, ibid. vol. ii. p. 255, and 'American Naturalist', 1873, p. 672. The English Night-jar likewise makes in the spring a curious noise during its rapid flight.

Fig. 41. Outer tail-feather of Scolopax gallinago (from 'Proc. Zool. Soc.' 1858).

Fig. 42. Outer tail-feather of Scolopax frenata.

Fig. 43. Outer tail-feather of Scolopax javensis.

and after zig-zagging about for a time descends to the earth in a curved line, with outspread tail and quivering pinions, and surprising velocity. The sound is emitted only during this rapid descent. No one was able to explain the cause, until M. Meves observed that on each side of the tail the outer feathers are peculiarly formed (fig. 41), having a stiff sabre-shaped shaft with the oblique barbs of unusual length, the outer webs being strongly bound together. He found that by blowing on these feathers, or by fastening them to a long thin stick and waving them rapidly through the air, he could reproduce the drumming noise made by the living bird. Both sexes are furnished with these feathers, but they are generally larger in the male than in the female, and emit a deeper note. In some species, as in *S. frenata* (fig. 42), four feathers, and in *S. javensis* (fig. 43), no less than eight on each side of the tail are greatly modified. Different tones are emitted by the feathers of the different species when waved through the air; and the *Scolopax Wilsonii* of the United States makes a switching noise whilst descending rapidly to the earth.[53]

53. See M. Meves' interesting paper in 'Proc. Zool. Soc.', 1858, p. 199. For the habits of the snipe, Macgillivray, 'Hist. British Birds', vol. iv. p. 371. For the American snipe, Capt. Blakiston, 'Ibis', vol. v. 1863, p. 131.

In the male of the *Chamaepetes unicolor* (a large gallinaceous bird of America) the first primary wing-feather is arched to-wards the tip and is much more attenu-ated than in the female. In an allied bird, the *Penelope nigra*, Mr Salvin observed a male, which, whilst it flew downwards 'with outstretched wings, gave forth a kind of crashing rushing noise', like the falling of a tree.[54] The male alone of one of the Indian bustards (*Sypheotides auritus*) has its primary wing-feathers greatly acuminated; and the male of an allied species is known to make a hum-ming noise whilst courting the female.[55]

Fig. 44 Primary wing-feather of a Humming-bird, the *Selasphorus platycercus* (from a sketch by Mr Salvin). Upper figure, that of male; lower figure, corresponding feather of female.

In a widely different group of birds, namely Humming-birds, the males alone of certain kinds have either the shafts of their primary wing-feathers broadly dilated, or the webs abruptly ex-cised towards the extremity. The male, for instance, of *Selasphorus platycercus*, when adult, has the first primary wing-feather (fig. 44), thus excised. Whilst flying from flower to flower he makes 'a shrill, almost whistling noise';[56] but it did not appear to Mr Salvin that the noise was intentionally made.

Lastly, in several species of a sub-genus of Pipra or Mana-kin, the males, as described by Mr Sclater, have their *secondary* wing-feathers modified in a still more remarkable manner. In the brilliantly-coloured *P. deliciosa* the first three secondaries are thick-stemmed and curved towards the body; in the fourth and fifth (fig. 45, *a*) the change is greater; and in the sixth and seventh

54. Mr Salvin, in 'Proc. Zool. Soc.', 1867, p. 160. I am much indebted to this distin-guished ornithologist for sketches of the feathers of the Chamaepetes, and for other information.
55. Jerdon, 'Birds of India', vol. iii. pp. 618, 621.
56. Gould, 'Introduction to the Trochilidae', 1861, p. 49. Salvin, 'Proc. Zoolog. Soc.', 1867, p. 160.

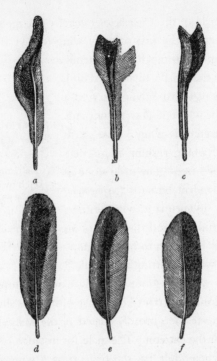

Fig. 45. Secondary wing-feathers of *Pipra deliciosa* (from Mr Scalter, in 'Proc. Zool. Soc.', 1860). The three upper feathers, *a, b, c,* from the male; the three lower corresponding feathers, *d, e, f,* from the female. *a* and *d,* fifth secondary wing-feather of male and female, upper surface. *b* and *e,* sixth secondary, upper surface. *c* and *f,* seventh secondary, lower surface.

(*b, c*) the shaft 'is thickened to an extraordinary degree, forming a solid horny lump'. The barbs also are greatly changed in shape, in comparison with the corresponding feathers (*d, e, f*) in the female. Even the bones of the wing, which support these singular feathers in the male, are said by Mr Fraser to be much thickened. These little birds make an extraordinary noise, the first 'sharp note being not unlike the crack of a whip'.[57]

57. Sclater, in 'Proc. Zool. Soc.', 1860, p. 90, and in 'Ibis', vol. iv. 1862, p. 175. Also Salvin, in 'Ibis', 1860, p. 37.

The diversity of the sounds, both vocal and instrumental, made by the males of many birds during the breeding-season, and the diversity of the means for producing such sounds, are highly remarkable. We thus gain a high idea of their importance for sexual purposes, and are reminded of the conclusion arrived at as to insects. It is not difficult to imagine the steps by which the notes of a bird, primarily used as a mere call or for some other purpose, might have been improved into a melodious love song. In the case of the modified feathers, by which the drumming, whistling, or roaring noises are produced, we know that some birds during their courtship flutter, shake, or rattle their unmodified feathers together; and if the females were led to select the best performers, the males which possessed the strongest or thickest, or most attenuated feathers, situated on any part of the body, would be the most successful; and thus by slow degrees the feathers might be modified to almost any extent. The females, of course, would not notice each slight successive alteration in shape, but only the sounds thus produced. It is a curious fact that in the same class of animals, sounds so different as the drumming of the snipe's tail, the tapping of the woodpecker's beak, the harsh trumpet-like cry of certain water-fowl, the cooing of the turtle-dove, and the song of the nightingale, should all be pleasing to the females of the several species. But we must not judge of the tastes of distinct species by a uniform standard; nor must we judge by the standard of man's taste. Even with man, we should remember what discordant noises, the beating of tom-toms and the shrill notes of reeds, please the ears of savages. Sir S. Baker remarks,[58] that 'as the stomach of the Arab prefers the raw meat and reeking liver taken hot from the animal, so does his ear prefer his equally coarse and discordant music to all other'.

58. 'The Nile Tributaries of Abyssinia', 1867, p. 203.

Love-Antics and Dances—The curious love gestures of some birds have already been incidentally noticed; so that little need here be added. In Northern America, large numbers of a grouse, the *Tetrao phasianellus*, meet every morning during the breeding-season on a selected level spot, and here they run round and round in a circle of about fifteen or twenty feet in diameter, so that the ground is worn quite bare, like a fairy-ring. In these Partridge-dances, as they are called by the hunters, the birds assume the strangest attitudes, and run round, some to the left and some to the right. Audubon describes the males of a heron (*Ardea herodias*) as walking about on their long legs with great dignity before the females, bidding defiance to their rivals. With one of the disgusting carrion-vultures (*Cathartes jota*) the same naturalist states that 'the gesticulations and parade of the males at the beginning of the love-season are extremely ludicrous'. Certain birds perform their love antics on the wing, as we have seen with the black African weaver, instead of on the ground. During the spring our little white-throat (*Sylvia cinerea*) often rises a few feet or yards in the air above some bush, and 'flutters with a fitful and fantastic motion, singing all the while, and then drops to its perch'. The great English bustard throws himself into indescribably odd attitudes whilst courting the female, as has been figured by Wolf. An allied Indian bustard (*Otis bengalensis*) at such times 'rises perpendicularly into the air with a hurried flapping of his wings, raising his crest and puffing out the feathers of his neck and breast, and then drops to the ground'; he repeats this manoeuvre several times, at the same time humming in a peculiar tone. Such females as happen to be near 'obey this saltatory summons', and when they approach he trails his wings and spreads his tail like a turkey-cock.[59]

59. For *Tetrao phasianellus*, see Richardson, 'Fauna, Bor. America', p. 361, and for further particulars Capt. Blakiston, 'Ibis', 1863, p. 125. For the Cathartes and Ardea, Audubon, 'Ornith. Biography', vol. ii. p. 51, and vol. iii. p. 89. On the White-throat, Macgillivray, 'Hist. British Birds', vol. ii. p. 354. On the Indian Bustard, Jerdon, 'Birds of India', vol. iii. p. 618.

Fig. 46. Bower-bird, *Chlamydera maculata*, with bower (from Brehm).

But the most curious case is afforded by three allied genera of Australian birds, the famous Bower-birds—no doubt the co-descendants of some ancient species which first acquired the strange instinct of constructing bowers for performing their love-antics. The bowers (fig. 46), which, as we shall hereafter see, are decorated with feathers, shells, bones, and leaves, are built on the ground for the sole purpose of courtship, for their nests are formed in trees. Both sexes assist in the erection of the bowers, but the male is the principal workman. So strong is this instinct that it is practised under confinement, and Mr Strange has described[60] the habits of some Satin Bower-birds which he kept in an aviary in New South Wales. 'At times the male will chase the female all over the aviary, then go to the bower, pick up a

60. Gould, 'Handbook to the Birds of Australia', vol. 1. pp. 444, 449, 455. The bower of the Satin Bower-bird may be seen in the Zoological Society's Gardens, Regent's Park.

gay feather or a large leaf, utter a curious kind of note, set all his feathers erect, run round the bower and become so excited that his eyes appear ready to start from his head; he continues opening first one wing then the other, uttering a low, whistling note, and, like the domestic cock, seems to be picking up something from the ground until at last the female goes gently towards him'. Captain Stokes has described the habits and 'play-houses' of another species, the Great Bower-bird, which was seen 'amusing itself by flying backwards and forwards, taking a shell alternately from each side, and carrying it through the archway in its mouth'. These curious structures, formed solely as halls of assemblage, where both sexes amuse themselves and pay their court, must cost the birds much labour. The bower, for instance, of the Fawn-breasted species, is nearly four feet in length, eighteen inches in height, and is raised on a thick platform of sticks.

Decoration—I will first discuss the cases in which the males are ornamented either exclusively or in a much higher degree than the females, and in a succeeding chapter those in which both sexes are equally ornamented, and finally the rare cases in which the female is somewhat more brightly-coloured than the male. As with the artificial ornaments used by savage and civilised men, so with the natural ornaments of birds, the head is the chief seat of decoration.[61] The ornaments, as mentioned at the commencement of this chapter, are wonderfully diversified. The plumes on the front or back of the head consist of variously-shaped feathers, sometimes capable of erection or expansion, by which their beautiful colours are fully displayed. Elegant ear-tufts (see fig. 39, ante) are occasionally present. The head is sometimes covered with velvety down, as with the pheasant; or is naked and viv-

61. See remarks to this effect, on the 'Feeling of Beauty among Animals', by Mr J. Shaw, in the 'Athenaeum', Nov. 24th, 1866, p. 681.

idly coloured. The throat, also, is sometimes ornamented with a beard, wattles, or caruncles. Such appendages are generally brightly-coloured, and no doubt serve as ornaments, though not always ornamental in our eyes; for whilst the male is in the act of courting the female, they often swell and assume vivid tints, as in the male turkey. At such times the fleshy appendages about the head of the male Tragopan pheasant (*Ceriornis Temminckii*) swell into a large lappet on the throat and into two horns, one on each side of the splendid top-knot; and these are then coloured of the most intense blue which I have ever beheld.[62] The African hornbill (*Bucorax abyssinicus*) inflates the scarlet bladder-like wattle on its neck, and with its wings drooping and tail expanded 'makes quite a grand appearance'.[63] Even the iris of the eye is sometimes more brightly-coloured in the male than in the female; and this is frequently the case with the beak, for instance, in our common blackbird. In *Buceros corrugatus*, the whole beak and immense casque are coloured more conspicuously in the male than in the female; and 'the oblique grooves upon the sides of the lower mandible are peculiar to the male sex'.[64]

The head, again, often supports fleshy appendages, filaments, and solid protuberances. These, if not common to both sexes, are always confined to the males. The solid protuberances have been described in detail by Dr W. Marshall,[65] who shews that they are formed either of cancellated bone coated with skin, or of dermal and other tissues. With mammals true horns are always supported on the frontal bones, but with birds various bones have been modified for this purpose; and in species of the same group the protuberances may have cores of bone, or be quite destitute

---

62. See Dr Murie's account with coloured figures in 'Proc. Zoolog. Soc.', 1872, p. 730.
63. Mr Monteiro, 'Ibis', vol. iv. 1862, p. 339.
64. 'Land and Water', 1868, p. 217.
65. 'Ueber die Schädelhöcker', &c., 'Niederländischen Arhiv für Zoologie', B. I. Heft. 2, 1872.

of them, with intermediate gradations connecting these two extremes. Hence, as Dr Marshall justly remarks, variations of the most different kinds have served for the development through sexual selection of these ornamental appendages. Elongated feathers or plumes spring from almost every part of the body. The feathers on the throat and breast are sometimes developed into beautiful ruffs and collars. The tail-feathers are frequently increased in length; as we see in the tail-coverts of the peacock, and in the tail itself of the Argus pheasant. With the peacock even the bones of the tail have been modified to support the heavy tail-coverts.[66] The body of the Argus is not larger than that of a fowl; yet the length from the end of the beak to the extremity of the tail is no less than five feet three inches,[67] and that of the beautifully ocellated secondary wing-feathers nearly three feet. In a small African night-jar (*Cosmetornis vexillarius*) one of the primary wing-feathers, during the breeding-season, attains a length of twenty-six inches, whilst the bird itself is only ten inches in length. In another closely-allied genus of night-jars, the shafts of the elongated wing-feathers are naked, except at the extremity, where there is a disc.[68] Again, in another genus of night-jars, the tail-feathers are even still more prodigiously developed. In general the feathers of the tail are more often elongated than those of the wings, as any great elongation of the latter impedes flight. We thus see that in closely-allied birds ornaments of the same kind have been gained by the males through the development of widely different feathers.

It is a curious fact that the feathers of species belonging to very distinct groups have been modified in almost exactly the same peculiar manner. Thus the wing-feathers in one of the above-

---

66. Dr W. Marshall, 'Über den Vogelschwanz', ibid. B. I. Heft 2, 1872.
67. Jardine's 'Naturalist Library: Birds', vol. xiv. p. 166.
68. Sclater, in the 'Ibis', vol. vi. 1864, p. 114. Livingstone, 'Expedition to the Zambesi', 1865, p. 66.

mentioned night-jars are bare along the shaft, and terminate in a disc; or are, as they are sometimes called, spoon or racket-shaped. Feathers of this kind occur in the tail of a motmot (*Eumomota superciliaris*), of a king-fisher, finch, humming-bird, parrot, several Indian drongos (Dicrurus and Edolius, in one of which the disc stands vertically), and in the tail of certain birds of paradise. In these latter birds, similar feathers, beautifully ocellated, ornament the head, as is likewise the case with some gallinaceous birds. In an Indian bustard (*Sypheotides auritus*) the feathers forming the ear-tufts, which are about four inches in length, also terminate in discs.[69] It is a most singular fact that the motmots, as Mr Salvin has clearly shewn,[70] give to their tail feathers the racket-shape by biting off the barbs, and, further, that this continued mutilation has produced a certain amount of inherited effect.

Again, the barbs of the feathers in various widely-distinct birds are filamentous or plumose, as with some herons, ibises, birds of paradise, and Gallinaceae. In other cases the barbs disappear, leaving the shafts bare from end to end; and these in the tail of the *Paradisea apoda* attain a length of thirty-four inches:[71] in *P. Papuana* (fig. 47) they are much shorter and thin. Smaller feathers when thus denuded appear like bristles, as on the breast of the turkey-cock. As any fleeting fashion in dress comes to be admired by man, so with birds a change of almost any kind in the structure or colouring of the feathers in the male appears to have been admired by the female. The fact of the feathers in widely distinct groups, having been modified in an analogous manner, no doubt depends primarily on all the feathers having nearly the same structure and manner of development, and consequently tending to vary in the same manner.

---

69. Jerdon, 'Birds of India', vol. iii. p. 620.
70. 'Proc. Zoolog. Soc.', 1873, p. 429.
71. Wallace, in 'Annals and Mag. of Nat. Hist.', vol. xx. 1857, p. 416, and in his 'Malay Archipelago', vol. ii. 1869, p. 390.

Fig. 47. Paradisea Papuana (T. W. Wood).

We often see a tendency to analogous variability in the plumage of our domestic breeds belonging to distinct species. Thus top-knots have appeared in several species. In an extinct variety of the turkey, the top-knot consisted of bare quills surmounted with plumes of down, so that they somewhat resembled the racket-shaped feathers above described. In certain breeds of the pigeon and fowl the feathers are plumose, with some tendency in the shafts to be naked. In the Sebastopol goose the scapular feathers are greatly elongated, curled, or even spirally twisted, with the margins plumose.[72]

In regard to colour hardly anything need here be said, for every one knows how splendid are the tints of many birds, and how harmoniously they are combined. The colours are often metallic and iridescent. Circular spots are sometimes surrounded by one or more differently shaded zones, and are thus converted into ocelli. Nor need much be said on the wonderful difference between the sexes of many birds. The common peacock offers a striking instance. Female birds of paradise are obscurely coloured and destitute of all ornaments, whilst the males are probably the most highly decorated of all birds, and in so many different ways, that they must be seen to be appreciated. The elongated and golden-orange plumes which spring from beneath the wings of the *Paradisea apoda*, when vertically erected and made to vibrate, are described as forming a sort of halo, in the centre of which the head 'looks like a little emerald sun with its rays formed by the two plumes.'[73] In another most beautiful species the head is bald, 'and of a rich cobalt blue, crossed by several lines of black velvety feathers'.[74]

Male humming-birds (figs. 48 and 49) almost vie with birds of paradise in their beauty, as every one will admit who has seen Mr Gould's splendid volumes, or his rich collection. It is very remarkable in how many different ways these birds are or-namented. Almost every part of their plumage has been taken advantage of, and modified; and the modifications have been carried, as Mr Gould shewed me, to a wonderful extreme in some species belonging to nearly every sub-group. Such cases

72. See my work on 'The Variation of Animals and Plants under Domestication', vol. i. pp. 289, 293.
73. Quoted from M. de Lafresnaye in 'Annals and Mag. of Nat. Hist.', vol. xiii. 1854, p. 157: see also Mr Wallace's much fuller account in vol. xx. 1857, p. 412, and in his 'Malay Archipelago'.
74. Wallace, 'The Malay Archipelago', vol. ii. 1869, p. 405.

Fig. 48. Lophorins ornatus, male and female (from Brehm).

are curiously like those which we see in our fancy breeds, reared by man for the sake of ornament: certain individuals originally varied in one character, and other individuals of the same species in other characters; and these have been seized on by man and much augmented—as shewn by the tail of the fantail-pigeon, the hood of the jacobin, the beak and wattle of the carrier, and so forth. The sole difference between these cases is that in the one, the result is due to man's selection,

Fig. 49. Spathura underwoodi, male and female (from Brehm).

whilst in the other, as with humming-birds, birds of paradise, &c., it is due to the selection by the females of the more beautiful males.

I will mention only one other bird, remarkable from the extreme contrast in colour between the sexes, namely the famous bell-bird (*Chasmorhynchus niveus*) of S. America, the note of which can be distinguished at the distance of nearly three miles, and astonishes every one when first hearing it. The male is pure white, whilst the female is dusky-green; and white is a very rare colour in terrestrial species of moderate size and inoffensive habits. The male, also, as described by Waterton, has a spiral tube, nearly three inches in length, which rises from the base of the beak. It is jet-black, dotted over with minute downy feathers. This tube can be inflated with air, through a communication with the palate; and when not inflated hangs down on one side. The genus consists of four species, the males of which are very distinct, whilst the females, as described by Mr Sclater in a very interesting paper, closely resemble each other, thus offering an excellent instance of the common rule that within the same group the males differ much more from each other than do the females. In a second species (*C. nudicollis*) the male is likewise snow-white, with the exception of a large space of naked skin on the throat and round the eyes, which during the breeding-season is of a fine green colour. In a third species (*C. tricarunculatus*) the head and neck alone of the male are white, the rest of the body being chesnut-brown, and the male of this species is provided with three filamentous projections half as long as the body—one rising from the base of the beak, and the two others from the corners of the mouth.[75]

The coloured plumage and certain other ornaments of the adult males are either retained for life, or are periodically renewed during the summer and breeding-season. At this same season the beak and naked skin about the head frequently change colour, as with some herons, ibises, gulls, one of the bell-birds just noticed, &c. In the white ibis, the cheeks, the inflatable skin of the throat, and the basal portion of the beak then become crimson.[76] In one of the rails, *Gallicrex cristatus*, a

75. Mr Sclater, 'Intellectual Observer', Jan. 1867. 'Waterton's Wanderings', p. 118. See also Mr Salvin's interesting paper, with a plate, in the 'Ibis', 1865, p. 90.
76. 'Land and Water', 1867, p. 394.

large red caruncle is developed during this period on the head of the male. So it is with a thin horny crest on the beak of one of the pelicans, *P. erythrorhynchus*; for after the breeding-season, these horny crests are shed, like horns from the heads of stags, and the shore of an island in a lake in Nevada was found covered with these curious exuviae.[77]

Changes of colour in the plumage according to the season depend, firstly on a double annual moult, secondly on an actual change of colour in the feathers themselves, and thirdly on their dull-coloured margins being periodically shed, or on these three processes more or less combined. The shedding of the deciduary margins may be compared with the shedding of their down by very young birds; for the down in most cases arises from the summits of the first true feathers.[78]

With respect to the birds which annually undergo a double moult, there are, firstly, some kinds, for instance snipes, swallow-plovers (Glareolae), and curlews, in which the two sexes resemble each other, and do not change colour at any season. I do not know whether the winter plumage is thicker and warmer than the summer plumage, but warmth seems the most probable end attained of a double moult, where there is no change of colour. Secondly, there are birds, for instance, certain species of Totanus and other Grallatores, the sexes of which resemble each other, but in which the summer and winter plumage differ slightly in colour. The difference, however, in these cases is so small that it can hardly be an advantage to them; and it may, perhaps, be attributed to the direct action of the different conditions to which the birds are exposed during the two seasons. Thirdly, there are many other birds the sexes of which are alike, but which are widely different in their summer and winter plumage. Fourthly,

---

77. Mr D. G. Elliot, in 'Proc. Zool. Soc.', 1869, p. 589.
78. Nitzsch's 'Pterylography', edited by P. L. Sclater. Ray Soc. 1867, p. 14.

there are birds, the sexes of which differ from each other in colour; but the females, though moulting twice, retain the same colours throughout the year, whilst the males undergo a change of colour, sometimes a great one, as with certain bustards. Fifthly and lastly, there are birds the sexes of which differ from each other in both their summer and winter plumage; but the male undergoes a greater amount of change at each recurrent season than the female—of which the ruff (*Machetes pugnax*) offers a good instance.

With respect to the cause or purpose of the differences in colour between the summer and winter plumage, this may in some instances, as with the ptarmigan,[79] serve during both seasons as a protection. When the difference between the two plumages is slight, it may perhaps be attributed, as already remarked, to the direct action of the conditions of life. But with many birds there can hardly be a doubt that the summer plumage is ornamental, even when both sexes are alike. We may conclude that this is the case with many herons, egrets, &c., for they acquire their beautiful plumes only during the breeding-season. Moreover, such plumes, top-knots, &c., though possessed by both sexes, are occasionally a little more developed in the male than in the female; and they resemble the plumes and ornaments possessed by the males alone of other birds. It is also known that confinement, by affecting the reproductive system of male birds, frequently checks the development of their secondary sexual characters, but has no immediate influence on any other characters; and I am informed by Mr Bartlett that eight or nine specimens of the Knot (*Tringa canutus*) retained their unadorned winter plumage in the Zoological Gardens

79. The brown mottled summer plumage of the ptarmigan is of as much importance to it, as a protection, as the white winter plumage; for in Scandinavia, during the spring, when the snow has disappeared, this bird is known to suffer greatly from birds of prey, before it has acquired its summer dress: see Wilhelm von Wright, in Lloyd, 'Game Birds of Sweden', 1867, p. 125.

throughout the year, from which fact we may infer that the summer plumage though common to both sexes partakes of the nature of the exclusively masculine plumage of many other birds.[80]

From the foregoing facts, more especially from neither sex of certain birds changing colour during either annual moult, or changing so slightly that the change can hardly be of any service to them, and from the females of other species moulting twice yet retaining the same colours throughout the year, we may conclude that the habit of annually moulting twice has not been acquired in order that the male should assume an ornamental character during the breeding-season; but that the double moult, having been originally acquired for some distinct purpose, has subsequently been taken advantage of in certain cases for gaining a nuptial plumage.

It appears at first sight a surprising circumstance that some closely-allied species should regularly undergo a double annual moult, and others only a single one. The ptarmigan, for instance, moults twice or even thrice in the year, and the black-cock only once: some of the splendidly coloured honey-suckers (Nectariniae) of India and some sub-genera of obscurely coloured pipits (Anthus) have a double, whilst others have only a single annual moult.[81] But the gradations in the manner of moulting, which are known to occur with various birds, shew us how species, or whole groups, might have originally acquired their double annual moult, or having once gained the habit, have again lost it. With certain bus-

---

80. In regard to the previous statements on moulting, see, on snipes, &c., Macgillivray, 'Hist. Brit. Birds', vol. iv. p. 371; on Glareolae, curlews, and bustards, Jerdon, 'Birds of India', vol. iii. pp. 615, 630, 683; on Totanus, ibid. p. 700; on the plumes of herons, ibid. p. 738, and Macgillivray, vol. iv. pp. 435 and 444, and Mr Stafford Allen, in the 'Ibis', vol. v. 1863, p. 33.

81. On the moulting of the ptarmigan, see Gould's 'Birds of Great Britain'. On the honey-suckers, Jerdon, 'Birds of India', vol. i. pp. 359, 365, 369. On the moulting of Anthus, see Blyth, in 'Ibis', 1867, p. 32.

tards and plovers the vernal moult is far from complete, some feathers being renewed, and some changed in colour. There is also reason to believe that with certain bustards and rail-like birds, which properly undergo a double moult, some of the older males retain their nuptial plumage throughout the year. A few highly modified feathers may merely be added during the spring to the plumage, as occurs with the disc-formed tail-feathers of certain drongos (*Bhringa*) in India, and with the elongated feathers on the back, neck, and crest of certain herons. By such steps as these, the vernal moult might be rendered more and more complete, until a perfect double moult was acquired. Some of the birds of paradise retain their nuptial feathers throughout the year, and thus have only a single moult; others cast them directly after the breeding-season, and thus have a double moult; and others again cast them at this season during the first year, but not afterwards; so that these latter species are intermediate in their manner of moulting. There is also a great difference with many birds in the length of time during which the two annual plumages are retained; so that the one might come to be retained for the whole year, and the other completely lost. Thus in the spring *Machetes pugnax* retains his ruff for barely two months. In Natal the male widow-bird (*Chera progne*) acquires his fine plumage and long tail-feathers in December or January, and loses them in March; so that they are retained only for about three months. Most species, which undergo a double moult, keep their ornamental feathers for about six months. The male, however, of the wild *Gallus bankiva* retains his neck-hackles for nine or ten months; and when these are cast off, the underlying black feathers on the neck are fully exposed to view. But with the domesticated descendant of this species, the neckhackles of the male are immediately replaced by new ones; so that we here see, as to part of the plumage, a double moult changed under domestication into a single moult.[82]

The common drake (*Anas boschas*) after the breeding-season is well known to lose his male plumage for a period of three months, during which time he assumes that of the female. The male pintail-duck (*Anas acuta*) loses his plumage for the shorter period of six weeks or two months; and Montagu remarks that 'this double moult within so short a time is a most extraordinary circumstance, that seems to bid defiance to all human reasoning'. But the believer in the gradual modification of species will be far from feeling surprise at finding gradations of all kinds. If the male pintail were to acquire his new plumage within a still shorter period, the new male feathers would almost necessarily be mingled with the old, and both with some proper to the female; and this apparently is the case with the male of a not distantly-allied bird, namely the *Merganser serrator*, for the males are said to 'undergo a change of plumage, which assimilates them in some measure to the female'. By a little further acceleration in the process, the double moult would be completely lost.[83]

Some male birds, as before stated, become more brightly coloured in the spring, not by a vernal moult, but either by an actual change of colour in the feathers, or by their obscurely-coloured

---

82. For the foregoing statements in regard to partial moults, and on old males retaining their nuptial plumage, see Jerdon, on bustards and plovers, in 'Birds of India', vol. iii. pp. 617, 637, 709, 711. Also Blyth in 'Land and Water', 1867, p. 84. On the moulting of Paradisea, see an interesting article by Dr W. Marshall, 'Archives Neerlandaises', tom. vi. 1871. On the Vidua, 'Ibis', vol. iii. 1861, p. 133. On the Drongo-shrikes, Jerdon, ibid. vol. i. p. 435. On the vernal moult of the *Herodias bubulcus*, Mr S. S. Allen, in 'Ibis', 1863, p. 33. On *Gallus bankiva*, Blyth, in 'Annals and Mag. of Nat. Hist.', vol. i. 1848, p. 455; see, also, on this subject, my 'Variation of Animals under Domestication', vol. i. p. 236.

83. See Macgillivray, 'Hist. British Birds' (vol. v. pp. 34, 70, and 223), on the moulting of the Anatidae, with quotations from Waterton and Montagu. Also Yarrell, 'Hist. of British Birds', vol. iii. p. 243.

deciduary margins being shed. Changes of colour thus caused may last for a longer or shorter time. In the *Pelecanus onocrotalus* a beautiful rosy tint, with lemon-coloured marks on the breast, overspreads the whole plumage in the spring; but these tints, as Mr Sclater states, 'do not last long, disappearing generally in about six weeks or two months after they have been attained'. Certain finches shed the margins of their feathers in the spring, and then become brighter coloured, while other finches undergo no such change. Thus the *Fringilla tristis* of the United States (as well as many other American species) exhibits its bright colours only when the winter is past, whilst our goldfinch, which exactly represents this bird in habits, and our siskin, which represents it still more closely in structure, undergo no such annual change. But a difference of this kind in the plumage of allied species is not surprising, for with the common linnet, which belongs to the same family, the crimson forehead and breast are displayed only during the summer in England, whilst in Madeira these colours are retained throughout the year.[84]

Display by Male Birds of their Plumage—Ornaments of all kinds, whether permanently or temporarily gained, are sedulously displayed by the males, and apparently serve to excite, attract, or fascinate the females. But the males will sometimes display their ornaments, when not in the presence of the females, as occasionally occurs with grouse at their balz-places, and as may be noticed with the peacock; this latter bird, however, evidently wishes for a spectator of some kind, and, as I have often seen, will shew off his finery before poultry, or even pigs.[85] All naturalists who have

---

84. On the pelican, see Sclater, in 'Proc. Zool. Soc.', 1868, p. 265. On the American finches, see Audubon, 'Ornith. Biography', vol. i. pp. 174, 221, and Jerdon, 'Birds of India', vol. ii. p. 383. On the *Fringilla cannabina* of Madeira, Mr E. Vernon Harcourt, 'Ibis', vol. v. 1863, p. 230.

85. See also 'Ornamental Poultry', by Rev. E. S. Dixon, 1848, p. 8.

closely attended to the habits of birds, whether in a state of nature or under confinement, are unanimously of opinion that the males take delight in displaying their beauty. Audubon frequently speaks of the male as endeavouring in various ways to charm the female. Mr Gould, after describing some peculiarities in a male humming-bird, says he has no doubt that it has the power of displaying them to the greatest advantage before the female. Dr Jerdon[86] insists that the beautiful plumage of the male serves 'to fascinate and attract the female'. Mr Bartlett, at the Zoological Gardens, expressed himself to me in the strongest terms to the same effect.

It must be a grand sight in the forests of India 'to come sud-

Fig. 50. Rupicola crocea, male (T. W. Wood).

86. 'Birds of India', introduct. vol. i. p. xxiv.; on the peacock, vol. iii. p. 507. See Gould's 'Introduction to the Trochilidae', 1861, pp. 15 and 111.

denly on twenty or thirty pea-fowl, the males displaying their gorgeous trains, and strutting about in all the pomp of pride before the gratified females'. The wild turkey-cock erects his glittering plumage, expands his finely-zoned tail and barred wing-feathers, and altogether, with his crimson and blue wattles, makes a superb, though to our eyes, grotesque appearance. Similar facts have already been given with respect to grouse of various kinds. Turning to another Order. The male *Rupicola crocea* (fig. 50) is one of the most beautiful birds in the world, being of a splendid orange, with some of the feathers curiously truncated and plumose. The female is brownish-green, shaded with red, and has a much smaller crest. Sir R. Schomburgk has described their courtship; he found one of their meeting-places where ten males and two females were present. The space was from four to five feet in diameter, and appeared to have been cleared of every blade of grass and smoothed as if by human hands. A male 'was capering, to the apparent delight of several others. Now spreading its wings, throwing up its head, or opening its tail like a fan; now strutting about with a hopping gait until tired, when it gabbled some kind of note, and was relieved by another. Thus three of them successively took the field, and then, with self-approbation, withdrew to rest.' The Indians, in order to obtain their skins, wait at one of the meeting-places till the birds are eagerly engaged in dancing, and then are able to kill with their poisoned arrows four or five males, one after the other.[87] With birds of paradise a dozen or more full-plumaged males congregate in a tree to hold a dancing-party, as it is called by the natives: and here they fly about, raise their wings, elevate their exquisite plumes, and make them vibrate, and the whole tree seems, as Mr Wallace remarks, to be filled with waving plumes. When thus engaged, they become so absorbed that a skilful archer may shoot nearly the whole party. These birds, when

87. 'Journal of R. Geograph. Soc.', vol. x. 1840, p. 236.

Fig. 51. Polyplectron chinquis, male (T. W. Wood).

kept in confinement in the Malay Archipelago, are said to take
much care in keeping their feathers clean; often spreading them
out, examining them, and removing every speck of dirt. One observer, who kept several pairs alive, did not doubt that the display
of the male was intended to please the female.[88]

88. 'Annals and Mag. of Nat. Hist.', vol. xiii. 1854, p. 157; also Wallace, ibid. vol. xx.
    1857, p. 412, and 'The Malay Archipelago', vol. ii. 1869, p. 252. Also Dr Bennett,
    as quoted by Brehm, 'Thierleben', B. iii. s. 326.

The Gold and Amherst pheasants during their courtship not only expand and raise their splendid frills, but twist them, as I have myself seen, obliquely towards the female on whichever side she may be standing, obviously in order that a large surface may be displayed before her.[89] They likewise turn their beautiful tails and tail-coverts a little towards the same side. Mr Bartlett has observed a male Polyplectron (fig. 51) in the act of courtship, and has shewn me a specimen stuffed in the attitude then assumed. The tail and wing-feathers of this bird are ornamented with beautiful ocelli, like those on the peacock's train. Now when the peacock displays himself, he expands and erects his tail transversely to his body, for he stands in front of the female, and has to shew off, at the same time, his rich blue throat and breast. But the breast of the Polyplectron is obscurely coloured, and the ocelli are not confined to the tail-feathers. Consequently the Polyplectron does not stand in front of the female; but he erects and expands his tail-feathers a little obliquely, lowering the expanded wing on the same side, and raising that on the opposite side. In this attitude the ocelli over the whole body are exposed at the same time before the eyes of the admiring female in one grand bespangled expanse. To whichever side she may turn, the expanded wings and the obliquely-held tail are turned towards her. The male Tragopan pheasant acts in nearly the same manner, for he raises the feathers of the body, though not the wing itself, on the side which is opposite to the female, and which would otherwise be concealed, so that nearly all the beautifully spotted feathers are exhibited at the same time.

The Argus pheasant affords a much more remarkable case. The immensely developed secondary wing-feathers are confined to the male; and each is ornamented with a row of from

89. Mr T. W. Wood has given ('The Student', April 1870, p. 115) a full account of this manner of display, by the Gold pheasant and by the Japanese pheasant, *Ph. versicolor*; and he calls it the lateral or one-sided display.

twenty to twenty-three ocelli, above an inch in diameter. These feathers are also elegantly marked with oblique stripes and rows of spots of a dark colour, like those on the skin of a tiger and leopard combined. These beautiful ornaments are hidden until the male shews himself off before the female. He then erects his tail, and expands his wing-feathers into a great, almost upright, circular fan or shield, which is carried in front of the body. The neck and head are held on one side, so that they are concealed by the fan; but the bird in order to see the female, before whom he is displaying himself, sometimes pushes his head between two of the long wing-feathers (as Mr Bartlett has seen), and then presents a grotesque appearance. This must be a frequent habit with the bird in a state of nature, for Mr Bartlett and his son on examining some perfect skins sent from the East, found a place between two of the feathers, which was much frayed, as if the head had here frequently been pushed through. Mr Wood thinks that the male can also peep at the female on one side, beyond the margin of the fan.

The ocelli on the wing-feathers are wonderful objects; for they are so shaded that, as the Duke of Argyll remarks,[90] they stand out like balls lying loosely within sockets. When I looked at the specimen in the British Museum, which is mounted with the wings expanded and trailing downwards, I was however greatly disappointed, for the ocelli appeared flat, or even concave. But Mr Gould soon made the case clear to me, for he held the feathers erect, in the position in which they would naturally be displayed, and now, from the light shining on them from above, each ocellus at once resembled the ornament called a ball and socket. These feathers have been shewn to several artists, and all have expressed their admiration at the perfect shading. It may well be asked, could such artistically shaded ornaments have

90. 'The Reign of Law', 1867, p. 203.

been formed by means of sexual selection? But it will be convenient to defer giving an answer to this question, until we treat in the next chapter of the principle of gradation.

The foregoing remarks relate to the secondary wing-feathers, but the primary wing-feathers, which in most gallinaceous birds are uniformly coloured, are in the Argus pheasant equally wonderful. They are of a soft brown tint with numerous dark spots, each of which consists of two or three black dots with a surrounding dark zone. But the chief ornament is a space parallel to the dark-blue shaft, which in outline forms a perfect second feather lying within the true feather. This inner part is coloured of a lighter chestnut, and is thickly dotted with minute white points. I have shewn this feather to several persons, and many have admired it even more than the ball and socket feathers, and have declared that it was more like a work of art than of nature. Now these feathers are quite hidden on all ordinary occasions, but are fully displayed, together with the long secondary feathers, when they are all expanded together so as to form the great fan or shield.

The case of the male Argus pheasant is eminently interesting, because it affords good evidence that the most refined beauty may serve as a sexual charm, and for no other purpose. We must conclude that this is the case, as the secondary and primary wing-feathers are not at all displayed, and the ball and socket ornaments are not exhibited in full perfection, until the male assumes the attitude of courtship. The Argus pheasant does not possess brilliant colours, so that his success in love appears to depend on the great size of his plumes, and on the elaboration of the most elegant patterns. Many will declare that it is utterly incredible that a female bird should be able to appreciate fine shading and exquisite patterns. It is undoubtedly a marvellous fact that she should possess this almost human degree of taste. He who thinks that he can safely gauge the discrimination and taste of the lower

Fig. 52. Side view of male Argus pheasant, whilst displaying before the female.
Observed and sketched from nature by Mr T. W. Wood.

animals may deny that the female Argus pheasant can appreciate such refined beauty; but he will then be compelled to admit that the extraordinary attitudes assumed by the male during the act of courtship, by which the wonderful beauty of his plumage is fully displayed, are purposeless; and this is a conclusion which I for one will never admit.

Although so many pheasants and allied gallinaceous birds

carefully display their plumage before the females, it is remarkable, as Mr Bartlett informs me, that this is not the case with the dull-coloured Eared and Cheer pheasants (*Crossoptilon auritum* and *Phasianus wallichii*); so that these birds seem conscious that they have little beauty to display. Mr Bartlett has never seen the males of either of these species fighting together, though he has not had such good opportunities for observing the Cheer as the Eared pheasant. Mr Jenner Weir, also, finds that all male birds with rich or strongly-characterised plumage are more quarrelsome than the dull-coloured species belonging to the same groups. The goldfinch, for instance, is far more pugnacious than the linnet, and the blackbird than the thrush. Those birds which undergo a seasonal change of plumage likewise become much more pugnacious at the period when they are most gaily ornamented. No doubt the males of some obscurely-coloured birds fight desperately together, but it appears that when sexual selection has been highly influential, and has given bright colours to the males of any species, it has also very often given a strong tendency to pugnacity. We shall meet with nearly analogous cases when we treat of mammals. On the other hand, with birds the power of song and brilliant colours have rarely been both acquired by the males of the same species; but in this case, the advantage gained would have been the same, namely, success in charming the female. Nevertheless it must be owned that the males of several brilliantly coloured birds have had their feathers specially modified for the sake of producing instrumental music, though the beauty of this cannot be compared, at least according to our taste, with that of the vocal music of many songsters.

We will now turn to male birds which are not ornamented in any high degree, but which nevertheless display during their courtship whatever attractions they may possess. These cases are in some respects more curious than the foregoing, and have been but little noticed. I owe the following facts to Mr Weir, who has long kept confined birds of many kinds, including all the British Fringillidae and Emberizidae. The facts have been selected from a large body of valuable notes kindly sent me by him. The bullfinch makes his advances in front of the female, and then puffs out his breast, so that many more of the crimson feathers are seen at once than otherwise would be the case. At the same time he twists and bows his black tail from side to side in a ludicrous manner. The male chaffinch also stands in front of the female, thus shewing his red breast and 'blue bell', as the fanciers call his head; the wings at the same time being slightly expanded, with the pure white bands on the shoulders thus rendered conspicuous. The common linnet distends his rosy breast, slightly expands his brown wings and tail, so as to make the best of them by exhibiting their white edgings. We must, however, be cautious in concluding that the wings are spread out solely for display, as some birds do so whose wings are not beautiful. This is the case with the domestic cock, but it is always the wing on the side opposite to the female which is expanded, and at the same time scraped on the ground. The male goldfinch behaves differently from all other finches: his wings are beautiful, the shoulders being black, with the dark-tipped wing-feathers spotted with white and edged with golden yellow. When he courts the female, he sways his body from side to side, and quickly turns his slightly expanded wings first to one side, then to the other, with a golden flashing effect. Mr Weir informs me that no other British finch turns thus from side to side during his courtship, not even the closely-allied male siskin, for he would not thus add to his beauty.

Most of the British Buntings are plain coloured birds; but in the spring the feathers on the head of the male reed-bunting (*Emberiza schoeniculus*) acquire a fine black colour by the abrasion of the dusky tips; and these are erected during the act of courtship.

Mr Weir has kept two species of Amadina from Australia: the *A. castanotis* is a very small and chastely coloured finch, with a dark tail, white rump, and jet-black upper tail-coverts, each of the latter being marked with three large conspicuous oval spots of white.[91] This species, when courting the female, slightly spreads out and vibrates these parti-coloured tail-coverts in a very peculiar manner. The male *Amadina Lathami* behaves very differently, exhibiting before the female his brilliantly spotted breast, scarlet rump, and scarlet upper tail-coverts. I may here add from Dr Jerdon that the Indian bulbul (*Pycnonotus haemorrhous*) has its under tail-coverts of a crimson colour, and these, it might be thought, could never be well exhibited; but the bird 'when excited often spreads them out laterally, so that they can be seen even from above'.[92] The crimson under tail-coverts of some other birds, as with one of the woodpeckers, *Picus major*, can be seen without any such display. The common pigeon has iridescent feathers on the breast, and every one must have seen how the male inflates his breast whilst courting the female, thus shewing them off to the best advantage. One of the beautiful bronze-winged pigeons of Australia (*Ocyphaps lophotes*) behaves, as described to me by Mr Weir, very differently: the male, whilst standing before the female, lowers his head almost to the ground, spreads out and raises his tail, and half expands his wings. He then alternately and slowly raises and depresses his body, so that the iridescent metallic feathers are all seen at once, and glitter in the sun.

Sufficient facts have now been given to shew with what care male birds display their various charms, and this they do with the utmost skill. Whilst preening their feathers, they have frequent opportunities for admiring themselves, and of study-

91. For the description of these birds, see Gould's 'Handbook to the Birds of Australia', vol. i. 1865, p. 417.
92. 'Birds of India', vol. ii. p. 96.

ing how best to exhibit their beauty. But as all the males of the same species display themselves in exactly the same manner, it appears that actions, at first perhaps intentional, have become instinctive. If so, we ought not to accuse birds of conscious vanity; yet when we see a peacock strutting about, with expanded and quivering tail-feathers, he seems the very emblem of pride and vanity.

The various ornaments possessed by the males are certainly of the highest importance to them, for in some cases they have been acquired at the expense of greatly impeded powers of flight or of running. The African night-jar (Cosmetornis), which during the pairing-season has one of its primary wing-feathers developed into a streamer of very great length, is thereby much retarded in its flight, although at other times remarkable for its swiftness. The 'unwieldy size' of the secondary wing-feathers of the male Argus pheasant are said 'almost entirely to deprive the bird of flight'. The fine plumes of male birds of paradise trouble them during a high wind. The extremely long tailfeathers of the male widow-birds (Vidua) of Southern Africa render 'their flight heavy'; but as soon as these are cast off they fly as well as the females. As birds always breed when food is abundant, the males probably do not suffer much inconvenience in searching for food from their impeded powers of movement; but there can hardly be a doubt that they must be much more liable to be struck down by birds of prey. Nor can we doubt that the long train of the peacock and the long tail and wing-feathers of the Argus pheasant must render them an easier prey to any prowling tiger-cat, than would otherwise be the case. Even the bright colours of many male birds cannot fail to make them conspicuous to their enemies of all kinds. Hence, as Mr Gould has remarked, it probably is that such birds are generally of a shy disposition, as if conscious that their beauty was a source of danger, and are much more difficult to discover or approach, than the sombre coloured and comparatively tame females, or than the young and as yet unadorned males.[93]

It is a more curious fact that the males of some birds which are provided with special weapons for battle, and which in a state of nature are so pugnacious that they often kill each other, suffer from possessing certain ornaments. Cock-fighters trim the hackles and cut off the combs and gills of their cocks; and the birds are then said to be dubbed. An undubbed bird, as Mr Tegetmeier insists, 'is at a fearful disadvantage; the comb and gills offer an easy hold to his adversary's beak, and as a cock always strikes where he holds, when once he has seized his foe, he has him entirely in his power. Even supposing that the bird is not killed, the loss of blood suffered by an undubbed cock is much greater than that sustained by one that has been trimmed.'[94] Young turkey-cocks in fighting always seize hold of each other's wattles; and I presume that the old birds fight in the same manner. It may perhaps be objected that the comb and wattles are not ornamental, and cannot be of service to the birds in this way; but even to our eyes, the beauty of the glossy black Spanish cock is much enhanced by his white face and crimson comb; and no one who has ever seen the splendid blue wattles of the male Tragopan pheasant, distended in courtship, can for a moment doubt that beauty is the object gained. From the foregoing facts we clearly see that the plumes and other ornaments of the males must be of the highest importance to them; and we further see that beauty is even sometimes more important than success in battle.

93. On the Cosmetornis, see Livingstone's 'Expedition to the Zambesi', 1865, p. 66. On the Argus pheasant, Jardine's 'Nat. Hist. Lib.: Birds', vol. xiv. p. 167. On Birds of Paradise, Lesson, quoted by Brehm, 'Thierleben', B. iii. s. 325. On the widow-bird, Barrow's 'Travels in Africa', vol. i. p. 243, and 'Ibis', vol. iii. 1861, p. 133. Mr Gould, on the shyness of male birds, 'Handbook to Birds of Australia', vol. i. 1865, pp. 210, 457.

94. Tegetmeier, 'The Poultry Book', 1866, p. 139.

From Chapter 16: Birds Concluded

Summary of the Four Chapters on Birds—Most male birds are highly pugnacious during the breeding-season, and some possess weapons adapted for fighting with their rivals. But the most pugnacious and the best armed males rarely or never depend for success solely on their power to drive away or kill their rivals, but have special means for charming the female. With some it is the power of song, or of giving forth strange cries, or instrumental music, and the males in consequence differ from the females in their vocal organs, or in the structure of certain feathers. From the curiously diversified means for producing various sounds, we gain a high idea of the importance of this means of courtship. Many birds endeavour to charm the females by love-dances or antics, performed on the ground or in the air, and sometimes at prepared places. But ornaments of many kinds, the most brilliant tints, combs and wattles, beautiful plumes, elongated feathers, top-knots, and so forth, are by far the commonest means. In some cases mere novelty appears to have acted as a charm. The ornaments of the males must be highly important to them, for they have been acquired in not a few cases at the cost of increased danger from enemies, and even at some loss of power in fighting with their rivals. The males of very many species do not assume their ornamental dress until they arrive at maturity, or they assume it only during the breeding-season, or the tints then become more vivid. Certain ornamental appendages become enlarged, turgid, and brightly coloured during the act of courtship. The males display their charms with elaborate care and to the best effect; and this is done in the presence of the females. The courtship is sometimes a prolonged affair, and many males and females congregate at an appointed place. To suppose that the females do not appreciate the beauty of the males, is to admit that their splendid decorations, all their pomp and display, are useless; and this is incredible. Birds have fine powers of discrimination, and in some few instances it can be shewn that they have a taste for the beautiful. The females, moreover, are known occasionally to exhibit a marked preference or antipathy for certain individual males.

If it be admitted that the females prefer, or are unconsciously excited by the more beautiful males, then the males would slowly but surely be rendered more and more attractive through sexual selection. That it is this sex which has been chiefly modified, we may infer from the fact that, in almost every genus where the sexes differ, the males differ much more from one another than do the females; this is well shewn in certain closely-allied representative species, in which the females can hardly be distinguished, whilst the males are quite distinct. Birds in a state of nature offer individual differences which would amply suffice for the work of sexual selection; but we have seen that they occasionally present more strongly-marked variations which recur so frequently that they would immediately be fixed, if they served to allure the female. The laws of variation must determine the nature of the initial changes, and will have largely influenced the final result. The gradations, which may be observed between the males of allied species, indicate the nature of the steps through which they have passed. They explain also in the most interesting manner how certain characters have originated, such as the indented ocelli on the tail-feathers of the peacock, and the ball and socket ocelli on the wing-feathers of the Argus pheasant. It is evident that the brilliant colours, top-knots, fine plumes, &c., of many male birds cannot have been acquired as a protection; indeed, they sometimes lead to danger. That they are not due to the direct and definite action of the conditions of life, we may feel assured, because the females have been exposed to the same conditions, and yet often differ from the males to an extreme degree. Although it is probable that changed conditions acting during a lengthened period have in some cases produced a definite effect on both sexes, or sometimes on one sex alone, the more important result will have been an increased tendency to vary or to present more strongly marked individual differences; and such differences will have afforded an excellent groundwork for the action of sexual selection.

The laws of inheritance, irrespectively of selection, appear to have determined whether the characters acquired by the males for the sake of ornament, for producing various sounds, and for fighting together, have been transmitted to the males alone or to both sexes, either permanently, or periodically during certain seasons of the year. Why various characters should have been transmitted sometimes in one way and sometimes in another, is not in most cases known; but the period of variability seems often to have been the determining cause. When the two sexes have inherited all characters in common they necessarily resemble each other; but as the successive variations may be differently transmitted, every possible gradation may be found, even within the same genus, from the closest similarity to the widest dissimilarity between the sexes. With many closely-allied species, following nearly the same habits of life, the males have come to differ from each other chiefly through the action of sexual selection; whilst the females have come to differ chiefly from partaking more or less of the characters thus acquired by the males. The effects, moreover, of the definite action of the conditions of life, will not have been masked in the females, as in the males, by the accumulation through sexual selection of strongly-pronounced colours and other ornaments. The individuals of both sexes, however affected, will have been kept at each successive period nearly uniform by the free intercrossing of many individuals.

With species, in which the sexes differ in colour, it is possible or probable that some of the successive variations often tended to be transmitted equally to both sexes; but that when this occurred the females were prevented from acquiring the bright colours of the males, by the destruction which they suffered during incubation. There is no evidence that it is possible by natural selection to convert one form of transmission into another. But there would not be the least difficulty in rendering a female dull-coloured, the male being still kept bright-coloured, by the selection of successive variations, which were from the first limited in their transmission to the same sex. Whether the females of many species have actually been thus modified, must at present remain doubtful. When, through the law of the equal transmission of characters to both sexes, the females were rendered as conspicuously coloured as the males, their instincts appear often to have been modified so that they were led to build domed or concealed nests.

In one small and curious class of cases the characters and habits of the two sexes have been completely transposed, for the females are larger, stronger, more vociferous and brighter coloured than the males. They have, also, become so quarrelsome that they often fight together for the possession of the males, like the males of other pugnacious species for the possession of the females. If, as seems probable, such females habitually drive away their rivals, and by the display of their bright colours or other charms endeavour to attract the males, we can understand how it is that they have gradually been rendered, by sexual selection and sexually-limited transmission, more beautiful than the males—the latter being left unmodified or only slightly modified.

Whenever the law of inheritance at corresponding ages prevails but not that of sexually-limited transmission, then if the parents vary late in life—and we know that this constantly occurs with our poultry, and occasionally with other birds—the young will be left unaffected, whilst the adults of both sexes will be modified. If both these laws of inheritance prevail and either sex varies late in life, that sex alone will be modified, the other sex and the young being unaffected. When variations in brightness or in other conspicuous characters occur early in life, as no doubt often happens, they will not be acted on through sexual selection until the period of reproduction arrives, consequently if dangerous to the young, they will be eliminated through natural selection. Thus we can understand how it is that variations arising late in life have so often been preserved for the ornamentation of the males; the females and the young being left almost unaffected, and therefore like each other. With species having a distinct summer and winter plumage, the males of which either resemble or differ from the females during both seasons or during the summer alone, the degrees and kinds of resemblance between the young and the old are exceedingly complex; and this complexity apparently depends on characters, first acquired by the males, being transmitted in various ways and degrees, as limited by age, sex, and season.

As the young of so many species have been but little modified in colour and in other ornaments, we are enabled to form some judgment with respect to the plumage of their early progenitors; and we may infer that the beauty of our existing species, if we look to the whole class, has been largely increased since that period, of which the immature plumage gives us an indirect record. Many birds, especially those which live much on the ground, have undoubtedly been obscurely coloured for the sake of protection. In some instances the upper exposed surface of the plumage has been thus coloured in both sexes, whilst the lower surface in the males alone has been variously ornamented through sexual selection. Finally, from the facts given in these four chapters, we may conclude that weapons for battle, organs for producing sound, ornaments of many kinds, bright and conspicuous colours, have generally been acquired by the males through variation and sexual selection, and have been transmitted in various ways according to the several laws of inheritance—the females and the young being left comparatively but little modified.[95]

95. I am greatly indebted to the kindness of Mr Sclater for having looked over these four chapters on birds, and the two following ones on mammals. In this way I have been saved from making mistakes about the names of the species, and from stating anything as a fact which is known to this distinguished naturalist to be erroneous. But of course he is not at all answerable for the accuracy of the statements quoted by me from various authorities.

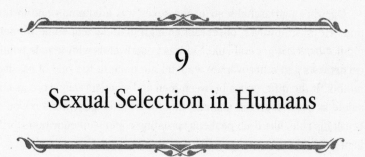

9

Sexual Selection in Humans

Darwin concludes his discussion of sexual selection by turning back to our own species. Having surveyed the differences between male and female animals, he catalogs differences—both real and perceived—between men and women. Men, Darwin points out, are larger than women; the two sexes are also distinguished by other secondary sexual characteristics such as beards and buttocks. Along with these physical differences, Darwin claimed mental ones. "The chief distinction in the intellectual powers of the two sexes," he wrote, "is shown by man's attaining to a higher eminence, in whatever he takes up, than can woman—whether requiring deep thought, reason, or imagination, or merely the use of the senses and hands."

Darwin ascribed these differences to sexual selection. As men competed for women, those with the most physical and mental prowess tended to win. Women's minds were not shaped so keenly by sexual selection. In effect, they remained at the bottom of the ladder of humanity. "It is generally admitted that with woman the powers of intuition, of rapid perception, and perhaps of imitation, are more characteristic of the lower races, and therefore of a past and lower state of civilization," he writes.

Yet women also had a profound effect on human evolution according to Darwin. By preferring certain sorts of beauty over others, women had produced an intense sexual selection that might have created humanity's range of skin colors and other differences.

Darwin's patriarchal view of the sexes was a common one in his time. He was, however, observant enough to know that some women did not meet his general rule. He was good friends with female political activists and other women who did not quite fit the role of passive mother. If the differences between men and women really were as absolute as those found in animals, then these intelligent women would break his rule, like drab peahens sprouting a plume of enormous, brilliant tail feathers.

Darwin tried to avoid this contradiction by summoning up Lamarck's ghost. In a letter he wrote in 1882 to Caroline Kennard, an American correspondent, he declared,

> I certainly think that women, though generally superior to men in moral qualities, are inferior intellectually: and this seems to me to get a grant from the laws of inheritance (if I understand the laws right) in their becoming the intellectual equals of men. On the other hand, there is some reason to believe that aboriginally (& to the present day in the case of savages) men & women were equal in this respect & this would greatly favour their recovering this equality. But to do this, as I believe, women must become as regular "bread-winners" as are men; & we may suspect that the easy education of our children, not to mention the happiness of our homes, would in this case greatly suffer.

In other words, it was possible for women to be equal to men, but Darwin feared the social disruption that would entail.

Society has changed in many ways since Darwin's time. Women serve as prime ministers, as physicists, as lawyers, as doctors, as ministers, as soldiers. They can vote, divorce their husbands, and hold their own property. Some of these changes have happened with relative speed. Take anthropology, the topic Darwin explores so thoroughly in *The Descent of Man*. In the United States in 1973, 26 percent of the people earning PhDs in anthropology were women. By 1998, 55 percent were. With the proper encouragement and training, women can achieve a greatness that is hard to find in *The Descent of Man*. Clearly, sexual selection has not impaired their brains.

And yet those female anthropologists and prime ministers and doctors continue generally to look quite different from men. Women also face different risks for diseases, and women typically live longer than men. Culture may be able to reshape the sexes to some degree, but we are not made of pure putty. And these differences even extend to the brain. Some structures are relatively larger in the brains of women than in those of men, and vice versa. In some regions of the brain, women have much higher densities of neurons. Male and female brains may be sculpted by hormones in different ways.

It is always tempting to read differences in behavior into differences in brain structure. In the nineteenth century, phrenologists tried to ascribe every imaginable quality to a particular bump or curve of the brain. Brains are far too complex to be understood with such a simplistic approach. Psychologists study brains at a higher level, observing the behaviors they produce. And they find some tantalizing clues that some differences between the sexes really are innate. One-year-old girls pay more attention to faces than do boys, who will prefer a film of a car instead. Even so, it is easy to overinterpret these sorts of differences. When faced with certain kinds of mental challenges, men and women use different parts of their brains—and end up performing equally well.

These findings don't fit comfortably in Darwin's stark portrait of the sexes. On the other hand, one can still find evidence of sexual selection in human behavior, particularly when it comes to how the sexes behave toward one another. Across the world's cultures, women tend to prefer marrying men who are older than they are, while the reverse is true for men. These preferences may reflect ancient strategies for reproductive success: young women are more fertile than old ones, and older men can provide more assistance—food, protection, and so on—for a woman's children. As women go through their ovulation cycle, their attraction to different sorts of male faces changes significantly. When women are most fertile, they are attracted to men with large brows and chins, and deep voices—all traits influenced by the male hormone testosterone, and potential signs of good genes. Just as sexual selection in animals turned out to be subtler and richer than Darwin realized, scientists today are finding surprises in humans.

From CHAPTER 19:
Secondary Sexual Characters of Man;
and CHAPTER 20: *Secondary Sexual Characters of Man—continued*

WITH MANKIND THE differences between the sexes are greater than in most of the Quadrumana, but not so great as in some, for instance, the mandrill. Man on an average is considerably taller, heavier, and stronger than woman, with squarer shoulders and more plainly-pronounced muscles. Owing to the relation which exists between muscular development and the projection of the brows,[1] the superciliary ridge is generally more marked in man than in woman. His body, and especially his face, is more hairy, and his voice has a different and more powerful tone. In certain races the women are said to differ slightly in tint from the men. For instance, Schweinfurth, in speaking of a negress belonging to the Monbuttoos, who inhabit the interior of Africa a few degrees north of the Equator, says, 'Like all her race, she had a skin several shades lighter than her husband's, being something of the colour of half-roasted coffee.'[2] As the women labour in the fields and are quite unclothed, it is not likely that they differ in colour from the men owing to less exposure to the weather. European women are perhaps the brighter coloured of the two sexes, as may be seen when both have been equally exposed.

---

1. Schaaffhausen, translation in 'Anthropological Review', Oct. 1868, pp. 419, 420, 427.
2. 'The Heart of Africa', English Transl. 1873, vol. i. p. 544.

Man is more courageous, pugnacious and energetic than woman, and has a more inventive genius. His brain is absolutely larger, but whether or not proportionately to his larger body, has not, I believe, been fully ascertained. In woman the face is rounder; the jaws and the base of the skull smaller; the outlines of the body rounder, in parts more prominent; and her pelvis is broader than in man;[3] but this latter character may perhaps be considered rather as a primary than a secondary sexual character. She comes to maturity at an earlier age than man.

As with animals of all classes, so with man, the distinctive characters of the male sex are not fully developed until he is nearly mature; and if emasculated they never appear. The beard, for instance, is a secondary sexual character, and male children are beardless, though at an early age they have abundant hair on the head. It is probably due to the rather late appearance in life of the successive variations whereby man has acquired his masculine characters, that they are transmitted to the male sex alone. Male and female children resemble each other closely, like the young of so many other animals in which the adult sexes differ widely; they likewise resemble the mature female much more closely than the mature male. The female, however, ultimately assumes certain distinctive characters, and in the formation of her skull, is said to be intermediate between the child and the man.[4] Again, as the young of closely allied though distinct species do not differ nearly so much from each other as do the adults, so it is with the children of the different races of man. Some have even maintained that race-differences cannot be detected in the infantile skull.[5] In regard to colour, the new-born negro child is

3. Ecker, translation in 'Anthropological Review', Oct. 1868, pp. 351–356. The comparison of the form of the skull in men and women has been followed out with much care by Welcker.
4. Ecker and Welcker, ibid. pp. 352, 355; Vogt, 'Lectures on Man', Eng. translat. p. 81.
5. Schaaffhausen, 'Anthropolog. Review', ibid. p. 429.

reddish nut-brown, which soon becomes slaty-grey; the black colour being fully developed within a year in the Soudan, but not until three years in Egypt. The eyes of the negro are at first blue, and the hair chesnut-brown rather than black, being curled only at the ends. The children of the Australians immediately after birth are yellowish-brown, and become dark at a later age. Those of the Guaranys of Paraguay are whitish-yellow, but they acquire in the course of a few weeks the yellowish-brown tint of their parents. Similar observations have been made in other parts of America.[6]

I have specified the foregoing differences between the male and female sex in mankind, because they are curiously like those of the Quadrumana. With these animals the female is mature at an earlier age than the male; at least this is certainly the case in the *Cebus azarae*.[7] The males of most species are larger and stronger than the females, of which fact the gorilla affords a well-known instance. Even in so trifling a character as the greater prominence of the superciliary ridge, the males of certain monkeys differ from the females,[8] and agree in this respect with mankind. In the gorilla and certain other monkeys, the cranium of the adult male presents a strongly-marked sagittal crest, which is absent in the female; and Ecker found a trace of a similar difference between the two sexes in the Australians.[9] With monkeys when there is any difference in the voice, that of the male is the more powerful. We have seen that certain male monkeys have

6. Pruner-Bey, on negro infants as quoted by Vogt, 'Lectures on Man', Eng. translat. 1864, p. 189: for further facts on negro infants, as quoted from Winterbottom and Camper, see Lawrence, 'Lectures on Physiology', &c. 1822, p. 451. For the infants of the Guaranys, see Rengger, 'Säugethiere', &c. s. 3. See also Godron, 'De l'Espèce', tom. ii. 1859, p. 253. For the Australians, Waitz, 'Introduct. to Anthropology', Eng. translat. 1863, p. 99.

7. Rengger, 'Säugethiere', &c. 1830, s. 49.

8. As in *Macacus cynomolgus* (Desmarest, 'Mammalogie', p. 65), and in *Hylobates agilis* (Geoffroy St-Hilaire and F. Cuvier, 'Hist. Nat. des Mamm.' 1824, tom. i. p. 2).

9. 'Anthropological Review', Oct. 1868, p. 353.

a well-developed beard, which is quite deficient, or much less developed in the female. No instance is known of the beard, whiskers, or moustache being larger in the female than in the male monkey. Even in the colour of the beard there is a curious parallelism between man and the Quadrumana, for with man when the beard differs in colour from the hair of the head, as is commonly the case, it is, I believe, almost always of a lighter tint, being often reddish. I have repeatedly observed this fact in England; but two gentlemen have lately written to me, saying that they form an exception to the rule. One of these gentlemen accounts for the fact by the wide difference in colour of the hair on the paternal and maternal sides of his family. Both had been long aware of this peculiarity (one of them having often been accused of dyeing his beard), and had been thus led to observe other men, and were convinced that the exceptions were very rare. Dr Hooker attended to this little point for me in Russia, and found no exception to the rule. In Calcutta, Mr J. Scott, of the Botanic Gardens, was so kind as to observe the many races of men to be seen there, as well as in some other parts of India, namely, two races in Sikhim, the Bhoteas, Hindoos, Burmese, and Chinese, most of which races have very little hair on the face and he always found that when there was any difference in colour between the hair of the head and the beard, the latter was invariably lighter. Now with monkeys, as has already been stated, the beard frequently differs strikingly in colour from the hair of the head, and in such cases it is always of a lighter hue, being often pure white, sometimes yellow or reddish.[10]

10. Mr Blyth informs me that he has only seen one instance of the beard, whiskers, &c., in a monkey becoming white with old age, as is so commonly the case with us. This, however, occurred in an aged *Macacus cynomolgus*, kept in confinement, whose moustaches were 'remarkably long and human-like'. Altogether this old monkey presented a ludicrous resemblance to one of the reigning monarchs of Europe, after whom he was universally nick-named. In certain races of man the hair on the head hardly ever becomes grey; thus Mr D. Forbes has never, as he informs me, seen an instance with the Aymaras and Quichuas of S. America.

In regard to the general hairiness of the body, the women in all races are less hairy than the men; and in some few Quadrumana the under side of the body of the female is less hairy than that of the male.[11] Lastly, male monkeys, like men, are bolder and fiercer than the females. They lead the troop, and when there is danger, come to the front. We thus see how close is the parallelism between the sexual differences of man and the Quadrumana. With some few species, however, as with certain baboons, the orang and the gorilla, there is a considerably greater difference between the sexes, as in the size of the canine teeth, in the development and colour of the hair, and especially in the colour of the naked parts of the skin, than in mankind.

All the secondary sexual characters of man are highly variable, even within the limits of the same race; and they differ much in the several races. These two rules hold good generally throughout the animal kingdom. In the excellent observations made on board the *Novara*,[12] the male Australians were found to exceed the females by only 65 millim. in height, whilst with the Javans the average excess was 218 millim.; so that in this latter race the difference in height between the sexes is more than thrice as great as with the Australians. Numerous measurements were carefully made of the stature, the circumference of the neck and chest, the length of the back-bone and of the arms, in various races; and nearly all these measurements shew that the males differ much more from one another than do the females. This fact indicates that, as far as these characters are concerned, it is the male which has been chiefly modified, since the several races diverged from their common stock.

11. This is the case with the females of several species of Hylobates, see Geoffroy St-Hiliare and F. Cuvier, 'Hist. Nat. des Mamm.', tom. i. See, also, on *H. lar.* 'Penny Cyclopedia', vol. ii. pp. 149, 150.
12. The results were deduced by Dr Weisbach from the measurements made by Drs K. Scherzer and Schwarz, see 'Reise der Novara: Anthropolog. Theil', 1867, ss. 216, 231, 234, 236, 239, 269.

The development of the beard and the hairiness of the body differ remarkably in the men of distinct races, and even in different tribes or families of the same race. We Europeans see this amongst ourselves. In the Island of St Kilda, according to Martin,[13] the men do not acquire beards until the age of thirty or upwards, and even then the beards are very thin. On the Europaeo-Asiatic continent, beards prevail until we pass beyond India; though with the natives of Ceylon they are often absent, as was noticed in ancient times by Diodorus.[14] Eastward of India beards disappear, as with the Siamese, Malays, Kalmucks, Chinese, and Japanese; nevertheless the Ainos,[15] who inhabit the northernmost islands of the Japan Archipelago, are the hairiest men in the world. With negroes the beard is scanty or wanting, and they rarely have whiskers; in both sexes the body is frequently almost destitute of fine down.[16] On the other hand, the Papuans of the Malay Archipelago, who are nearly as black as negroes, possess well-developed beards.[17] In the Pacific Ocean the inhabitants of the Fiji Archipelago have large bushy beards, whilst those of the not distant archipelagoes of Tonga and Samoa are beardless; but these men belong to distinct races. In the Ellice group all the inhabitants belong to the same race; yet on one island alone, namely Nunemaya, 'the men have splendid beards'; whilst on the other islands 'they have, as a rule, a dozen straggling hairs for a beard'.[18]

13. 'Voyage to St Kilda' (3rd edit. 1753), p. 37.
14. Sir J. E. Tennent, 'Ceylon', vol. ii. 1859, p. 107.
15. Quatrefages, 'Revue des Cours Scientifiques', Aug. 29, 1868, p. 630; Vogt 'Lectures on Man', Eng. translat. p. 127.
16. On the beards of negroes, Vogt, 'Lectures', &c. p. 127; Waitz, 'Introduct. to Anthropology', Engl. translat. 1863, vol. i. p. 96. It is remarkable that in the United States ('Investigations in Military and Anthropological Statistics of American Soldiers', 1869, p. 569) the pure negroes and their crossed offspring seem to have bodies almost as hairy as Europeans.
17. Wallace, 'The Malay Arch.', vol. ii. 1869, p. 178.
18. Dr J. Barnard Davis On Oceanic Races, in 'Anthropolog. Review', April, 1870, pp. 185, 191.

Throughout the great American continent the men may be said to be beardless; but in almost all the tribes a few short hairs are apt to appear on the face, especially in old age. With the tribes of North America, Catlin estimates that eighteen out of twenty men are completely destitute by nature of a beard; but occasionally there may be seen a man, who has neglected to pluck out the hairs at puberty, with a soft beard an inch or two in length. The Guaranys of Paraguay differ from all the surrounding tribes in having a small beard, and even some hair on the body, but no whiskers.[19] I am informed by Mr D. Forbes, who particularly attended to this point, that the Aymaras and Quichuas of the Cordillera are remarkably hairless, yet in old age a few straggling hairs occasionally appear on the chin. The men of these two tribes have very little hair on the various parts of the body where hair grows abundantly in Europeans, and the women have none on the corresponding parts. The hair on the head, however, attains an extraordinary length in both sexes, often reaching almost to the ground; and this is likewise the case with some of the N. American tribes. In the amount of hair, and in the general shape of the body, the sexes of the American aborigines do not differ so much from each other, as in most other races.[20] This fact is analogous with what occurs with some closely allied monkeys; thus the sexes of the chimpanzee are not as different as those of the orang or gorilla.[21]

In the previous chapters we have seen that with mammals, birds, fishes, insects, &c., many characters, which there is every reason to believe were primarily gained through sexual selection

---

19. Catlin, 'North American Indians', 3rd edit. 1842, vol. ii. p. 227. On the Guaranys, see Azara, 'Voyages dans l'Amérique Mérid.', tom. ii. 1809, p. 58; also Rengger, 'Säugethiere von Paraguay', s. 3.
20. Prof. and Mrs Agassiz ('Journey in Brazil', p. 530) remark that the sexes of the American Indians differ less than those of the negroes and of the higher races. See also Rengger, ibid. p. 3, on the Guaranys.
21. Rütimeyer, 'Die Grenzen der Thierwelt; eine Betrachtung zu Darwin's Lehre', 1868, s. 54.

by one sex, have been transferred to the other. As this same form of transmission has apparently prevailed much with mankind, it will save useless repetition if we discuss the origin of characters peculiar to the male sex together with certain other characters common to both sexes.

Law of Battle—With savages, for instance the Australians, the women are the constant cause of war both between members of the same tribe and between distinct tribes. So no doubt it was in ancient times; 'nam fuit ante Helenam mulier teterrima belli causa'. ['For even before Helen (of Troy) a woman was a most hideous cause of war'.] With some of the North American Indians, the contest is reduced to a system. That excellent observer, Hearne,[22] says:—'It has ever been the custom among these people for the men to wrestle for any woman to whom they are attached; and, of course, the strongest party always carries off the prize. A weak man, unless he be a good hunter, and well-beloved, is seldom permitted to keep a wife that a stronger man thinks worth his notice. This custom prevails throughout all the tribes, and causes a great spirit of emulation among their youth, who are upon all occasions, from their childhood, trying their strength and skill in wrestling.' With the Guanas of South America, Azara states that the men rarely marry till twenty years old or more, as before that age they cannot conquer their rivals.

Other similar facts could be given; but even if we had no evidence on this head, we might feel almost sure, from the analogy of the higher Quadrumana,[23] that the law of battle had prevailed with man during the early stages of his development. The oc-

---

22. 'A Journey from Prince of Wales Fort.' 8 vo. edit. Dublin, 1796, p. 104. Sir J. Lubbock ('Origin of Civilisation', 1870, p. 69) gives other and similar cases in North America. For the Guanas of S. America see Azara, 'Voyages', &c. tom. ii. p. 94.
23. On the fighting of the male gorillas, see Dr Savage, in 'Boston Journal of Nat. Hist.', vol. v. 1847, p. 423. On *Presbytis entellus*, see the 'Indian Field', 1859, p. 146.

casional appearance at the present day of canine teeth which project above the others, with traces of a diastema or open space for the reception of the opposite canines, is in all probability a case of reversion to a former state, when the progenitors of man were provided with these weapons, like so many existing male Quadrumana. It was remarked in a former chapter that as man gradually became erect, and continually used his hands and arms for fighting with sticks and stones, as well as for the other purposes of life, he would have used his jaws and teeth less and less. The jaws, together with their muscles, would then have been reduced through disuse, as would the teeth through the not well understood principles of correlation and economy of growth; for we everywhere see that parts, which are no longer of service, are reduced in size. By such steps the original inequality between the jaws and teeth in the two sexes of mankind would ultimately have been obliterated. The case is almost parallel with that of many male Ruminants, in which the canine teeth have been reduced to mere rudiments, or have disappeared, apparently in consequence of the development of horns. As the prodigious difference between the skulls of the two sexes in the orang and gorilla stands in close relation with the development of the immense canine teeth in the males, we may infer that the reduction of the jaws and teeth in the early male progenitors of man must have led to a most striking and favourable change in his appearance.

There can be little doubt that the greater size and strength of man, in comparison with woman, together with his broader shoulders, more developed muscles, rugged outline of body, his greater courage and pugnacity, are all due in chief part to inheritance from his half-human male ancestors. These characters would, however, have been preserved or even augmented during the long ages of man's savagery, by the success of the strongest and boldest men, both in the general struggle for life and in their contests for wives; a success which would have ensured their leaving a more numerous progeny than their less favoured brethren. It is not probable that the greater strength of man was primarily acquired through the inherited effects of his having worked harder than woman for his own subsistence and that of his family; for the women in all barbarous nations are compelled to work at least as hard as the men. With civilised people the arbitrament of battle for the possession of the women has long ceased; on the other hand, the men, as a general rule, have to work harder than the women for their joint subsistence, and thus their greater strength will have been kept up.

Difference in the Mental Powers of the two Sexes—With respect to differences of this nature between man and woman, it is probable that sexual selection has played a highly important part. I am aware that some writers doubt whether there is any such inherent difference; but this is at least probable from the analogy of the lower animals which present other secondary sexual characters. No one disputes that the bull differs in disposition from the cow, the wild-boar from the sow, the stallion from the mare, and, as is well known to the keepers of menageries, the males of the larger apes from the females. Woman seems to differ from man in mental disposition, chiefly in her greater tenderness and less selfishness; and this holds good even with savages, as shewn by a well-known passage in Mungo Park's Travels, and by statements made by many other travellers. Woman, owing to her maternal instincts, displays these qualities towards her infants in an eminent degree; therefore it is likely that she would often extend them towards her fellow-creatures. Man is the rival of other men; he delights in competition, and this leads to ambition which passes too easily into selfishness. These latter qualities seem to be his natural and unfortunate birthright. It is generally admitted that with woman the powers of intuition, of rapid perception, and perhaps of imitation, are more strongly marked than in man; but some, at least, of these faculties are characteristic of the lower races, and therefore of a past and lower state of civilisation.

The chief distinction in the intellectual powers of the two sexes is shewn by man's attaining to a higher eminence, in whatever he takes up, than can woman—whether requiring deep thought, reason, or imagination, or merely the use of the senses and hands. If two lists were made of the most eminent men and women in poetry, painting, sculpture, music (inclusive both of composition and performance), history, science, and philosophy, with half-a-dozen names under each subject, the two lists would not bear comparison. We may also infer, from the law of the deviation from averages, so well illustrated by Mr Galton, in his work on 'Hereditary Genius', that if men are capable of a decided pre-eminence over women in many subjects, the average of mental power in man must be above that of woman.

Amongst the half-human progenitors of man, and amongst savages, there have been struggles between the males during many generations for the possession of the females. But mere bodily strength and size would do little for victory, unless associated with courage, perseverance, and determined energy. With social animals, the young males have to pass through many a contest before they win a female, and the older males have to retain their females by renewed battles. They have, also, in the case of mankind, to defend their females, as well as their young, from enemies of all kinds, and to hunt for their joint subsistence. But to avoid enemies or to attack them with success, to capture wild animals, and to fashion weapons, requires the aid of the higher mental faculties, namely, observation, reason, invention, or imagination. These various faculties will thus have been continually put to the test and selected during manhood; they will, moreover, have been strengthened by use during this same period of life. Consequently, in accordance with the principle often alluded to, we might expect that they would at least tend to be transmitted chiefly to the male offspring at the corresponding period of manhood.

Now, when two men are put into competition, or a man with a woman, both possessed of every mental quality in equal perfection, save that one has higher energy, perseverance, and courage, the latter will generally become more eminent in every pursuit, and will gain the ascendancy.[24] He may be said to possess genius—for genius has been declared by a great authority to be patience; and patience, in this sense, means unflinching, undaunted perseverance. But this view of genius is perhaps deficient; for without the higher powers of the imagination and reason, no eminent success can be gained in many subjects. These latter faculties, as well as the former, will have been developed in man, partly through sexual selection—that is, through the contest of rival males, and partly through natural selection—that is, from success in the general struggle for life; and as in both cases the struggle will have been during maturity, the characters gained will have been transmitted more fully to the male than to the female offspring. It accords in a striking manner with this view of the modification and re-inforcement of many of our mental faculties by sexual selection, that, firstly, they notoriously undergo a considerable change at puberty,[25] and, secondly, that eunuchs remain throughout life inferior in these same qualities. Thus man has ultimately become superior to woman. It is, indeed, fortunate that the law of the equal transmission of characters to both sexes prevails with mammals; otherwise it is probable that man would have become as superior in mental endowment to woman, as the peacock is in ornamental plumage to the peahen.

24. J. Stuart Mill remarks ('The Subjection of Women', 1869, p. 122), 'The things in which man most excels woman are those which require most plodding, and long hammering at single thoughts.' What is this but energy and perseverance?
25. Maudsley, 'Mind and Body', p. 31.

It must be borne in mind that the tendency in characters acquired by either sex late in life, to be transmitted to the same sex at the same age, and of early acquired characters to be transmitted to both sexes, are rules which, though general, do not always hold. If they always held good, we might conclude (but I here exceed my proper bounds) that the inherited effects of the early education of boys and girls would be transmitted equally to both sexes; so that the present inequality in mental power between the sexes would not be effaced by a similar course of early training; nor can it have been caused by their dissimilar early training. In order that woman should reach the same standard as man, she ought, when nearly adult, to be trained to energy and perseverance, and to have her reason and imagination exercised to the highest point; and then she would probably transmit these qualities chiefly to her adult daughters. All women, however, could not be thus raised, unless during many generations those who excelled in the above robust virtues were married, and produced offspring in larger numbers than other women. As before remarked of bodily strength, although men do not now fight for their wives, and this form of selection has passed away, yet during manhood, they generally undergo a severe struggle in order to maintain themselves and their families; and this will tend to keep up or even increase their mental powers, and, as a consequence, the present inequality between the sexes.[26]

Having made these preliminary remarks on the admiration felt by savages for various ornaments, and for deformities most un-

---

26. An observation by Vogt bears on this subject: he says, 'It is a remarkable circumstance, that the difference between the sexes, as regards the cranial cavity, increases with the development of the race, so that the male European excels much more the female, than the negro the negress. Welcker confirms this statement of Huschke from his measurements of negro and German skulls.' But Vogt admits ('Lectures on Man', Eng. translat. 1864, p. 81) that more observations are requisite on this point.

sightly in our eyes, let us see how far the men are attracted by the appearance of their women, and what are their ideas of beauty. I have heard it maintained that savages are quite indifferent about the beauty of their women, valuing them solely as slaves; it may therefore be well to observe that this conclusion does not at all agree with the care which the women take in ornamenting themselves, or with their vanity. Burchell[27] gives an amusing account of a Bush-woman who used as much grease, red ochre, and shining powder 'as would have ruined any but a very rich husband'. She displayed also 'much vanity and too evident a consciousness of her superiority'. Mr Winwood Reade informs me that the negroes of the West Coast often discuss the beauty of their women. Some competent observers have attributed the fearfully common practice of infanticide partly to the desire felt by the women to retain their good looks.[28] In several regions the women wear charms and use love-philters to gain the affections of the men; and Mr Brown enumerates four plants used for this purpose by the women of North-Western America.[29]

Hearne,[30] an excellent observer, who lived many years with the American Indians, says, in speaking of the women, 'Ask a Northern Indian what is beauty, and he will answer, a broad flat face, small eyes, high cheek-bones, three or four broad black lines across each cheek, a low forehead, a large broad chin, a clumsy hook nose, a tawny hide, and breasts hanging down to the belt.' Pallas, who visited the northern parts of the Chinese empire, says 'those women are preferred who have the Mandschú form; that is to say, a broad face, high cheekbones, very broad noses,

---

27. 'Travels in S. Africa', 1824, vol. i. p. 41
28. See, for references, Gerland 'Ueber das Aussterben der Naturvölker', 1868, s. 51, 53, 55; also Azara, 'Voyages', &c., tom. ii. p. 116.
29. On the vegetable productions used by the North-Western American Indians, 'Pharmaceutical Journal', vol. x.
30. 'A Journey from Prince of Wales Fort', 8 vo. edit. 1796, p. 89.

and enormous ears';[31] and Vogt remarks that the obliquity of the eye, which is proper to the Chinese and Japanese, is exaggerated in their pictures for the purpose, as it 'seems, of exhibiting its beauty, as contrasted with the eye of the red-haired barbarians'. It is well known, as Huc repeatedly remarks, that the Chinese of the interior think Europeans hideous, with their white skins and prominent noses. The nose is far from being too prominent, according to our ideas, in the natives of Ceylon; yet 'the Chinese in the seventh century, accustomed to the flat features of the Mongol races, were surprised at the prominent noses of the Cingalese; and Thsang described them as having "the beak of a bird, with the body of man".'

Finlayson, after minutely describing the people of Cochin China, says that their rounded heads and faces are their chief characteristics; and, he adds, 'the roundness of the whole countenance is more striking in the women, who are reckoned beautiful in proportion as they display this form of face'. The Siamese have small noses with divergent nostrils, a wide mouth, rather thick lips, a remarkably large face, with very high and broad cheekbones. It is, therefore, not wonderful that 'beauty, according to our notion is a stranger to them. Yet they consider their own females to be much more beautiful than those of Europe.'[32]

It is well known that with many Hottentot women the posterior part of the body projects in a wonderful manner; they are steatopygous; and Sir Andrew Smith is certain that this peculiarity is greatly admired by the men.[33] He once saw a woman who

31. Quoted by Prichard, 'Phys. Hist. of Mankind', 3rd edit. vol. iv. 1844, p. 519; Vogt, 'Lectures on Man', Eng. translat. p. 129. On the opinion of the Chinese on the Cingalese, E. Tennent, 'Ceylon', 1859, vol. ii. p. 107.

32. Prichard, as taken from Crawfurd and Finlayson, 'Phys. Hist. of Mankind', vol. iv. pp. 534, 535.

33. Idem illustrissimus viator dixit mihi praecinctorium vel tabulam foeminae, quod nobis teterrimum est, quondam permagno aestimari ab hominibus in hâc gente. Nunc res mutata est, et censent talem conformationem minime optandam esse. ['The famous explorer told me that the very girdle or protuberance on women

was considered a beauty, and she was so immensely developed behind, that when seated on level ground she could not rise, and had to push herself along until she came to a slope. Some of the women in various negro tribes have the same peculiarity; and, according to Burton, the Somal men 'are said to choose their wives by ranging them in a line, and by picking her out who projects farthest *a tergo*. Nothing can be more hateful to a negro than the opposite form.'[34]

With respect to colour, the negroes rallied Mungo Park on the whiteness of his skin and the prominence of his nose, both of which they considered as 'unsightly and unnatural conformations'. He in return praised the glossy jet of their skins and the lovely depression of their noses; this they said was, 'honeymouth', nevertheless they gave him food. The African Moors, also, 'knitted their brows and seemed to shudder' at the whiteness of his skin. On the eastern coast, the negro boys when they saw Burton, cried out 'Look at the white man; does he not look like a white ape?' On the western coast, as Mr Winwood Reade informs me, the negroes admire a very black skin more than one of a lighter tint. But their horror of whiteness may be attributed, according to this same traveller, partly to the belief held by most negroes that demons and spirits are white, and partly to their thinking it a sign of ill-health.

The Banyai of the more southern part of the continent are negroes, but 'a great many of them are of a light coffee-and-milk colour, and, indeed, this colour is considered handsome throughout the whole country'; so that here we have a different standard of taste. With the Kafirs, who differ much from negroes, 'the

which we see as repulsive is thought to be of considerable value by the men of this tribe. Now, though, the case has changed and they think that such a shape is by no means desirable.']

34. 'The Anthropological Review', November, 1864, p. 237. For additional references, see Waitz, 'Introduct. to Anthropology', Eng. translat. 1863, vol. i. p. 105.

skin, except among the tribes near Delagoa Bay, is not usually black, the prevailing colour being a mixture of black and red, the most common shade being chocolate. Dark complexions, as being most common are naturally held in the highest esteem. To be told that he is light-coloured, or like a white man, would be deemed a very poor compliment by a Kafir. I have heard of one unfortunate man who was so very fair that no girl would marry him.' One of the titles of the Zulu king is 'You who are black'.[35] Mr Galton, in speaking to me about the natives of S. Africa, remarked that their ideas of beauty seem very different from ours; for in one tribe two slim, slight, and pretty girls were not admired by the natives.

Turning to other quarters of the world; in Java, a yellow, not a white girl, is considered, according to Madame Pfeiffer, a beauty. A man of Cochin China 'spoke with contempt of the wife of the English Ambassador, that she had white teeth like a dog, and a rosy colour like that of potato-flowers'. We have seen that the Chinese dislike our white skin, and that the N. Americans admire 'a tawny hide'. In S. America, the Yuracaras, who inhabit the wooded, damp slopes of the eastern Cordillera, are remarkably pale-coloured, as their name in their own language expresses; nevertheless they consider European women as very inferior to their own.[36]

In several of the tribes of North America the hair on the head grows to a wonderful length; and Catlin gives a curious proof how much this is esteemed, for the chief of the Crows was elected to this office from having the longest hair of any

35. Mungo Park's 'Travels in Africa', 4to. 1816, pp. 53, 131. Burton's statement is quoted by Schaaffhausen, 'Archiv für Anthropolog.', 1866, s. 163. On the Banyai, Livingstone, 'Travels', p. 64. On the Kafirs, the Rev. J. Shooter, 'The Kafirs of Natal and the Zulu Country', 1857, p. 1.

36. For the Javans and Cochin-Chinese, see Waitz, 'Introduct. to Anthropology', Eng. translat. vol. i. p. 305. On the Yura-caras, A. d'Orbigny, as quoted in Prichard, 'Phys. Hist. of Mankind', vol. v. 3rd edit. p. 476.

man in the tribe, namely ten feet and seven inches. The Aymaras and Quichuas of S. America, likewise have very long hair;
and this, as Mr D. Forbes informs me, is so much valued as a
beauty, that cutting it off was the severest punishment which
he could inflict on them. In both the Northern and Southern
halves of the continent the natives sometimes increase the apparent length of their hair by weaving into it fibrous substances.
Although the hair on the head is thus cherished, that on the
face is considered by the North American Indians 'as very vul-
gar', and every hair is carefully eradicated. This practice prevails
throughout the American continent from Vancouver's Island in
the north to Tierra del Fuego in the south. When York Minster,
a Fuegian on board the 'Beagle', was taken back to his country,
the natives told him he ought to pull out the few short hairs on
his face. They also threatened a young missionary, who was left
for a time with them, to strip him naked, and pluck the hairs
from his face and body, yet he was far from being a hairy man.
This fashion is carried so far that the Indians of Paraguay eradicate their eyebrows and eyelashes, saying that they do not wish
to be like horses.[37]

It is remarkable that throughout the world the races which
are almost completely destitute of a beard, dislike hairs on the
face and body, and take pains to eradicate them. The Kalmucks
are beardless, and they are well known, like the Americans, to
pluck out all straggling hairs; and so it is with the Polynesians,
some of the Malays, and the Siamese. Mr Veitch states that the
Japanese ladies 'all objected to our whiskers, considering them
very ugly, and told us to cut them off, and be like Japanese men'.
The New Zealanders have short, curled beards; yet they formerly plucked out the hairs on the face. They had a saying that

37. 'North American Indians', by G. Catlin, 3rd edit. 1842, vol. i. p. 49; vol. ii. p. 227.
On the natives of Vancouver's Island, see Sproat, 'Scenes and Studies of Savage
Life', 1868, p. 25. On the Indians of Paraguay, Azara, 'Voyages', tom. ii. p. 105.

'there is no woman for a hairy man'; but it would appear that the fashion has changed in New Zealand, perhaps owing to the presence of Europeans, and I am assured that beards are now admired by the Maories.[38]

On the other hand, bearded races admire and greatly value their beards; among the Anglo-Saxons every part of the body had a recognised value; 'the loss of the beard being estimated at twenty shillings, while the breaking of a thigh was fixed at only twelve'.[39] In the East men swear solemnly by their beards. We have seen that Chinsurdi, the chief of the Makalolo in Africa, thought that beards were a great ornament. In the Pacific the Fijian's beard is 'profuse and bushy, and is his greatest pride'; whilst the inhabitants of the adjacent archipelagoes of Tonga and Samoa are 'beardless, and abhor a rough chin'. In one island alone of the Ellice group 'the men are heavily bearded, and not a little proud thereof'.[40]

We thus see how widely the different races of man differ in their taste for the beautiful. In every nation sufficiently advanced to have made effigies of their gods or of their deified rulers, the sculptors no doubt have endeavoured to express their highest ideal of beauty and grandeur.[41] Under this point of view it is well to compare in our mind the Jupiter or Apollo of the Greeks with the Egyptian or Assyrian statues; and these with the hideous bas-reliefs on the ruined buildings of Central America.

I have met with very few statements opposed to this conclusion. Mr Winwood Reade, however, who has had ample opportu-

---

38. On the Siamese, Prichard, ibid. vol. iv. p. 533. On the Japanese, Veitch in 'Gardeners' Chronicle', 1860, p. 1104. On the New Zealanders. Mantegazza, 'Viaggi e Studi', 1867, p. 526. For the other nations mentioned, see references in Lawrence, 'Lectures on Physiology', &c. 1822, p. 272.
39. Lubbock, 'Origin of Civilisation', 1870, p. 321.
40. Dr Barnard Davis quotes Mr Prichard and others for these facts in regard to the Polynesians, in 'Anthropological Review', April, 1870, pp. 185, 191.
41. Ch. Comte has remarks to this effect in his 'Traité de Législation', 3rd edit. 1837, p. 136.

nities for observation, not only with the negroes of the West Coast of Africa, but with those of the interior who have never associated with Europeans, is convinced that their ideas of beauty are *on the whole* the same as ours; and Dr Rohlfs writes to me to the same effect with respect to Bornu and the countries inhabited by the Pullo tribes. Mr Reade found that he agreed with the negroes in their estimation of the beauty of the native girls; and that their appreciation of the beauty of European women corresponded with ours. They admire long hair, and use artificial means to make it appear abundant; they admire also a beard, though themselves very scantily provided. Mr Reade feels doubtful what kind of nose is most appreciated: a girl has been heard to say, 'I do not want to marry him, he has got no nose', and this shews that a very flat nose is not admired. We should, however, bear in mind that the depressed, broad noses and projecting jaws of the negroes of the West Coast are exceptional types with the inhabitants of Africa. Notwithstanding the foregoing statements, Mr Reade admits that negroes 'do not like the colour of our skin; they look on blue eyes with aversion, and they think our noses too long and our lips too thin'. He does not think it probable that negroes would ever prefer the most beautiful European woman, on the mere grounds of physical admiration, to a good-looking negress.[42]

The general truth of the principle, long ago insisted on by Humboldt,[43] that man admires and often tries to exaggerate whatever characters nature may have given him, is shewn in

---

42. The 'African Sketch Book', vol. ii. 1873, pp. 253, 394, 521. The Fuegians, as I have been informed by a missionary who long resided with them, consider European women as extremely beautiful; but from what we have seen of the judgment of the other aborigines of America, I cannot but think that this must be a mistake, unless indeed the statement refers to the few Fuegians who have lived for some time with Europeans, and who must consider us as superior beings. I should add that a most experienced observer, Capt. Burton, believes that a woman whom we consider beautiful is admired throughout the world, 'Anthropological Review', March, 1864, p. 245.

43. 'Personal Narrative', Eng. translat. vol. iv. p. 518, and elsewhere. Mantegazza, in his 'Viaggi e Studi', 1867, strongly insists on this same principle.

many ways. The practice of beardless races extirpating every trace of a beard, and often all the hairs on the body, affords one illustration. The skull has been greatly modified during ancient and modern times by many nations; and there can be little doubt that this has been practised, especially in N. and S. America, in order to exaggerate some natural and admired peculiarity. Many American Indians are known to admire a head so extremely flattened as to appear to us idiotic. The natives on the north-western coast compress the head into a pointed cone; and it is their constant practice to gather the hair into a knot on the top of the head, for the sake, as Dr Wilson remarks, 'of increasing the apparent elevation of the favourite conoid form'. The inhabitants of Arakhan 'admire a broad, smooth forehead, and in order to produce it, they fasten a plate of lead on the heads of the new-born children'. On the other hand, 'a broad, well-rounded occiput is considered a great beauty' by the natives of the Fiji islands.[44]

As with the skull, so with the nose; the ancient Huns during the age of Attila were accustomed to flatten the noses of their infants with bandages, 'for the sake of exaggerating a natural conformation'. With the Tahitians, to be called *long-nose* is considered as an insult, and they compress the noses and foreheads of their children for the sake of beauty. The same holds with the Malays of Sumatra, the Hottentots, certain Negroes, and the natives of Brazil.[45] The Chinese have by nature unusually small

---

44. On the skulls of the American tribes, see Nott and Gliddon, 'Types of Mankind', 1854, p. 440; Prichard, 'Phys. Hist. of Mankind', vol. i. 3rd edit. p. 321; on the natives of Arakhan, ibid. vol. iv. p. 537. Wilson, 'Physical Ethnology', Smithsonian Institution, 1863, p. 288; on the Fijians, p. 290. Sir J. Lubbock ('Prehistoric Times', 2nd edit. 1869, p. 506) gives an excellent résumé on this subject.

45. On the Huns, Godron, 'De l'Espèce', tom. ii. 1859, p. 300. On the Tahitians, Waitz, 'Anthropolog.', Eng. translat. vol. i. p. 305. Marsden, quoted by Prichard, 'Phys. Hist. of Mankind', 3rd edit. vol. v. p. 67. Lawrence, 'Lectures on Physiology', p. 337.

feet;[46] and it is well known that the women of the upper classes distort their feet to make them still smaller. Lastly, Humboldt thinks that the American Indians prefer colouring their bodies with red paint in order to exaggerate their natural tint; and until recently European women added to their naturally bright colours by rouge and white cosmetics; but it may be doubted whether barbarous nations have generally had any such intention in painting themselves.

In the fashions of our own dress we see exactly the same principle and the same desire to carry every point to an extreme; we exhibit, also, the same spirit of emulation. But the fashions of savages are far more permanent than ours; and whenever their bodies are artificially modified, this is necessarily the case. The Arab women of the Upper Nile occupy about three days in dressing their hair; they never imitate other tribes, 'but simply vie with each other in the superlativeness of their own style'. Dr Wilson, in speaking of the compressed skulls of various American races, adds, 'such usages are among the least eradicable, and long survive the shock of revolutions that change dynasties and efface more important national peculiarities'.[47] The same principle comes into play in the art of breeding; and we can thus understand, as I have elsewhere explained,[48] the wonderful development of the many races of animals and plants, which have been kept merely for ornament. Fanciers always wish each character to be somewhat increased; they do not admire a medium standard; they certainly do not desire any great and abrupt change in the character of their breeds; they admire solely what they are accustomed to, but

46. This fact was ascertained in the 'Reise der Novara: Anthropolog. Thiel', Dr Weisbach, 1867, s. 265.
47. 'Smithsonian Institution', 1863, p. 289. On the fashions of Arab women, Sir S. Baker, 'The Nile Tributaries', 1867, p. 121.
48. 'The Variation of Animals and Plants under Domestication', vol. i. p. 214; vol. ii. p. 240.

they ardently desire to see each characteristic feature a little more developed.

The senses of man and of the lower animals seem to be so constituted that brilliant colours and certain forms, as well as harmonious and rhythmical sounds, give pleasure and are called beautiful; but why this should be so, we know not. It is certainly not true that there is in the mind of man any universal standard of beauty with respect to the human body. It is, however, possible that certain tastes may in the course of time become inherited, though there is no evidence in favour of this belief; and if so, each race would possess its own innate ideal standard of beauty. It has been argued[49] that ugliness consists in an approach to the structure of the lower animals, and no doubt this is partly true with the more civilised nations, in which intellect is highly appreciated; but this explanation will hardly apply to all forms of ugliness. The men of each race prefer what they are accustomed to; they cannot endure any great change; but they like variety, and admire each characteristic carried to a moderate extreme.[50] Men accustomed to a nearly oval face, to straight and regular features, and to bright colours, admire, as we Europeans know, these points when strongly developed. On the other hand, men accustomed to a broad face, with high cheek-bones, a depressed nose, and a black skin, admire these peculiarities when strongly marked. No doubt characters of all kinds may be too much developed for beauty. Hence a perfect beauty, which implies many characters modified in a particular manner, will be in every race a prodigy. As the great anatomist Bichat long ago said, if every one were cast in the same mould, there would be no such thing as beauty. If all our women were to become as beautiful as the

---

49. Schaaffhausen, 'Archiv für Anthropologie', 1866, s. 164.
50. Mr Bain has collected ('Mental and Moral Science', 1868, pp. 304–314) about a dozen more or less different theories of the idea of beauty; but none are quite the same as that here given.

Venus de' Medici, we should for a time be charmed; but we should soon wish for variety; and as soon as we had obtained variety, we should wish to see certain characters a little exaggerated beyond the then existing common standard.

From Chapter 20:
Secondary Sexual Characters of Man—continued

Colour of the Skin—The best kind of evidence that in man the colour of the skin has been modified through sexual selection is scanty; for in most races the sexes do not differ in this respect, and only slightly, as we have seen, in others. We know, however, from the many facts already given that the colour of the skin is regarded by the men of all races as a highly important element in their beauty; so that it is a character which would be likely to have been modified through selection, as has occurred in innumerable instances with the lower animals. It seems at first sight a monstrous supposition that the jet-blackness of the negro should have been gained through sexual selection; but this view is supported by various analogies, and we know that negroes admire their own colour. With mammals, when the sexes differ in colour, the male is often black or much darker than the female; and it depends merely on the form of inheritance whether this or any other tint is transmitted to both sexes or to one alone. The resemblance to a negro in miniature of *Pithecia satanas* with his jet black skin, white rolling eyeballs, and hair parted on the top of the head, is almost ludicrous.

The colour of the face differs much more widely in the various kinds of monkeys than it does in the races of man; and we have some reason to believe that the red, blue, orange, almost white and black tints of their skin, even when common to both sexes, as well as the bright colours of their fur, and the ornamental tufts about the head, have all been acquired through sexual selection. As the order of development during growth, generally indicates the order in which the characters of a species have been developed and modified during previous generations; and as the newly-born infants of the various races of man do not differ nearly as much in colour as do the adults, although their bodies are as completely destitute of hair, we have some slight evidence that the tints of the different races were acquired at a period subsequent to the removal of the hair, which must have occurred at a very early period in the history of man.

Summary—We may conclude that the greater size, strength, courage, pugnacity, and energy of man, in comparison with woman, were acquired during primeval times, and have subsequently been augmented, chiefly through the contests of rival males for the possession of the females. The greater intellectual vigour and power of invention in man is probably due to natural selection, combined with the inherited effects of habit, for the most able men will have succeeded best in defending and providing for themselves and for their wives and offspring. As far as the extreme intricacy of the subject permits us to judge, it appears that our male ape-like progenitors acquired their beards as an ornament to charm or excite the opposite sex, and transmitted them only to their male offspring. The females apparently first had their bodies denuded of hair, also as a sexual ornament; but they transmitted this character almost equally to both sexes. It is not improbable that the females were modified in other respects for the same purpose and by the same means; so that women have acquired sweeter voices and become more beautiful than men.

It deserves attention that with mankind the conditions were in many respects much more favourable for sexual selection, during a very early period, when man had only just attained to the rank of manhood, than during later times. For he would then, as we may safely conclude, have been guided more by his instinctive passions, and less by foresight or reason. He would have jealously guarded his wife or wives. He would not have practised infanticide; nor valued his wives merely as useful slaves; nor have been betrothed to them during infancy. Hence we may infer that the races of men were differentiated, as far as sexual selection is concerned, in chief part at a very remote epoch; and this conclusion throws light on the remarkable fact that at the most ancient period, of which we have as yet any record, the races of man had already come to differ nearly or quite as much as they do at the present day.

The views here advanced, on the part which sexual selection has played in the history of man, want scientific precision. He who does not admit this agency in the case of the lower animals, will disregard all that I have written in the later chapters on man. We cannot positively say that this character, but not that, has been thus modified; it has, however, been shewn that the races of man differ from each other and from their nearest allies, in certain characters which are of no service to them in their daily habits of life, and which it is extremely probable would have been modified through sexual selection. We have seen that with the lowest savages the people of each tribe admire their own characteristic qualities—the shape of the head and face, the squareness of the cheek-bones, the prominence or depression of the nose, the colour of the skin, the length of the hair on the head, the absence of hair on the face and body, or the presence of a great beard, and so forth. Hence these and other such points could hardly fail to be slowly and gradually exaggerated, from the more powerful and able men in each tribe, who would succeed in rearing the largest number of offspring, having selected during many generations for their wives the most strongly characterised and therefore most attractive women. For my own part I conclude that of all the causes which have led to the differences in external appearance between the races of man, and to a certain extent between man and the lower animals, sexual selection has been the most efficient.

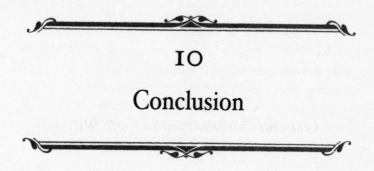

10

Conclusion

Despite the great length and wandering course of *The Descent of Man*, Darwin ends it with a cogent, concise summary. There is little need to introduce his conclusion, except perhaps to note that with the completion of *The Descent of Man*, Darwin had relatively little left to say about humans. In 1872, a year after *The Descent of Man* came out, he published *The Expression of Emotion in Man and Animals*. And then he was done. Darwin opened up a vast new science, the investigation of how we came to be, and then walked away. He went on to publish books on climbing vines, the fertilization of flowers, and earthworms. To Darwin, man was a fascinating species, but one of many.

CHAPTER 21:
General Summary and Conclusion

A BRIEF SUMMARY will be sufficient to recall to the reader's mind the more salient points in this work. Many of the views which have been advanced are highly speculative, and some no doubt will prove erroneous; but I have in every case given the reasons which have led me to one view rather than to another. It seemed worth while to try how far the principle of evolution would throw light on some of the more complex problems in the natural history of man. False facts are highly injurious to the progress of science, for they often endure long; but false views, if supported by some evidence, do little harm, for every one takes a salutary pleasure in proving their falseness; and when this is done, one path towards error is closed and the road to truth is often at the same time opened.

The main conclusion here arrived at, and now held by many naturalists who are well competent to form a sound judgment, is that man is descended from some less highly organised form. The grounds upon which this conclusion rests will never be shaken, for the close similarity between man and the lower animals in embryonic development, as well as in innumerable points of structure and constitution, both of high and of the most trifling importance—the rudiments which he retains, and the abnormal reversions to which he is occasionally liable—are facts which cannot be disputed. They have long been known, but until recently they told us nothing with respect to the origin of man. Now when viewed by the light of our knowledge of the whole organic world, their meaning is unmistakable. The great principle of evolution stands up clear and firm, when these groups of facts are considered in connection with others, such as the mutual affinities of the members of the same group, their geographical distribution in past and present times, and their geological succession. It is incredible that all these facts should speak falsely. He who is not content to look, like a savage, at the phenomena of nature as disconnected, cannot any longer believe that man is the work of a separate act of creation. He will be forced to admit that the close resemblance of the embryo of man to that, for instance, of a dog—the construction of his skull, limbs and whole frame on the same plan with that of other mammals, independently of the uses to which the parts may be put—the occasional reappearance of various structures, for instance of several muscles, which man does not normally possess, but which are common to the Quadrumana—and a crowd of analogous facts—all point in the plainest manner to the conclusion that man is the co-descendant with other mammals of a common progenitor.

We have seen that man incessantly presents individual differences in all parts of his body and in his mental faculties. These differences or variations seem to be induced by the same general causes, and to obey the same laws as with the the lower animals. In both cases similar laws of inheritance prevail. Man tends to increase at a greater rate than his means of subsistence; consequently he is occasionally subjected to a severe struggle for existence, and natural selection will have effected whatever lies within its scope. A succession of strongly-marked variations of a similar nature is by no means requisite; slight fluctuating differences in the individual suffice for the work of natural selection; not that we have any reason to suppose that in the same species, all parts of the organisation tend to vary to the same degree. We may feel assured that the inherited effects of the long-continued use or disuse of parts will have done much in the same direction with natural selection. Modifications formerly of importance, though no longer of any special use, are long-inherited. When one part is modified, other parts change through the principle of correlation, of which we have instances in many curious cases of correlated monstrosities. Something may be attributed to the direct and definite action of the surrounding conditions of life, such as abundant food, heat or moisture; and lastly, many characters of slight physiological importance, some indeed of considerable importance, have been gained through sexual selection.

No doubt man, as well as every other animal, presents structures, which seem to our limited knowledge, not to be now of any service to him, nor to have been so formerly, either for the general conditions of life, or in the relations of one sex to the other. Such structures cannot be accounted for by any form of selection, or by the inherited effects of the use and disuse of parts. We know, however, that many strange and strongly-marked peculiarites of structure occasionally appear in our domesticated productions, and if their unknown causes were to act more uniformly, they would probably become common to all the individuals of the species. We may hope hereafter to understand something about the causes of such occasional modifications, especially through the study of monstrosities: hence the labours of experimentalists, such as those of M. Camille Dareste, are full of promise for the future. In general we can only say that the cause of each slight variation and of each monstrosity lies much more in the constitution of the organism, than in the nature of the surrounding conditions; though new and changed conditions certainly play an important part in exciting organic changes of many kinds.

Through the means just specified, aided perhaps by others as yet undiscovered, man has been raised to his present state. But since he attained to the rank of manhood, he has diverged into distinct races, or as they may be more fitly called, sub-species. Some of these, such as the Negro and European, are so distinct that, if specimens had been brought to a naturalist without any further information, they would undoubtedly have been considered by him as good and true species. Nevertheless all the races agree in so many unimportant details of structure and in so many mental peculiarities, that these can be accounted for only by inheritance from a common progenitor; and a progenitor thus characterised would probably deserve to rank as man.

It must not be supposed that the divergence of each race from the other races, and of all from a common stock, can be traced back to any one pair of progenitors. On the contrary, at every stage in the process of modification, all the individuals which were in any way better fitted for their conditions of life, though in different degrees, would have survived in greater numbers than the less well-fitted. The process would have been like that followed by man, when he does not intentionally select particular individuals, but breeds from all the superior individuals, and neglects the inferior. He thus slowly but surely modifies his stock, and unconsciously forms a new strain. So with respect to modifications acquired independently of selection, and due to variations arising from the nature of the organism and the action of the surrounding conditions, or from changed habits of life, no single pair will have been modified much more than the other pairs inhabiting the same country, for all will have been continually blended through free intercrossing.

By considering the embryological structure of man—the homologies which he presents with the lower animals—the rudiments which he retains—and the reversions to which he is liable, we can partly recall in imagination the former condition of our early progenitors; and can approximately place them in their proper place in the zoological series. We thus learn that man is descended from a hairy, tailed quadruped, probably arboreal in its habits, and an inhabitant of the Old World. This creature, if its whole structure had been examined by a naturalist, would have been classed amongst the Quadrumana, as surely as the still more ancient progenitor of the Old and New World monkeys. The Quadrumana and all the higher mammals are probably derived from an ancient marsupial animal, and this through a long line of diversified forms, from some amphibian-like creature, and this again from some fish-like animal. In the dim obscurity of the past we can see that the early progenitor of all the Vertebrata must have been an aquatic animal, provided with branchiae, with the two sexes united in the same individual, and with the most important organs of the body (such as the brain and heart) imperfectly or not at all developed. This animal seems to have been more like the larvae of the existing marine Ascidians than any other known form.

The high standard of our intellectual powers and moral disposition is the greatest difficulty which presents itself, after we have been driven to this conclusion on the origin of man. But every one who admits the principle of evolution, must see that the mental powers of the higher animals, which are the same in kind with those of man, though so different in degree, are capable of advancement. Thus the interval between the mental powers of one of the higher apes and of a fish, or between those of an ant and scale-insect, is immense; yet their development does not offer any special difficulty; for with our domesticated animals, the mental faculties are certainly variable, and the variations are inherited. No one doubts that they are of the utmost importance to animals in a state of nature. Therefore the conditions are favourable for their development through natural selection. The same conclusion may be extended to man, the intellect must have been all-important to him, even at a very remote period, as enabling him to invent and use language, to make weapons, tools, traps, &c., whereby with the aid of his social habits, he long ago became the most dominant of all living creatures.

A great stride in the development of the intellect will have followed, as soon as the half-art and half-instinct of language came into use; for the continued use of language will have re-acted on the brain and produced an inherited effect; and this again will have reacted on the improvement of language. As Mr Chauncey Wright[1] has well remarked, the largeness of the brain in man relatively to his body, compared with the lower animals, may be attributed in chief part to the early use of some simple form of language—that wonderful engine which affixes signs to all sorts of objects and qualities, and excites trains of thought which would never arise from the mere impression of the senses, or if they did arise could not be followed out. The higher intellectual powers of man, such as those of ratiocination, abstraction, self-consciousness, &c., probably follow from the continued improvement and exercise of the other mental faculties.

The development of the moral qualities is a more interesting problem. The foundation lies in the social instincts, including under this term the family ties. These instincts are highly complex, and in the case of the lower animals give special tendencies towards certain definite actions; but the more important elements are love, and the distinct emotion of sympathy. Animals endowed with the social instincts take pleasure in one another's company, warn one another of danger, defend and

1. 'On the Limits of Natural Selection', in the 'North American Review', Oct. 1870, p. 295.

aid one another in many ways. These instincts do not extend to all the individuals of the species, but only to those of the same community. As they are highly beneficial to the species, they have in all probability been acquired through natural selection.

A moral being is one who is capable of reflecting on his past actions and their motives—of approving of some and disapproving of others; and the fact that man is the one being who certainly deserves this designation, is the greatest of all distinctions between him and the lower animals. But in the fourth chapter I have endeavoured to shew that the moral sense follows, firstly, from the enduring and ever-present nature of the social instincts; secondly, from man's appreciation of the approbation and disapprobation of his fellows; and thirdly, from the high activity of his mental faculties, with past impressions extremely vivid; and in these latter respects he differs from the lower animals. Owing to this condition of mind, man cannot avoid looking both backwards and forwards, and comparing past impressions. Hence after some temporary desire or passion has mastered his social instincts, he reflects and compares the now weakened impression of such past impulses with the ever-present social instincts; and he then feels that sense of dissatisfaction which all unsatisfied instincts leave behind them, he therefore resolves to act differently for the future—and this is conscience. Any instinct, permanently stronger or more enduring than another, gives rise to a feeling which we express by saying that it ought to be obeyed. A pointer dog, if able to reflect on his past conduct, would say to himself, I ought (as indeed we say of him) to have pointed at that hare and not have yielded to the passing temptation of hunting it.

Social animals are impelled partly by a wish to aid the members of their community in a general manner, but more commonly to perform certain definite actions. Man is impelled by the same general wish to aid his fellows; but has few or no special instincts. He differs also from the lower animals in the power of expressing his desires by words, which thus become a guide to the aid required and bestowed. The motive to give aid is likewise much modified in man: it no longer consists solely of a blind instinctive impulse, but is much influenced by the praise or blame of his fellows. The appreciation and the bestowal of praise and blame both rest on sympathy; and this emotion, as we have seen, is one of the most important elements of the social instincts. Sympathy, though gained as an instinct, is also much strengthened by exercise or habit. As all men desire their own happiness, praise or blame is bestowed on actions and motives, according as they lead to this end; and as happiness is an essential part of the general good, the greatest-happiness principle indirectly serves as a nearly safe standard of right and wrong. As the reasoning powers advance and experience is gained, the remoter effects of certain lines of conduct on the character of the individual, and on the general good, are perceived; and then the self-regarding virtues come within the scope of public opinion, and receive praise, and their opposites blame. But with the less civilised nations reason often errs, and many bad customs and base superstitions come within the same scope, and are then esteemed as high virtues, and their breach as heavy crimes.

The moral faculties are generally and justly esteemed as of higher value than the intellectual powers. But we should bear in mind that the activity of the mind in vividly recalling past impressions is one of the fundamental though secondary bases of conscience. This affords the strongest argument for educating and stimulating in all possible ways the intellectual faculties of every human being. No doubt a man with a torpid mind, if his social affections and sympathies are well developed, will be led to good actions, and may have a fairly sensitive conscience. But whatever renders the imagination more vivid and strengthens the habit of recalling and comparing past impressions, will make the conscience more sensitive, and may even somewhat compensate for weak social affections and sympathies.

The moral nature of man has reached its present standard, partly through the advancement of his reasoning powers and consequently of a just public opinion, but especially from his sympathies having been rendered more tender and widely diffused through the effects of habit, example, instruction, and reflection. It is not improbable that after long practice virtuous tendencies may be inherited. With the more civilised races, the conviction of the existence of an all-seeing Deity has had a potent influence on the advance of morality. Ultimately man does not accept the praise or blame of his fellows as his sole guide, though few escape this influence, but his habitual convictions, controlled by reason, afford him the safest rule. His conscience then becomes the supreme judge and monitor. Nevertheless the first foundation or origin of the moral sense lies in the social instincts, including sympathy; and these instincts no doubt were primarily gained, as in the case of the lower animals, through natural selection.

The belief in God has often been advanced as not only the greatest, but the most complete of all the distinctions between man and the lower animals. It is however impossible, as we have seen, to maintain that this belief is innate or instinctive in man. On the other hand a belief in all-pervading spiritual agencies seems to be universal; and apparently follows from a considerable advance in man's reason, and from a still greater advance in his faculties of imagination, curiosity and wonder. I am aware that the assumed instinctive belief in God has been used by many persons as an argument for His existence. But this is a rash argument, as we should thus be compelled to believe in the existence of many cruel and malignant spirits, only a little more powerful than man; for the belief in them is far more general than in a beneficent Deity. The idea of a universal and beneficent Creator does not seem to arise in the mind of man, until he has been elevated by long-continued culture.

He who believes in the advancement of man from some low organised form, will naturally ask how does this bear on the belief in the immortality of the soul. The barbarous races of man, as Sir J. Lubbock has shewn, possess no clear belief of this kind; but arguments derived from the primeval beliefs of savages are, as we have just seen, of little or no avail. Few persons feel any anxiety from the impossibility of determining at what precise period in the development of the individual, from the first trace of a minute germinal vesicle, man becomes an immortal being; and there is no greater cause for anxiety because the period cannot possibly be determined in the gradually ascending organic scale.[2]

I am aware that the conclusions arrived at in this work will be denounced by some as highly irreligious; but he who denounces them is bound to shew why it is more irreligious to explain the origin of man as a distinct species by descent from some lower form, through the laws of variation and natural selection, than to explain the birth of the individual through the laws of ordinary reproduction. The birth both of the species and of the individual are equally parts of that grand sequence of events, which our minds refuse to accept as the result of blind chance. The understanding revolts at such a conclusion, whether or not we are able to believe that every slight variation of structure—the union of each pair in marriage—the dissemination of each

---

2. The Rev. J. A. Picton gives a discussion to this effect in his 'New Theories and the Old Faith', 1870.

seed—and other such events, have all been ordained for some special purpose.

Sexual selection has been treated at great length in this work; for, as I have attempted to shew, it has played an important part in the history of the organic world. I am aware that much remains doubtful, but I have endeavoured to give a fair view of the whole case. In the lower divisions of the animal kingdom, sexual selection seems to have done nothing: such animals are often affixed for life to the same spot, or have the sexes combined in the same individual, or what is still more important, their perceptive and intellectual faculties are not sufficiently advanced to allow of the feelings of love and jealousy, or of the exertion of choice. When, however, we come to the Arthropoda and Vertebrata, even to the lowest classes in these two great Sub-Kingdoms, sexual selection has effected much.

In the several great classes of the animal kingdom—in mammals, birds, reptiles, fishes, insects, and even crustaceans—the differences between the sexes follow nearly the same rules. The males are almost always the wooers; and they alone are armed with special weapons for fighting with their rivals. They are generally stronger and larger than the females, and are endowed with the requisite qualities of courage and pugnacity. They are provided, either exclusively or in a much higher degree than the females, with organs for vocal or instrumental music, and with odoriferous glands. They are ornamented with infinitely diversified appendages, and with the most brilliant or conspicuous colours, often arranged in elegant patterns, whilst the females are unadorned. When the sexes differ in more important structures, it is the male which is provided with special sense-organs for discovering the female, with locomotive organs for reaching her, and often with prehensile organs for holding her. These various structures for charming or securing the female are often developed in the male during only part of the year, namely the breeding-season. They have in many cases been more or less transferred to the females; and in the latter case they often appear in her as mere rudiments. They are lost or never gained by the males after emasculation. Generally they are not developed in the male during early youth, but appear a short time before the age for reproduction. Hence in most cases the young of both sexes resemble each other; and the female somewhat resembles her young offspring throughout life. In almost every great class a few anomalous cases occur, where there has been an almost complete transposition of the characters proper to the two sexes; the females assuming characters which properly belong to the males. This surprising uniformity in the laws regulating the differences between the sexes in so many and such widely separated classes, is intelligible if we admit the action of one common cause, namely sexual selection.

Sexual selection depends on the success of certain individuals over others of the same sex, in relation to the propagation of the species; whilst natural selection depends on the success of both sexes, at all ages, in relation to the general conditions of life. The sexual struggle is of two kinds; in the one it is between the individuals of the same sex, generally the males, in order to drive away or kill their rivals, the females remaining passive; whilst in the other, the struggle is likewise between the individuals of the same sex, in order to excite or charm those of the opposite sex, generally the females, which no longer remain passive, but select the more agreeable partners. This latter kind of selection is closely analogous to that which man unintentionally, yet effectually, brings to bear on his domesticated productions, when he preserves during a long period the most pleasing or useful individuals, without any wish to modify the breed.

The laws of inheritance determine whether characters gained through sexual selection by either sex shall be transmitted to the same sex, or to both; as well as the age at which they shall be developed. It appears that variations arising late in life are commonly transmitted to one and the same sex. Variability is the necessary basis for the action of selection, and is wholly independent of it. It follows from this, that variations of the same general nature have often been taken advantage of and accumulated through sexual selection in relation to the propagation of the species, as well as through natural selection in relation to the general purposes of life. Hence secondary sexual characters, when equally transmitted to both sexes can be distinguished from ordinary specific characters only by the light of analogy. The modifications acquired through sexual selection are often so strongly pronounced that the two sexes have frequently been ranked as distinct species, or even as distinct genera. Such strongly-marked differences must be in some manner highly important; and we know that they have been acquired in some instances at the cost not only of inconvenience, but of exposure to actual danger.

The belief in the power of sexual selection rests chiefly on the following considerations. Certain characters are confined to one sex; and this alone renders it probable that in most cases they are connected with the act of reproduction. In innumerable instances these characters are fully developed only at maturity, and often during only a part of the year, which is always the breeding-season. The males (passing over a few exceptional cases) are the more active in courtship; they are the better armed, and are rendered the more attractive in various ways. It is to be especially observed that the males display their attractions with elaborate care in the presence of the females; and that they rarely or never display them excepting during the season of love. It is incredible that all this should be purposeless. Lastly we have distinct evidence with some quadrupeds and birds, that the individuals of one sex are capable of feeling a strong antipathy or preference for certain individuals of the other sex.

Bearing in mind these facts, and the marked results of man's unconscious selection, when applied to domesticated animals and cultivated plants, it seems to me almost certain that if the individuals of one sex were during a long series of generations to prefer pairing with certain individuals of the other sex, characterised in some peculiar manner, the offspring would slowly but surely become modified in this same manner. I have not attempted to conceal that, excepting when the males are more numerous than the females, or when polygamy prevails, it is doubtful how the more attractive males succeed in leaving a larger number of offspring to inherit their superiority in ornaments or other charms than the less attractive males; but I have shewn that this would probably follow from the females—especially the more vigorous ones, which would be the first to breed—preferring not only the more attractive but at the same time the more vigorous and victorious males.

Although we have some positive evidence that birds appreciate bright and beautiful objects, as with the bower-birds of Australia, and although they certainly appreciate the power of song, yet I fully admit that it is astonishing that the females of many birds and some mammals should be endowed with sufficient taste to appreciate ornaments, which we have reason to attribute to sexual selection; and this is even more astonishing in the case of reptiles, fish, and insects. But we really know little about the minds of the lower animals. It cannot be supposed, for instance, that male birds of paradise or peacocks should take such pains in erecting, spreading, and vibrating their beautiful plumes before the females for no purpose. We should remember the fact given on excellent authority in a former chapter, that several peahens, when debarred from an admired male, remained widows during a whole season rather than pair with another bird.

Nevertheless I know of no fact in natural history more wonderful than that the female Argus pheasant should appreciate the exquisite shading of the ball-and-socket ornaments and the elegant patterns on the wing-feathers of the male. He who thinks that the male was created as he now exists must admit that the great plumes, which prevent the wings from being used for flight, and which are displayed during courtship and at no other time in a manner quite peculiar to this one species, were given to him as an ornament. If so, he must likewise admit that the female was created and endowed with the capacity of appreciating such ornaments. I differ only in the conviction that the male Argus pheasant acquired his beauty gradually, through the preference of the females during many generations for the more highly ornamented males; the aesthetic capacity of the females having been advanced through exercise or habit, just as our own taste is gradually improved. In the male through the fortunate chance of a few feathers being left unchanged, we can distinctly trace how simple spots with a little fulvous shading on one side may have been developed by small steps into the wonderful ball-and-socket ornaments; and it is probable that they were actually thus developed.

Everyone who admits the principle of evolution, and yet feels great difficulty in admitting that female mammals, birds, reptiles, and fish, could have acquired the high taste implied by the beauty of the males, and which generally coincides with our own standard, should reflect that the nerve-cells of the brain in the highest as well as in the lowest members of the Vertebrate series, are derived from those of the common progenitor of this great Kingdom. For we can thus see how it has come to pass that certain mental faculties, in various and widely distinct groups of animals, have been developed in nearly the same manner and to nearly the same degree.

The reader who has taken the trouble to go through the

several chapters devoted to sexual selection, will be able to judge how far the conclusions at which I have arrived are supported by sufficient evidence. If he accepts these conclusions he may, I think, safely extend them to mankind; but it would be superfluous here to repeat what I have so lately said on the manner in which sexual selection apparently has acted on man, both on the male and female side, causing the two sexes to differ in body and mind, and the several races to differ from each other in various characters, as well as from their ancient and lowly-organised progenitors.

He who admits the principle of sexual selection will be led to the remarkable conclusion that the nervous system not only regulates most of the existing functions of the body, but has indirectly influenced the progressive development of various bodily structures and of certain mental qualities. Courage, pugnacity, perseverance, strength and size of body, weapons of all kinds, musical organs, both vocal and instrumental, bright colours and ornamental appendages, have all been indirectly gained by the one sex or the other, through the exertion of choice, the influence of love and jealousy, and the appreciation of the beautiful in sound, colour or form; and these powers of the mind manifestly depend on the development of the brain.

Man scans with scrupulous care the character and pedigree of his horses, cattle, and dogs before he matches them; but when he comes to his own marriage he rarely, or never, takes any such care. He is impelled by nearly the same motives as the lower animals, when they are left to their own free choice, though he is in so far superior to them that he highly values mental charms and virtues. On the other hand he is strongly attracted by mere wealth or rank. Yet he might by selection do something not only for the bodily constitution and frame of his offspring, but for their intellectual and moral qualities. Both sexes ought to refrain from marriage if they are in any marked degree inferior in body or mind; but such hopes are Utopian and will never be even partially realised until the laws of inheritance are thoroughly known. Everyone does good service, who aids towards this end. When the principles of breeding and inheritance are better understood, we shall not hear ignorant members of our legislature rejecting with scorn a plan for ascertaining whether or not consanguineous marriages are injurious to man.

The advancement of the welfare of mankind is a most intricate problem: all ought to refrain from marriage who cannot avoid abject poverty for their children; for poverty is not only a great evil, but tends to its own increase by leading to recklessness in marriage. On the other hand, as Mr Galton has remarked, if the prudent avoid marriage, whilst the reckless marry, the inferior members tend to supplant the better members of society. Man, like every other animal, has no doubt advanced to his present high condition through a struggle for existence consequent on his rapid multiplication; and if he is to advance still higher, it is to be feared that he must remain subject to a severe struggle. Otherwise he would sink into indolence, and the more gifted men would not be more successful in the battle of life than the less gifted. Hence our natural rate of increase, though leading to many and obvious evils, must not be greatly diminished by any means. There should be open competition for all men; and the most able should not be prevented by laws or customs from succeeding best and rearing the largest number of offspring. Important as the struggle for existence has been and even still is, yet as far as the highest part of man's nature is concerned there are other agencies more important. For the moral qualities are advanced, either directly or indirectly, much more through the effects of habit, the reasoning powers, instruction, religion, &c., than through natural selection; though to this latter agency may be safely attributed the social instincts, which afforded the basis for the development of the moral sense.

The main conclusion arrived at in this work, namely that man is descended from some lowly organised form, will, I regret to think, be highly distasteful to many. But there can hardly be a doubt that we are descended from barbarians. The astonishment which I felt on first seeing a party of Feugians on a wild and broken shore will never be forgotten by me, for the reflection at once rushed into my mind—such were our ancestors. These men were absolutely naked and bedaubed with paint, their long hair was tangled, their mouths frothed with excitement, and their expression was wild, startled, and distrustful. They possessed hardly any arts, and like wild animals lived on what they could catch; they had no government, and were merciless to every one not of their own small tribe. He who has seen a savage in his native land will not feel much shame, if forced to acknowledge that the blood of some more humble creature flows in his veins. For my own part I would as soon be descended from that heroic little monkey, who braved his dreaded enemy in order to save the life of his keeper, or from that old baboon, who descending from the mountains, carried away in triumph his young comrade from a crowd of astonished dogs—as from a savage who delights to torture his enemies, offers up bloody sacrifices, practises infanticide without remorse, treats his wives like slaves, knows no decency, and is haunted by the grossest superstitions.

Man may be excused for feeling some pride at having risen, though not through his own exertions, to the very summit of the organic scale; and the fact of his having thus risen, instead of having been aboriginally placed there, may give him hope for a still higher destiny in the distant future. But we are not here concerned with hopes or fears, only with the truth as far as our reason permits us to discover it; and I have given the evidence to the best of my ability. We must, however, acknowledge, as it seems to me, that man with all his noble qualities, with sympathy which feels for the most debased, with benevolence which extends not only to other men but to the humblest living creature, with his god-like intellect which has penetrated into the movements and constitution of the solar system—with all these exalted powers—Man still bears in his bodily frame the indelible stamp of his lowly origin.

Further Reading

These books and articles are a good starting point for curious readers who want to explore the vast realms of scholarship about human evolution and Darwin's life. Janet Browne's two-volume biography of Darwin is magisterial and intimate, while Desmond and Morris's places Darwin in his social and historical context. *Evolution: The Triumph of an Idea* is a layman's introduction to the history of evolutionary biology and the current scientific consensus. The books by Dugatkin, de Waal, Wade, and Wood are excellent introductions to the latest ideas about human evolution, from the fossil evidence to insights gained from studies on chimpanzees. The papers by Evans et al., Green et al., and Niimura and Nei document some of the remarkable details of human evolution stored in genomes.

Browne, E. J. *Charles Darwin: A Biography*. New York: Knopf, 1995.
——.*Charles Darwin. Vol. 2, The Power of Place*. Princeton, NJ: Princeton University Press, 2003.
Desmond, Adrian J., and James R. Moore. *Darwin*. New York: Viking Penguin, 1991.
Dugatkin, Lee Alan. *The Altruism Equation: Seven Scientists Search for the Origins of Goodness*. Princeton, NJ: Princeton University Press, 2006.
Evans, P. D., N. Mekel-Bobrov, E. J. Vallender, R. R. Hudson, and B. T. Lahn. "Evidence that the adaptive allele of the brain-size gene

microcephalin introgressed into Homo sapiens from an archaic *Homo* lineage." *Proceedings of the National Academy of Sciences* 103(48):18178-83. 2006.

Gould, Stephen Jay. *The Mismeasure of Man.* Rev. and expanded. ed. New York: Norton, 1996.

Green, R. E., J. Krause, S. E. Ptak, A. W. Briggs, M. T. Ronan, J. F. Simons, L. Du, M. Egholm, J. M. Rothberg, M. Paunovic, and S. Paabo. "Analysis of One Million Base Pairs of Neanderthal DNA." *Nature* 444(7117):330-6, 2006.

Niimura, Y., and M. Nei. "Evolutionary Dynamics of Olfactory and Other Chemosensory Receptor Genes in Vertebrates." *Journal of Human Genetics* 51(6):505-17, 2006.

Van Riper, A. Bowdoin. *Men Among the Mammoths: Victorian Science and the Discovery of Human Prehistory, Science and Its Conceptual Foundations.* Chicago: University of Chicago Press, 1993.

Waal, F. B. M. de. *Our Inner Ape: A Leading Primatologist Explains Why We Are Who We Are.* New York: Riverhead Books, 2005.

Wade, Nicholas. *Before the Dawn: Recovering the Lost History of Our Ancestors.* New York: Penguin Press, 2006.

Wood, Bernard A. *Human Evolution: A Very Short Introduction.* Oxford: Oxford University Press, 2005.

Zimmer, Carl. *Evolution: The Triumph of An Idea.* 1st Harper Perennial ed. New York: HarperPerennial, 2006.

Index